Environment and Services

MITCHELL'S BUILDING CONSTRUCTION

The volumes of this series of standard text books have been rewritten, re-illustrated, and amplified in order to bring them into line with the present-day needs of students of Architecture, Building and Surveying. There are five related volumes:

ENVIRONMENT AND SERVICES
Peter Burberry Dip Arch MSc ARIBA

MATERIALS
Alan Everett ARIBA

STRUCTURE AND FABRIC Part 1
Jack Stroud Foster FRIBA

STRUCTURE AND FABRIC Part 2
*Jack Stroud Foster FRIBA
and Raymond Harington Dip Arch ARIBA ARIAS*

COMPONENTS AND FINISHES
*Harold King ARIBA
and Alan Everett ARIBA*

In keeping with the increasingly rapid technical developments in building and building design and the revised syllabuses for courses at all levels of the building industry, the volumes of *Mitchell's Building Construction* have been completely rewritten and re-illustrated. All quantities in these new volumes are expressed in SI units, and tables are given showing the main CI/SfB numbering and the relevant volumes and chapters in MBC.

Environment and Services gives an introduction to the technical aspects of building design and construction in the fields of physical environment and services installations. It covers, briefly, the principles involved, materials and equipment where these are relevant, and design methods and applications. It includes 305 diagrams, 78 tables and a comprehensive index. References to more detailed and advanced reading in the various aspects covered are quoted in the text.

In addition to giving an introduction to principles, the technical data and estimation methods which are included, mean that the book can be used in connection with appropriate exercises and technical applications forming essential parts of a number of courses in Building Design and Technology.

Some of the examinations for which the book could be used as a first and basic text are: RIBA examinations in Environmental Science, Services Engineering and Design Technology, and for the Ordinary and Higher National Certificates in Building. It will also provide an ideal text to some of the City and Guilds of London Institute courses.

In related courses, such as Building Technology, Structural Engineering, Environmental Engineering, Illuminating Engineering and Quantity Surveying, the book will be most useful in giving an overall picture of the related technologies, an aspect of education which is becoming important for the increasingly complex buildings now being designed and built.

Those already qualified in these and similar disciplines will also find the book helpful for the same reasons.

To architects the design data and methods, which are all in SI units, will provide office reference material of considerable usefulness.

The Author, Peter Burberry, was Head of the Department of Science and Technology at the Architectural Association School of Architecture and is now Senior Lecturer and Tutor to the MSc Course at Bristol University Department of Architecture.

MITCHELL'S BUILDING CONSTRUCTION

Environment and Services

Peter Burberry *Dip Arch, MSc, ARIBA*

B T Batsford Limited
London and Sydney

For copyright reasons, paperback copies of this book may not be issued to the public, on loan or otherwise, except in their original soft covers. A hardback edition is available for library use.

© *Text Peter Burberry 1970 and 1975*
© *Illustrations B T Batsford Limited 1970 and 1975*
New revised edition 1975
ISBN 0 7134 0514 7 (hard cover) 0 7134 0515 5 (paper back)

Printed in Great Britain by
Tinling (1973) Limited, Prescot, Merseyside
(a member of the Oxley Printing Group Ltd)
for the publishers B T Batsford Limited
4 Fitzhardinge Street, London W1

Contents

Acknowledgment	8	
SI units	9	

ENVIRONMENT

1 General 17
   Environmental factors
      and their significance 17
   Design considerations 19

2 Moisture 21
   Precipitation 21
   Snow 22
   Moisture migration through the
      fabric of the building 23
   Condensation 23
   Tabular method 28
   Dynamic condensation prediction 30

3 Air movement 32
   Wind 23
   Ventilation 36
   Ventilation criteria 37

4 Daylighting 39
   Light from the overcast sky 39
   Use of BRS sky component
      protractors 42
   Methods of analysis 42
   Detailed use of protractors 43
   Presentation of data 44
   Internal reflected component 45
   Waldram diagram for daylighting
      analysis 45
   Block spacing 50
   Model analysis 55

   Direct sunlight 55
   Solar heat gain 57
   Geometry of the sun's movement 57
   Sunlight criteria 66
   Field measurements of insolation 66

5 Heat 67
   Thermal comfort 67
   Thermal properties of buildings 71
   Construction and planning in
      relation to heat balance 73
   Thermal insulation 75
   Heat transfer 76
   Resistance of surfaces 77
   Resistance of cavities and air spaces 78
   Calculation of U-values 80
   Heat loss calculations 86
   Maximum hourly rate of heat loss 86
   Example of convential heat loss
      calculation in SI units 87
   Seasonal heat loss 87
   Internal surface temperatures 92
   Admittance method 92

6 Sound 109
   General 109
   Noise control 113
   Planning in relation to noise 113
   Constructional precautions to
      reduce noise 113
   External noise indices 119
   Service installations 119
   Room acoustics 120
   Direct sound 120
   Reverberation time 121
   Standing waves 123

ENVIRONMENTAL SERVICES

7 Thermal installations — 127
- Energy sources, heat transfer mechanisms — 129
- *Direct heaters* — 135
- *Central heating* — 139
- Control of heating installations — 149
- Flues — 151
- Ventilation systems — 155
- *Air trunkings* — 158
- Input and extract devices for air distribution systems — 158
- *Special types of ventilation systems* — 159
- *Variable volume systems* — 161
- *Extract through lighting fittings* — 161
- *Trunking noise* — 163
- *Ventilation of internal lavatories* — 164
- Ventilation trunking sizing — 166

8 Electric lighting — 172
- Terminology — 173
- Criteria for design — 173
- *Lighting fittings* — 175
- Other factors of significance in electric lighting design — 176
- *Heat gain* — 176
- Cost — 177
- Estimation of lighting levels — 177
- *Luminance design* — 181

UTILITY SERVICES

9 Water supply — 191
- Private water supplies — 191
- *Pumps* — 192
- *Purity* — 194
- *Treatment* — 195
- Mains water supply — 196
- Cold water storage and distribution — 199
- Supplies to buildings rising above the level of the mains head — 207
- Hot water supply — 209

10 Sanitary appliances — 216
- Materials — 216
- Appliances — 217
- WCs — 217
- *Slop hoppers and housemaids' closets* — 219
- Urinals — 219
- *Bidets* — 221
- *Baths and showers* — 221
- *Wash basins* — 224
- *Sinks* — 225
- *Taps and Valves* — 225

11 Pipes — 229
- Properties and applications of materials — 229
- *Pipe fixing* — 234
- Estimation of sizes for cold and hot water — 235
- *Storage for cold water supply* — 235
- *Storage for hot water supply* — 235
- *Pipe sizing* — 236

12 Drainage installations — 245
- General principles — 245
- *Performance standards* — 245
- *Hydraulic design* — 245
- *Design considerations* — 250
- *Common misconceptions* — 251
- *Design of drainage installations* — 251
- Drainage above ground — 252
- *Types of drainage system* — 252
- Waste and soil systems — 253
- *Sizes for waste and soil pipes* — 261
- Rain water pipes and gutters — 261
- *Drainage of paved areas* — 265
- Snow — 265
- Drainage below ground — 265
- *Planning* — 265
- *Construction* — 265
- *Drain construction* — 269
- Systems — 275
- *Recent specialised developments in drain layout* — 276

13	Sewage disposal	286		Entry to the building	302
	Foul	286		Distribution circuits	302
	Dilution	286		*Socket outlets*	303
	Conservancy	286		*Fixed apparatus*	304
	Treatment	287		Bathroom	304
	Local authority standard details	290		Emergency supplies	304
	Surface water	290		Telephones and other telecommunication systems	307
	Land drainage	290		Regulations	310
14	Refuse collection and storage	292		Electrical terms	310
	Bins	292		Lightning protection	310
	Paladins	292			
	Paper sacks	292	16	Gas	312
	Refuse chutes	293		Town and natural	312
	Garchey system	294		Explosion	312
	Grinders	294		Pipes	312
	Incinerators	295		Typical domestic installation	312
	Compactors	295		Flues	313
	Non-domestic premises	295	17	Mechanical conveyors	314
15	Electricity	296		Mechanical movement of people	314
	General principles	296		*Lifts*	314
	Insulation	297		*Planning for lift installations*	322
	Fusing	297		*Escalators*	324
	Switch polarity	297		*Paternosters*	325
	Earthing	297		Tube conveyors	326
	Bathrooms	297			
	Basic wiring systems	298	18	Firefighting equipment	329
	Sheathed	298		Hand extinguishers	329
	Conduit	298		Fixed apparatus	331
	MICC cables	299		Fire alarms	333
	Wiring systems for larger installations	299		Equipment for use by fire brigade	333
	Cable trunking	299		*Fire brigade hydrants*	333
	Cable tap system	299		*Access by fire brigade appliances*	333
	Busbars	299		*Emergency control services*	334
	Ducts for electrical distribution	299			
	Ductube	300	19	Ducted distribution of services	335
	Skirting trunking	301			
	Floor trunking	301	CI/SfB		339
	Overhead distribution	301	Index		343

Acknowledgment

The author and publishers thank the following for permission to use certain material or to quote from books or articles, and to use drawings as a basis for diagrams in this volume:

Architectural Press for table 11 from *Designing for Thermal Comfort* by Alex Hardy (AJ 20.11.58): British Standards Institution for table 6 from CP3 Chapter 1(c), table 60 from BS417 table 1, tables 61 and 64 from CP310: 1965, table 63 from BS1565 1566; Building Research Establishment for table 5 Beaufort Scale, table 13 from BRS CP14/71 (Crown Copyright 1971), tables 29 and 33 from BRS CP47/68 (Crown Copyright 1968), table 34 from BRE CP61/74 (Crown Copyright 1974) and diagram 47 from *Theoretical and Practical Aspects of Thermal Comfort*; HMSO for tables 2 and 3 from BRS Digest 110, table 7 from PWBS No. 30, tables 36, 37 and 38 from BRS Digest 102 and 103, tables 72 and 73 from BRS Digest 34, diagram 4 from BRS Digest 110, diagram 29 from *Sunlight and Daylight*, 'U values for floors' from BRS Digest 145, diagrams 24 and 25 from *Architectural Physics Lighting;* Institution of Heating and Ventilating Engineers for tables 30, 32 and 35 from the IHVE *Guide 1970*, also diagram 56 (redrawn) from figure A8.4 of the IHVE *Guide 1970*.

SI units

All quantities in this volume are given in SI units, which have been adopted by the United Kingdom for use throughout the construction industry as from 1971.

Traditionally, in this and other countries, systems of measurement have grown up employing many different units not rationally related and indeed often in numerical conflict when measuring the same thing. The use of bushels and pecks for volume measurement has declined in this country but pints and gallons, and cubic feet and cubic yards are still both simultaneously in use as systems of volume measurement, and conversions between the two must often be made. The sub-division of the traditional units vary widely: 8 pints equal 1 gallon, 27 cubic feet equal 1 cubic yard; 12 inches equal 1 foot; 16 ounces equal 1 pound; 14 pounds equal 1 stone, 8 stones equal 1 hundredweight. In more sophisticated fields the same problem existed. Energy could be measured in terms of foot pounds, British Thermal Units, horsepower, kilowatt hour, etc. Conversion between various units of national systems were necessary and complex, and between national systems even more so. Attempts to rationalise units have been made for several centuries. The most significant stages being:

The establishment of the decimal metric system during the French Revolution.

The adoption of the centimetre and gramme as basic units by the British Association for the Advancement of Science in 1873, which led to CGS system (centimetre, gramme, second).

The use after approximately 1900 of metres, kilograms, and seconds as basic units (MKS system).

The incorporation of electrical units between 1933 and 1950 giving metres, kilograms, seconds and amperes as basic units (MKSA system).

The establishment in 1954 of a rationalised and coherent system of units based on MKSA but also including temperature and light. This was given the title *Système International d'Unites* which is abbreviated to SI units.

The international discussions which have led to the development of the SI system take place under the auspices of the Conference General des Poids et Mesures (CGPM) which meets in Paris. Eleven meetings have been held since its constitution in 1875.

The United Kingdom has formally adopted the SI system and it will become, as in some 25 countries, the only legal system of measurement. Several European countries, while adopting the SI system, will also retain the old metric system as a legal alternative. The USA has not adopted the SI system.

The SI system is based on six basic units:

Quantity	Unit	Symbol
Length	metre	m
Mass	kilogramme	kg
Time	second	s
Electric current	ampere	A
Temperature	degree Kelvin	°K
Luminous intensity	candela	cd

The degree Kelvin will be used for absolute temperatures, for customary use the degree Celsius (°C) will still be used and for temperature intervals (difference between two temperatures) degrees Celsius (°C) will also be used. (273.15°K is 0°C and °K and deg C are identical in terms of temperature intervals.) In addition to the basic units there are two supplementary units:

Quantity	Unit	Symbol
Plane angle	radian	rad
Solid angle	steradian	sr

Degrees °, minutes ' and seconds " will also be used as part of the system.

From these basic and supplementary units the

SI UNITS

remainder of the units necessary for measurement are derived, eg:

Area derived from length /m².

Volume derived from length /m³.

Velocity derived from length and time /m/s.

Some derived units have special symbols:

Quantity	Unit	Symbol	Basic units involved
Frequency	hertz	Hz	1 Hz = 1/sec (1 cycle per sec)
Force energy, work, quantity of heat	newton	N	1 N = 1 kg m/s²
	joule	J	1 J = 1 Nm
Power	watt	W	1 W = 1 J/s
Luminous flux	lumen	lm	1 lm = 1 cd sr
Illumination	lux	lx	1 lx = 1 lm/m²

Multiples and submultiples of SI are all formed in the same way and all are decimally related to the basic units. It is recommended that the factor 1000 should be consistently employed as the change point from unit to multiple or from one multiple to another. The following table gives the names and symbols of the multiples. When using multiples the description or the symbol is combined with the basic SI unit eg, kilojoule kJ.

Factor			Prefix	
			Name	Symbol
one million million (billion)		10^{12}	tera	T
one thousand million		10^9	giga	G
one million		10^6	mega	M
one thousand		10^3	kilo	k
one thousandth		10^{-3}	milli	m
one millionth		10^{-6}	micro	
one thousand millionth		10^{-9}	nano	n
one million millionth		10^{-12}	pico	p

It will be noted that the kilogram departs from the general SI rule with respect to multiples, being already 100 g. Where more than three significant figures are used it has been United Kingdom practice to group the digits into three and separate the groups with commas.

This could lead to confusion with calculation from other countries where the comma is used as a decimal point. It is recommended therefore that groups of three digits should be used separated by spaces, not commas. In the United Kingdom the decimal point can still, however, be represented by a point either on or above the bottom line.

DEFINITIONS

Basic SI units

The metre (m), unit of length. The metre is the length equal to 1 650 763.73 wave lengths in vacuum of the radiation corresponding to the transition between the levels $2p_{10}$ and $5d_5$ of the krypton-86 atom (Eleventh General Conference of Weights and Measures, 1960; XI CGPM, 1960).

The kilogramme (kg), unit of mass. The kilogramme is equal to the mass of the international prototype of the kilogramme (III CGPM, 1901).

> NOTE The international prototype is in the custody of the Bureau International des Poids et Mesures (BIPM), Sèvres, near Paris.

The second (s), unit of time-interval. The second is the duration of 9 192 631 770 periods of the radiation corresponding to the transition between the two hyperfine levels of the ground state of the caesium-133 atom (XII CGPM, 1967).

The ampere (A); unit of electric current. The empere is that constant current which, if maintained in two straight parallel conductors of infinite length, of negligible circular cross-section, and placed 1 metre apart in a vacuum, would produce between these conductors a force equal to 2×10^{-7} newton per metre of length (IX CGPM, 1948).

The kelvin (K); unit of thermodynamic temperature. The kelvin is the fraction 1/273.16 of the thermodynamic temperature of the triple point of water (XII CGPM, 1967).

> NOTES The temperature of the ice point is 273.15 K. The units of kelvin and Celsius temperature interval are identical.

The candela (cd); unit of luminous intensity. The candela is the luminous intensity, in the perpendicular direction, of a surface of 1/600 000 square

SI UNITS

metre of a black body at the temperature of freezing platinum under a pressure of 101 325 newtons per square metre (XIII CGPM, 1967).

Supplementary units

The radian (rad); unit of plane angle. The angle between two radii of a circle which cut off on the circumference an arc equal in length to the radius.

The steradian (sr); unit of solid angle. The solid angle which, having its vertex in the centre of a sphere, cuts off an area of the surface of the sphere equal to that of a square having sides of length equal to the radius of the sphere.

SOUND

The main units remain unchanged (eg, 1 hertz = 1 cycle per second, decibels and absorption coefficients, being ratios, remain unchanged). It is important to note, however, when using pre-metric tables that where absorption coefficients are quoted per unit (eg, per person or per seat) an area in square feet is implicit in the figure and a conversion to square metres should be made (square feet x 0.09 = square metres). The constant for use in the sabine formula becomes 0.16 (see text page 122).

ELECTRICITY

Units remain unchanged.

CONVERSIONS IN THIS VOLUME

Sizes defined by legislation in imperial units have not yet been redefined in SI units. A 4 in. diameter is approximately 101.6 mm. It does not seem likely that fractions of a millimetre or even single millimetres will be used when these dimensions are redefined and in this volume the sizes have been rounded off to the nearest convenient value. 4 in. diameters are expressed as 100 mm and 2 ft 6 in. depth of cover for buried water pipes as 0.75 m instead of 0.762 m.

Until new legislation is prepared, however, and new sizes established, the imperial sizes will be legally required. The Building Regulations 1965, *Metric Values, Consultative Proposals,* by the Ministry of Housing and Local Government, Welsh Office, published by HMSO, gives the metric equivalent of the present values and the proposed new metric values to be contained in the metric version of the Building Regulations. The proposals are subject to amendment following consultation.

CONVERSION FACTORS

Quantity or application	SI Unit		Present Unit (PU)	Conversion Factor (CF) PU × CF = SI value
	Description	Symbol		
SPACE length	metre	m	foot	0.31
			inch	0.025
area	square metre	m^2	square yard	0.84
			square foot	0.09
	square millimetre	mm^2	square inch	0.000 65
	square millimetre	mm^2	square inch	645
volume	cubic metre	m^3	cubic yard	0.76
			cubic foot	0.028
			gallon	0.0045
	litre	$\frac{m^3}{1000}$	gallon	4.55
			pint	0.57
MASS mass	kilogramme	kg	pound	0.45
			ton	1016
density	kilogramme per cubic metre	kg/m^3	pounds per cubic foot	1.6
			pounds per gallon	9.98
MOTION velocity	metre/second	m/s	feet per second	0.31
			feet per minute	0.0051
			miles per hour	0.45
FLOW RATE volume flow	cubic metre per second	m^3/s	cubic foot per second	0.028
	litres per second	l/s	gallons per minute	0.076
PRESSURE pressure	newtons per square metre	N/m^2	foot water gauge	2890
TEMPERATURE customary temperature (level)	degree Celsius	°C	degree Fahrenheit	$\frac{5(°F - 32)}{9}$
temperature interval (range or difference)	degree	deg C	degrees Fahrenheit	0.56
HEAT quantity (energy)	joule	J	British Thermal Unit	1055 *
flow rate (power)	watt	W	BTU per hour	0.29
			ton of refrigeration	3516
intensity of heat flow rate	watts per square metre	W/m^2	BTU per hour per square foot	3.16

12

CONVERSION FACTORS

	Quantity or application	SI Unit Description	SI Unit Symbol	Present Unit (PU)	Conversion Factor PU × CF = SI value
HEAT (contd)	thermal conductivity	watts per metre degree Celsius	W/m deg C	BTU inch/hour foot^2 deg F	0.14 †
	thermal conductance	watts per square metre per degree Celsius	W/m^2 deg C	BTU per hour per square foot per degree F	5.68
	thermal resistivity	metre degree Celsius per watt	m deg C/W	square foot hour degree F per BTU inch	6.93 †
	thermal resistance	square metre degree Celsius per watt	m^2 deg C/W	square foot hour degree F/BTU	0.18
	thermal diffusivity	square metre per second	m^2/s	square foot per hour	0.000 026
	thermal capacity per unit mass (specific heat)	kilojoule per kilogramme degree Celsius	kJ/kg deg C	BTU/pound °F	4.19
	thermal capacity per unit volume	kilojoule per cubic metre degree Celsius	kJ/m^3 deg C	BTU/cubic foot deg F	67.1
	calorific value (weight basis)	kilojoules per kilogramme	kJ/kg	BTU per pound	2.32
	calorific value (volume basis)	kilojoules per cubic metre	kJ/m^3	BTU per cubic foot	37.26
	latent heat	kilojoules per kilogramme	kJ/kg	BTU per pound	2.32
	refrigeration	watts	W	ton	3516
	moisture content	grammes per kilogramme	g/kg	grains per pound	14.28
MOISTURE	vapour permeability	kilogramme metre per newton second	kgm/Ns	pound foot per hour pound force	0.000 008 6
	illumination	lux	lx	foot candles lumen per square foot	0.107
LIGHT	luminance	candela per square metre	cd/m^2	foot lambert	3.43

* In practice 1 BTU is taken as equarent to 1 kilojoule.
† The apparent discrepancy between imperial and SI units may be made resolved by expressing the SI units in basic terms before cancellation of terms, eg:

Jm/m^{2}s deg C becomes W/m deg C $\left(\frac{m}{m^2} = m \text{ and } J/s = W\right)$ m^{2}s deg C/Jm similarly becomes m deg C/W

ENVIRONMENT

1 General

ENVIRONMENTAL FACTORS AND THEIR SIGNIFICANCE

Over much of the world's surface the natural environment does not provide, except momentarily, satisfactory conditions for human comfort. Dwellings and buildings have always set out to provide 'protection from the elements' for their occupants. Historically designers of buildings followed traditionally established precedents and relied for environmental decisions on their own experience of traditional solutions to local conditions. Thus the best combinations of materials and forms were established slowly by trial and error and indigenous building forms suited to local climate and materials developed over long periods. The acceleration of technical development which is now in progress in the field of building presents ever increasing ranges of building materials, products, functional problems and forms. New environmental problems and possibilities are involved together with more exacting demands from occupants for satisfactory solutions to environmental problems. In some cases electronic equipment or manufacturing processes demand more closely controlled conditions than are necessary for human comfort. Experience of previous practice is of only limited value in tackling the new situations which arise. Fortunately technical progress has also given rise to research which helps to make clear the fundamental principles underlying many environmental phenomena, to provide standards to guide designers and methods of analysis and means for the prediction of conditions to be expected in new buildings. This process has not begun to reach the point where established methods are available to meet most problems but it is possible to approach this aspect of building design rationally and with reasonable confidence that experienced designers, understanding the governing principles, familiar with accepted environmental standards and able to use analytical techniques where necessary to predict the performance of their projected building at appropriate stages in design will be able to produce well conceived answers to environmental problems.

Historically the main aim of environmental control has been human comfort. Provision for the storage of goods usually remained limited to providing roof and walls. In many modern buildings the main occupant is a piece of equipment or a medical or manufacturing process which may require more rigorously controlled conditions of temperature, humidity or particles in the air than are needed for human comfort. In some cases the mechanical equipment produces so much heat, noise or vibration that without control the building would become uninhabitable for humans. The main aspects of environment are the same whether equipment or people are considered. The preponderance of cases where human needs are paramount is such that the physiology and psychology of human sensation and perception must be regarded as one of the bases of environmental work. Indeed, when considered together with the relevant aspects of classical physics, the term psycho-physics has been used to describe fundamental work on environmental comfort. Detailed consideration of fundamental work on sensation and perception is beyond the scope of this book.

There is however an important general consideration which must be taken into account when considering the reaction of building occupants to environmental conditions. It is exemplified by the progressively increasing environmental standards demanded. At any given time performance standards with specific quantitative values exist in regulations, codes of practice and research recommendations. Over a period of time these standards can be observed to become more and more exacting. This is not entirely due to research on better standards. It is clear that an important environmental requirement in a new building is not merely to reach a particular currently accepted standard but to provide a quality of environment which is better than the occupants of the new building really expected.

Mechanical installations for the control of environment were used initially and for a considerable time thereafter in buildings designed and built in just the same way that was current before the mechanical installations became available. At this stage any improvement in environmental conditions was regarded as a worthwhile achievement. The considerable effect which the building itself has on its own internal environment was not consciously considered nor was it rewarding to do so since the range of building materials and forms was circumscribed and the types of environmental installations very limited. The best balance between use of the building, its form and construction and the installation, presented no special problems since the possible choices were very restricted and usually differed little in environmental terms. New materials and possible building forms, developing environmental standards, a wide range of different types of installation and increased consciousness of economy now present a very different situation. If satisfactory results are now to be achieved, it is vital to take into account the pattern of use of the building, the way in which the fabric of the building will react to the natural environmental conditions and to the operation of environmental congrol installations, the form and space occupied by the installations themselves and the form of the building and location of plant and distribution ducts. The environmental design must therefore form an integral part of the design of the building from the earliest stages.

Not only must environmental considerations be taken into account in the fundamental design of the building, they must also be balanced one with another. Larger windows giving better daylighting may result in excessive heat loss in winter and overheating due to the sun's rays in summer; windows and radiators may compete for particular wall space; better artificial lighting may make natural ventilation impossible since mechanical extract will be required to prevent too much heat from entering the rooms; heavy walls and floors, good for sound insulation, may, in a building with intermittent use, make it impossible to make the fuel savings which a fast reacting heating installation could have made possible. In addition these environmental factors interact with other aspects of interior design and the elegance, convenience and maintenance of rooms may be made or marred by the selection and positioning of environmental control equipment. It is clear therefore that the approach to environmental design must be an integrated one. Historically it has been fragmented between engineers skilled in specific types of energy distribution rather than integrated in relation to human needs or the decisions needed for the design of buildings. At present courses are being set up and professional groups formed dealing with environment as a unified concept related to human needs.

The apparent human environmental requirements are for light, air and warmth. In addition sound must be considered since it is indispensable from most activities, and humidity and hygiene are often relevant.

For comfort and efficiency the human body requires to be maintained within a limited range of environmental conditions which are only a small section of the full scope found in nature. The variation of conditions on the earth's surface with place and particularly with time is such that for most aspects it is necessary to consider means of augmentation and reduction, sometimes in quick succession.

In order to achieve effective environmental design judgment must be exercised at very early stages in the design of buildings. The possibilities for daylighting and natural ventilation are clearly implicit in any concept of shape and size. It is not generally appreciated that these factors have, throughout history, governed the maximum widths and forms of buildings. Modern light sources and techniques for mechanical ventilation mean that much deeper buildings are possible and architects are faced with a decision to be made. It is a decision however, which, whether consciously or by default, is made before the roughest outline is drawn and long before windows are considered. The width and shapes sketched at the outset of design will either allow or make impossible complete natural lighting of the interior. Similar considerations apply to air movement. Acoustic and thermal performance, and particularly solar heat gain, are substantially influenced by initial concepts although, in these cases, subsequent decisions about constructional materials, insulation, windows and screening can have significant effects.

It will be apparent that environmental considerations cannot be left until other aspects of the design are settled but must be borne in the designer's mind, together with their implications for building form, from the very outset of his work. It is also clear that prediction techniques for environmental aspects should, wherever possible, be able to be used in the early stages of design and should yield results in forms which will aid architects to shape and construct buildings appropriately. This has not always been appreciated by research workers and many existing techniques for environmental analysis can only be used after the decisions which they should influence have been made. There is an urgent need for research effort to be directed towards developing environmental design methods which can be used at the times, early in the design, when they will be effective.

DESIGN CONSIDERATIONS

Whatever the methods available to him the environmental and services designer must attempt to take part in the basic development of the design of buildings with which he is concerned. He must realise that balances have to be made between the various aspects involved. It is rarely possible, within the constraints of practical design and finance, to achieve the ideal solution to all the problems presented. Some, indeed, may be mutually opposed, improvement of one resulting inevitably in deterioration of the other. A specialist consultant must be able, not only to advocate the best solution from his point of view, but also to take a constructive part in reaching an optimum overall solution which, while not perfect from many specialist viewpoints, gives nevertheless the best result for the building as a whole.

Specialist consultants must therefore be able, not only to carry out the final precise analysis which will be required at the end of the design process, but also to take part in the establishment of initial concepts of form and fabric which will aid the subsequent progress of design, and to be able to provide quick, even if approximate, answers which will play a useful part in the early formative stages of design.

An ideal pattern of working, which would be followed by specialists and by architects dealing with these aspects of design themselves, may be postulated:

	Stage in process of design (part of overall architectural design)	*Simple example of specialist designers' work*
1	Establish the *objectives* to be achieved (in terms that are clear to all parties to the design including the client)	Good daylight in office
2	Establish the *criteria* which will enable the success, or otherwise of achieving the objectives to be assessed (this is usually a numerical technical standard which may be based upon statutory and recommended recommended standards	1% daylight factor 4 m from external wall of office
3	Propose *concept* of building from which will make achievement of the criteria feasible	eg width overall of building not to exceed approx 13 m (allows central corridor at lower standard) and glazing area not less than 40%
4	The architect, taking into account 3 above and all the other technical factors, will develop an overall *design solution* which will define precisely the size of the building and the windows	spot checks of lighting performance of stages in sketch design if requested by architect
5	An *appraisal* of all aspects of the design will be made by the architect, the client and other appropriate bodies	analysis of daylighting distribution in the design
6	The *decision* to proceed or, if design is not acceptable, to return to stage 3 must be made. (In some cases no satisfactory solution will be possible and modification of criteria and even objectives may be needed)	Does analysis demonstrate that daylighting is up to the criteria set, or if not sufficiently near to be acceptable?

GENERAL

Diagram 1 sums up schematically the pattern of this design process.

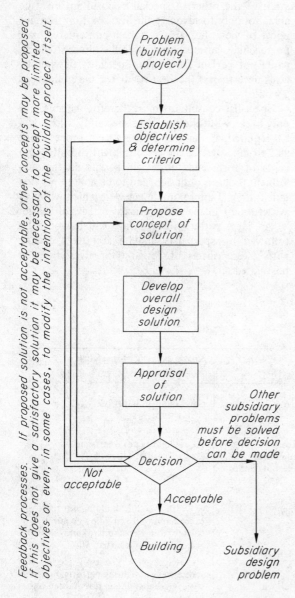

1   Schematic diagram of the basic design process

2 Moisture

Water in its free liquid form must be totally excluded from the interiors of buildings and its presence as vapour in the air or as moisture content of building materials must be controlled to within acceptable limits. Moisture in buildings comes from three main sources: precipitation as rain or snow; damp rising from the ground through building materials by capillary attraction or as vapour; and condensation upon cold surfaces of humidity from the air. In addition condensed droplets in the air (fog) can be carried by wind to wet the surfaces of the building and to penetrate joints and materials. The relative humidity of air in buildings can be a critical factor in governing thermal comfort since it influences the ability of the body to lose heat.

PRECIPITATION

It is usual to think of precipitation in terms of the volume of water which falls during the course of a year. This is not normally significant for buildings where critical aspects are the frequency, duration and intensity of the worst rainstorms since it is these factors that govern the quantity of water which must be carried away from the buildings by gutters, downpipes, channels, gullies and drains. Diagram 2 shows the relationship between the intensity, duration and frequency of British rainstorms. Detailed rainfall records are published in the annual *British Rainfall* but in Britain it is desirable to consult diagram 2 in order to establish a design standard rather than local records since very heavy falls may well occur even if not so far recorded at that spot. The intensity selected for design of gutters, downpipes and drains will depend on the consequences of overflowing and is discussed in chapter 11. The standards which are normally in use at present are a rainfall intensity of 75 mm per hour when designing gutters and downpipes and 50 mm per hour for drains although in both cases precautions are necessary to ensure that overflow from very heavy storms does not cause damage.

The effect of rain upon buildings can be influ-

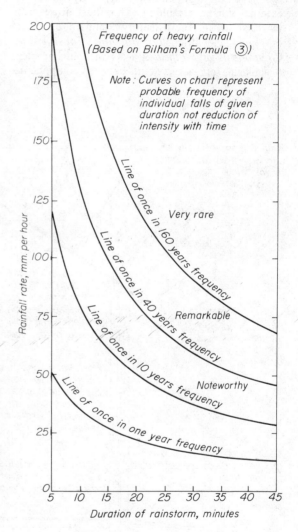

2 Frequency of heavy rainfall

enced by wind. The amount falling upon pitched roofs may be affected, and significant quantities may fall on, run down and have to be collected and discharged from the elevations of tall buildings. Although intended to deal with the wetting of walls

MOISTURE

rather than rainwater run-off the BRS *Index of Driving Rain* provides a useful insight into the combined behaviour of wind and rain.

In some cases due to bad joints water may flow or condensed droplets of water may be carried by wind through one or more layers of the external skin of a building. This must however be regarded as an unsatisfactory standard of design or building. *Mitchell's Building Construction: Components and Finishes* deals with the problem of achieving satisfactory joints between components.

SNOW

The melting of snow does not normally place such an acute load on the gutters and drains of individual buildings as does heavy rainfall and consequently is not a critical problem from this point of view. *MBC: Components and Finishes* volume deals with the precautions desirable to prevent the penetration of wind-driven snow into buildings and damp penetration due to melting. Care must be taken in gutter layout to avoid damage to the building

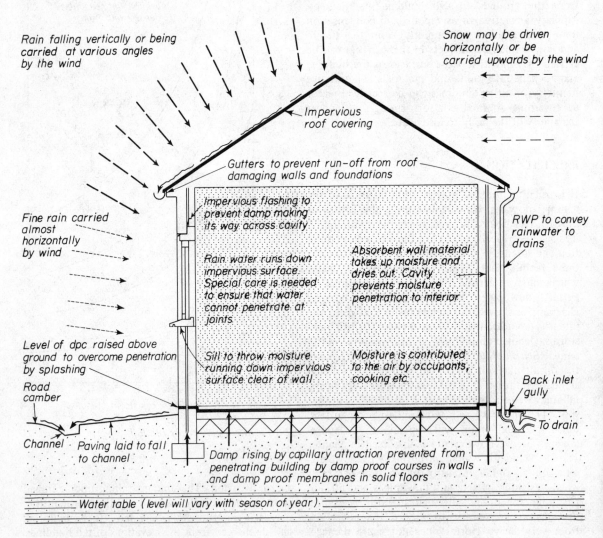

3 Diagram showing the ways in which moisture penetration is controlled in buildings

from overflows during snow thaw, and snow boards may be needed to prevent snow falls on to glazed areas or entrances.

Design of buildings and surrounding features in relation to snow drifting is not normal in this country.

MOISTURE MIGRATION THROUGH THE FABRIC OF THE BUILDING

Movement of moisture through the materials of which the building is constructed can occur through capillary attraction or by diffusion of vapour through the pores of the material. The sources of moisture which must be considered in this context include not only precipitation on external surfaces but also damp from the ground on which the building stands. Roofs are traditionally made of impervious materials but walls and ground floors are not often made in this way. Since, in the case of walls, the exposure to moisture is intermittent it is possible to use thick solid walls of porous materials which absorb moisture into their external faces and then allow it to evaporate without giving rise to serious internal penetration. This type of wall generally has to be very much thicker than is required for structural needs and in modern buildings, means other than thickness have to be used to limit moisture penetration.

Two methods of control are normally employed. Either a barrier of impervious material or an air gap is placed to intercept the movement. Flashings (capillary movement only), damp proof courses, damp proof membranes (capillary and vapour diffusion) and vapour barriers (vapour only) are examples of impervious materials being used as barriers. Cavities, provided that they are not bridged by moisture transmitting features, are a very effective means of arresting capillary movement. They are not necessarily effective against vapour diffusion since unless well ventilated it will be possible for the vapour to pass across. Suspended ground floors formed a good example of a cavity formed to prevent capillary and vapour movement which to be effective in preventing vapour reaching the other side had to be adequately ventilated. This type of construction has now largely gone out of use because of the high heat losses which resulted from the ventilation. A floor resting solidly on the ground and having a continuous damp proof membrane gives a more satisfactory overall result.

CONDENSATION

Condensation arises from the variation with temperature of the capacity of air to hold moisture in the form of water vapour. As temperatures increase the capacity increases. For each temperature, however, there is a saturation level of moisture. When air which has absorbed moisture is cooled to such a temperature that the moisture content exceeds the saturation point, the excess moisture will be deposited as water. In buildings, the air from outside is taken in and warmed and moisture is added from occupants and processes. If cold surfaces such as windows or badly insulated walls exist, they can cool the air immediately adjacent to them so that moisture is deposited on the surface in the form of condensation. This is both unsightly and likely to cause damage to the contents and finishes of rooms and also to the wall, ceiling and floor materials.

In recent years condensation, particularly in dwellings, appears to have been increasingly troublesome. A number of factors could account for this. The change from open fires to other forms of heating is one. The open fire promoted high rates of ventilation which disposed of water vapour in the air, and the heat output was partly radiant which helped to keep internal surfaces above dewpoint. Associated with the change from open fires is the improved sealing of buildings (eg weather stripping, elimination of air bricks and flues) which also contributes to reduced rates of venitlation. Increasing use of electric, gas and oil local heating appliances which can easily be turned on and off encourages fuel saving by intermittent use of the apparatus. In these conditions the wall surfaces may take some time to warm up during which period they may well be below dewpoint.

Some gas and oil fired room heaters do not have flues and then products of combustion can contribute significant amounts of water vapour to the atmosphere. The consequences of condensation are, in many modern constructions, more acute than was the case in the past. In masonry and brick constructions significant quantities of water can condense and subsequently evaporate without being apparent or causing serious deterioration. Many

MOISTURE

modern insulating materials are vegetable based and deteriorate rapidly under the conditions described, while in other cases impervious surfaces concentrate the condensation.

Risk of condensation can be minimised by increased ventilation rates. This, however, is not in practice feasible with naturally ventilated buildings in winter. The most effective method is to ensure that the internal insulation of the wall is sufficeintly high to keep the surface temperature above dewpoint. A method of estimating the surface temperature of walls (dry bulb) for design purposes is given on page 92. Diagram 4 is a psychometric chart which enables the saturation temperature, or dewpoint of air with varying moisture contents, to be determined. This chart is a vital tool in condensation prediction.

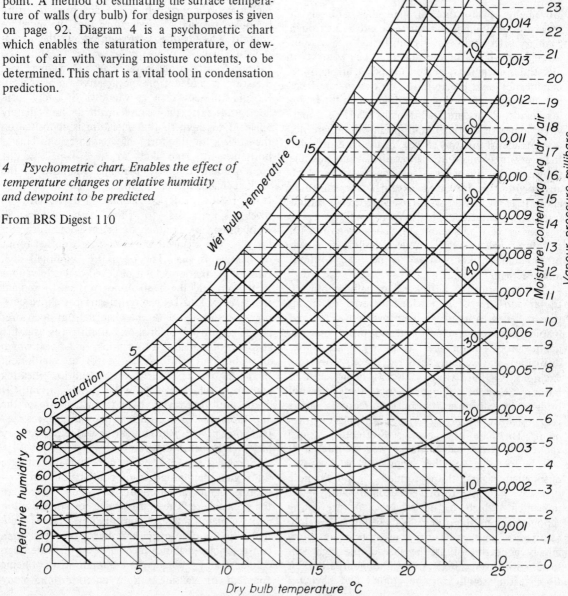

4 Psychometric chart. Enables the effect of temperature changes or relative humidity and dewpoint to be predicted

From BRS Digest 110

Crown Copyright reserved. Reproduced by permission of H M Stationery Office

Use of psychrometric chart (diagram 4)
The chart expresses the relationships between the quantity of water in the atmosphere (expressed as mixing ratio), the vapour pressure, the relative humidity (and dewpoint) and the wet and dry bulb temperature.

With normal meteorological or environmental data on dry bulb temperature and relative humidity, the appropriate vapour pressure can be determined for condensation calculations (eg 20°C 60 per cent RH gives 14.2 mb vapour pressure).

The variation of relative humidity with temperature for air with a given quantity of water can be established (eg consider in the case of the above a reduction in air temperature to 15°C, the vapour pressure and mixing ratio remaining constant. As a result of the temperature change the relative humidity changes to 83 per cent).

Dewpoint temperature may be predicted (eg taking the same case, 100 per cent RH (dewpoint) will occur at 12.5°C dry bulb).

The effect of adding moisture to the air may also be established by adding the vapour introduced to the mixing ratio (eg consider external air at 0°C 90 per cent RH warmed on entering a dwelling to 20°C and with 3.4g of moisture per kg of dry air added. The mixing ratio at 0°C 90 per cent RH is 3.4 g/kg. Adding 3.4 g/kg results in a total mixing ratio of 6.8 g/kg. This corresponds to an RH of 46 per cent at 20 °C and a dewpoint of 8.5 °C).

The actual moisture content of air in buildings depends on the original content externally and the additional moisture introduced by occupants and processes. Meteorological data can be used to establish a design value for external conditions (in England very moist conditions often exist). This information is usually in terms of wet and dry bulb temperatures, or dry bulb temperature and relative humidity. In either case a point representing the conditions can be found on diagram 4 and the quantity of moisture in the atmosphere per kg of dry air read off from the right hand scale.

CP3 Chapter 11: 1970, Thermal Insulation suggests dry bulb design temperatures of 0°C outside and 18 °C to 22 °C inside. For condensation prediction it is usual to assume that the outside air is saturated at 100% relative humidity. Where radiation to the cold night sky must be taken into account a reduction of the outside temperature to -6°C is suggested.

Internal conditions must also be determined.

Table 1 Sources and approximate quantities of vapour input to buildings

Source	Approximate quantity
People: Sedentary Active	0.05 kg per person per hour 0.2 kg per person per hour
Gas: Cooking or any flueless heaters or appliances	0.81 per m³ of town gas
Paraffin: Flueless heaters	1 kg per litre approximately (1 litre represents approx 4MJ heat output)
Cooking, bathing and showers	0.03 to 0.06 kg per hour
Clothes washing and drying	Too variable for figure to be given but very substantial quantities of water can be put into the atmosphere and special precautions against general penetration of this moisture into buildings are required

Many activities contribute moisture to the atmosphere inside buildings, including the moisture given off by the occupants themselves. Table 1 summarises some of the main sources in domestic buildings.

It is difficult to use this information directly to predict moisture conditions but fortunately the Building Research Establishment has made recommendations to form a basis for design which take into account typical ventilation rates and vapour inputs. Table 2 shows the amount of moisture, in g per kg of dry air, to be added to the mixing ratio of the outside air, taken from diagram 4. The resulting total mixing ratio enables the vapour pressure, or the relative humidity at any particular temperature to be determined.

It is now possible to select a new point on the psychometric chart, diagram 4, based on the total internal moisture content (mixing ratio scale) and on the internal surface temperature of the wall (dry bulb scale). Projection across and up respectively from these scales enables the relative humidity at the wall surface to be established. If this is near to or beyond the 100% relative humidity line condensation will occur.

MOISTURE

Table 2 Excess moisture content of air (over outdoor content) for various building types

Building type	Excess content g moisture per kg of dry air
Shops, offices, classrooms, public buildings, dry industrial processes	1.7
Dwellings	3.4
Catering establishments or wet industrial processes	6.8

It is possible for vapour to diffuse through walls so that condensation can take place within the construction itself. The problem is likely to be particularly acute where a very permeable inner layer, such as insulation board, also provides a large part of the internal insulation. This results in high moisture content combined with low temperatures at points within the construction. In these circumstances a vapour barrier such as polythene film (see *MBC: Materials* for properties, and *Structure and Fabric* for installation details) to govern the movement of moisture must be employed.

Troublesome condensation can also occur on cold water pipes, particularly service pipes, and on windows. The temperature of mains water can fall to 3° or 4°C during very cold weather. The method described can be used to check whether condensation will occur. If so, lagging, with an external impervious layer (eg gloss paint), should be provided or the pipe re-routed.

Condensation is almost inevitable on windows in modern dwellings and condensation grooves and drainage channels should be provided (see *MBC: Components and Finishes*). Where formation of condensation on windows must be avoided (shop windows) it is possible to increase the surface temperature by using double glazing, in which case the new conditions can be estimated in the way described, or by heating at the bottom of the window, or by blowing air across the window.

In many cases it may be thought desirable to consider the risk of interstitial condensation. Such a check would be particularly valuable where thermal insulation is used as an internal lining for walls,

since the type of lining often reduces the structural temperature in the wall without reducing the vapour pressure, as in the case of flat roofs which can present a critical condensation risk.

BRS Digest 110 gives a design data and a method for a steady state check of surface and interstitial condensation. It is also possible to carry out the calculation by graphical means and this approach has a number of advantages. A suitable procedure is:

Graphical method for condensation prediction

Consider the external wall of a dwelling house.

Stage 1

Establish external and internal air temperatures and vapour pressures. External air temperature 0°C and internal air temperature 20°C, Assuming the external air is saturated, the vapour pressure and mixing ratio may be determined from the psychrometric chart, diagram 4. (VP = 6 mb. MR = 3.9 g/kg.) From Table 2 the excess moisture content for dwellings is given as 3.4 g/kg. Add this to the external moisture content. The total is 7.3 g/kg at 20°C dry bulb which represents a RH of 50 per cent and a vapour pressure of 11.7 mb.

Stage 2

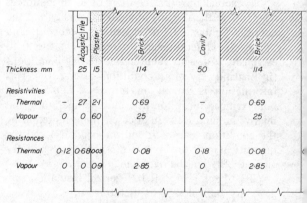

5A

Draw a section through the construction.
Mark on the section thicknesses and resistivities (thermal and vapour). Calculate and mark resistances (resistivity × thickness = resistance). Table 3 gives typical thermal and vapour resistivities.

CONDENSATION

Stage 3

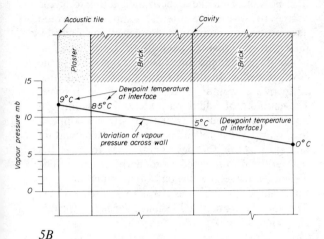

5B

In a similar way prepare scale diagram of the relative vapour resistance of each element (note that acoustic tile, cavity and surface resistances have negligible vapour resistance).

Construct scale of vapour pressure, mark internal and external vapour pressures on inner and outer faces.

Join points by straight line which represents vapour pressure variations to scale.

Establish vapour pressure at each interface and, using the psychrometric chart, determine the dew-dew-point temperature.

Stage 4

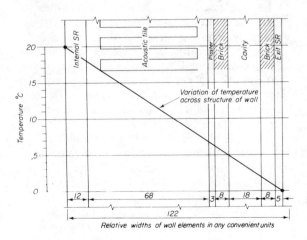

5C

Prepare scale diagram of the thermal properties of the wall. (Widths drawn for each material are in proportion to the relative thermal resistance of each element).

Construct convenient scale for temperature as shown.

Mark internal and external temperatures on appropriate faces.

Join points by straight line which represents temperature variation.

Stage 5

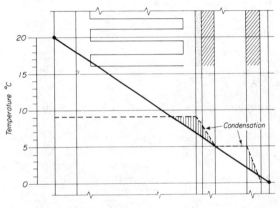

5D

On stage 4 diagram mark the dewpoint temperatures at each face and interface and draw line joining these points. Where the dewpoint line rises above the structural temperature line condensation will occur.

In the example, condensation is extensive and steps will have to be taken to reduce it. A vapour barrier at the back of the acoustic tile will eliminate condensation in the plaster and brickwork but will have no effect on condensation in the acoustic tile itself. An alternative thermal insulation having a higher vapour resistivity could provide a solution. (In this case it is interesting to note that leaving out the acoustic tile will eliminate condensation.)

Caution must be employed when vapour barriers are used to reduce the penetration of vapour into a wall or roof. Any bad workmanship or jointing may give rise to a patch of condensation. There have been many failures of vapour barriers applied in sheet or panel form.

MOISTURE

Table 3   Typical values of heat and vapour resistance

	Thermal resistance (r_t) m^2 deg C/W	Vapour resistance (r_v) MN s/g
Surfaces		
Wall surface—inside	0.12	
—outside	0.05	
Roof (or ceiling)		
surface—inside	0.11	
outside	00.04	
Internal airspace	0.18	
Membranes		
Average gloss paint film		7.5–40
Polythene sheet (0.06 mm)		250
Aluminium foil		4000

	Thermal resistivity m deg C/W	Vapour resistivity* MN s/g m
Materials		
Brickwork	0.69–1.38	25–100
Concrete	0.69	30–100
Rendering	0.83	100
Plaster	2.08	60
Timber	6.93	45–75
Plywood	6.93	1500–6000
Fibre building board	15.2–18.7	15–60
Hardboard	6.93	450–750
Plasterboard	6.24	45–60
Compressed strawboard	9.7–11.8	45–75
Wood-wool slab	8.66	15–40
Expanded polystyrene	27.72	100–600
Foamed urea-formaldehyde	27.72	20–30
Foamed polyurethane (open or closed cell)	27.72	30–100
Expanded ebonite	27.72	11000–60000

*Resistivity–1/diffusivity

Using the final (stage 5) condensation diagram it is possible to make a very quick estimate of the degree to which the internal temperature would have to be raised to avoid condensation simply by taking the external temperature (0°C) as fixed and drawing a new temperature line to run just above the dew point line. In this case the increase in internal temperature would be too great to be feasible. The amount of additional external insulation can also be established by taking the internal temperature as the fixed point and projecting the temperature line (just above the dew point line) to meet the 0°C grid line projected beyond the diagram. The distance outside the diagram of the junction, measured along the grid line gives, to the scale of the diagram, the amount of insulation needed (in this example 0.26 m² °C/W).

Tabular method

Anyone not able to carry out the simple graphical analysis of condensation may use the arithmetical procedure. If so it is desirable to set out the calculation as shown in table 4 (which shows the same calculation as the graphical method) and to check the column totals as shown to provide a check on the arithmetic. Comparison of the structural temperature and dew point temperature columns will reveal incidence of condensation although it is usual to plot both dew point temperature and structural temperature variations across a scale diagram of the construction.

In some cases, such as lecture rooms or theatres, where the moisture load varies with the pattern of occupancy, it may be desirable to make an estimate of the peak moisture levels. Apart from condensation risks moisture levels in rooms of this type may well rise above the 70% regarded as the maximum for comfort. Information about moisture levels is needed therefore to aid decisions about ventilation rates as well as condensation precautions.

A simple graphical method may be used to give an answer sufficiently accurate for design purposes. Diagram 6 shows this method. The period being investigated is divided into convenient time units, in this case half yours. In each half hour period the total weight of water present and input into the room is found and divided by the total weight of air present and input. The result is the mean mixing ratio, from which the vapour pressure, or the relative

Table 4 Tabular condensation analysis

Element	Thickness m	Thermal					Vapour					Dew point temp °C
		Resistivity	Resistance	% of total resistance	Temp drop across element	Structural temp at interfaces °C	Resistivity	Resistance	% of total resistance	Vapour pressure drop across element	Vapour pressure at interface mb	
Interior	–	–	–	–	–	18	–	–	–	–	10.3	–
Internal SR	–	–	0.12	10	1.8	16.2	–	–	–	–	10.3	–
Acoustic tile	0.025	27	0.68	56	10.0	6.2	0	0	0	0	10.3	8
Plaster	0.015	2.1	0.03	2	0.4	5.8	60	0.9	14	0.77	9.53	6.5
Brick	0.114	0.69	0.08	7	1.3	4.5	25	2.85	43	2.37	7.16	2.5
Cavity	0.050	–	0.18	15	2.2	2.3	–	0	0	0		
Brick	0.114	0.69	0.08	7	1.3	1.0	25	2.85	43	2.37	4.70	0
External Sr	–	–	0.05	6	1.1	-0.1	–	–				
Exterior	–	–	–		0		–				4.8	
			1.22	103	18.1			6.6	100	5.51		
				due to rounding off to whole numbers (should be 100)	due to rounding off (should be 18)					due to rounding off		

humidity at any given temperature can be established.

The example given is a classroom 50 m² in floor area, 3.5 m high. The internal temperature is to be 20°C. For daytime occupancy in term time an external temperature of 8°C with 100% relative humidity presents a possible selection for worst conditions. For an occupancy of 40 pupils a room of this volume should have a ventilation rate of 1120 m³ per hour or $6\frac{1}{2}$ air changes per hour (see chapter 3 p.38). In a building of this type, however, with window controlled natural ventilation it is unlikely that this rate would be achieved during the winter, particularly if blackout were used. 3 air changes might be a reasonable assumption. Occupancy is scheduled to be between 9.30 and 10.30, 11.00 and 12.30, and 14.00 to 16.30. Data required for the calculation is:

Volume of classroom 175 m³ ∴ weight of air 222 kg.
Volume of three air changes 525 m³, weight of three air changes 666 kg.
Moisture input from occupants 4) @ 50 gph = 2 kg per hour.
Mixing ratio for 100% RH at 8 °C = 6.5 g/kg ∴ weight of water in 175 m³ of air = 1.44 kg.

The calculation is made in the rows a-e in the top part of diagram 6 and plotted graphically.

1 Row a Enter weight of water in the room air (mixing ratio x weight of air in room). (Initially this will be based on the same mixing ratio as the external air. For subsequent calculations it will be the mixing

MOISTURE

1·44	1·44	1·44	1·84	2·00	2·06	1·69	1·94	2·04	2·08	1·7	1·54	1·48	1·86	2·01	2·07	a
2·16	2·16	2·16	2·16	2·16	2·16	2·16	2·16	2·16	2·16	2·16	2·16	2·16	2·16	2·16	2·16	b
—	—	1·00	1·00	1·00	—	1·00	1·00	1·00	—	—	—	1·00	1·00	1·00	1·00	c
3·60	3·60	4·60	5·00	5·16	4·22	4·85	5·10	5·20	4·24	3·86	3·70	4·64	5·02	5·17	5·23	d
6·5	6·5	8·29	9·01	9·30	7·60	8·74	9·19	9·37	7·64	6·95	6·67	8·36	9·05	9·32	9·42	e

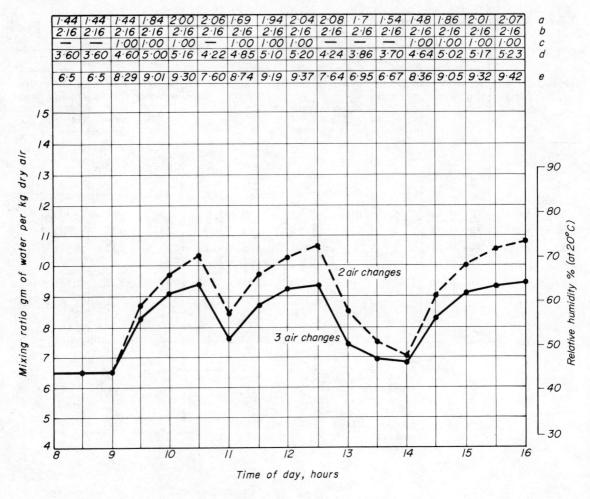

6 Graphical prediction of variation of moisture level with occupation

ratio resulting from the last half hour multiplied by the weight of air in room).

1 Row b Enter weight of water introduced by ventilation (external mixing ratio × weight of air introduced).

3 Row c Enter weight of water introduced by occupants.

4 Row d Sum total weight of water present and input during half-hour period.

5 Row e Enter resulting mean mixing ratio calculated by dividing weight of water from row d by the weight of air present and input during half-hour period.

6 Row a Using mixing ratio from 5 above multiplied by weight of air in room enter weight of water in the room air.

7 etc repeat for each time period.

The effects of two and three air changes are shown in the diagram.

Dynamic condensation prediction

The graphical and tabular prediction methods described deal with steady state conditions where it is assumed that internal and external temperatures and vapour pressures remain fixed. For sheet or lightweight constructions this assumption may be reasonable. For heavier constructions, however, the method may give very erroneous answers. If sufficient data is available to define internal and external vapour and temperature conditions over and around the period of the year under consideration then hybrid computing means may be used to obtain prediction which takes into account not only these variations but also thermal and vapour capacity and resistance.

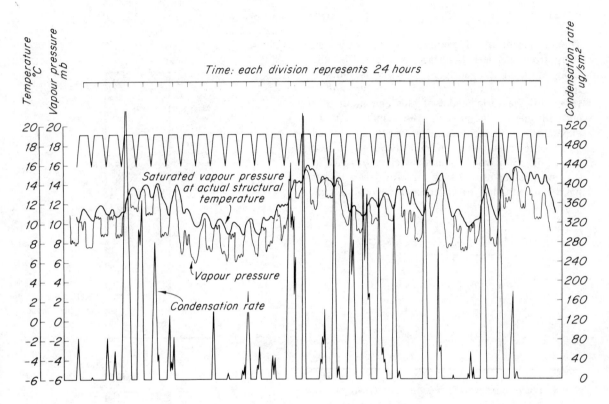

7 *Dynamic condensation prediction by means of electrical analogue. (Rate of condensation is shown at a particular point in the roof of a building)*

3 Air movement

WIND

The movement of air, taking place smoothly well away from the earth's surface, is affected by the topography of the surface. Velocity is reduced towards the ground, direction of movement is varied by the varied relief of hills and mountains and turbulence is created. The winds that we experience at surface level are not easily represented by specific speed values. Admiral Beaufort devised a scale in 1806 which is used to classify winds at sea and this has proved such a useful device that it has been extended to cover land conditions and,

8   A typical pattern of pressure distribution

9 and 10   *Air flow patterns round large buildings showing high velocity at parapet and through gap in building.*

recently at the Building Research Station, conditions for human comfort. Table 5 shows the Beaufort scale and the associated conditions for comfort. Ground areas where wind speeds exceed 5 m/s for significant periods are likely to be thought uncomfortable for human occupation, speeds greater than 10 m/s are positively unpleasant and speeds over 20 m/s can be dangerous both by the risk of being blown over and the risk of injury from broken branches and trees and falling pieces of building.

The traditional town gave substantial protection from high winds and discomfort was only rarely experienced. In recent years the construction of high buildings has caused a dramatic change. These

Table 5 Beaufort Scale

Beaufort number	Windspeed m/s*	Description	Sea condition	Land condition	Comfort
0	0-0.5	Calm	Mirror like surface	Smoke rises vertically	No noticeable wind
1	0.5-1.5	Light air	Ripples	Smoke drifts	
2	1.6-3.3	Light breeze	Small wavelets	Leaves rustle	Wind felt on face
3	3.4-5.4	Gentle breeze	Large wavelets scattered white horses	Wind extends flags	Hair disturbed clothing flaps
4	5.5-7.9	Moderate breeze	Small waves frequent white horses	Small branches in motion, raises dust and loose paper	Hair disarranged
5	8.0-10.7	Fresh breeze	Moderate waves many white horses chance of spray	Small trees in leaf begin to sway	Force of wind felt on body
6	10.8-13.8	Strong breeze	Large waves and white foam crests	Whistling in telegraph wires large branches in motion	Umbrellas used with difficulty Difficult to walk steadily Noise in ears
7	13.9-17.1	Near gale	Sea heaps up and foam begins to be blown in streaks	Whole trees in motion	Inconvenience in walking
8	17.2-20.7	Gale	High waves, spindrift and foam	Twigs broken from trees	Progress impeded Balance difficult in gusts
9	20.8-24.4	Strong gale	High waves with toppling crests Spray affects visibility	Slight structural damage (chimney pots and slates)	People blown over in gusts
10	24.4-28.5	Storm	Very high waves with long overhanging crests Surface white with foam visibility affected	Seldom experienced inland. Trees uprooted Considerable structural damage	

* Measured 10 m above sea or ground level

AIR MOVEMENT

buildings present a large area to the wind and considerable air pressure differentials result and consequently increased air velocities result, particularly at the corners of buildings. Diagram 8 shows a typical pattern of pressure distribution. The pressures themselves are not directly noticeable to people walking outside but diagrams 9 and 10 show the resulting velocity patterns which do affect comfort. It will be noted that particularly high velocities occur at the leading edges of the building both top and sides where high and low pressure zones approach closely. Buildings often have open areas at ground level and diagram 10 shows that high velocities of air movement are inevitable since these openings allow air flow between the high pressure area facing the wind and the low pressure areas on the other side. This feature of design usually presents air movement problems for people walking there.

The juxtaposition of high buildings with lower ones gives rise to special problems. Diagram 11 shows how a standing vortex may form between the high and low building giving very high velocities at ground level. Down wind of the high building down currents may develop, which are not likely to be noticeable directly, but which can carry smoke from chimneys downwards. Courtyards in buildings with blocks of different heights can suffer from the same problem.

The effects described above can have the effect of multiplying the original wind velocity. This contrasts with the traditional town pattern where velocities experienced at ground level were generally reduced. Pedestrians in new developments can, therefore, discover wind velocities four or five times as great as those to which they are accustomed. There have been examples where new layouts were quite unacceptable for pedestrians and ground areas have had to be roofed over to enable them to be

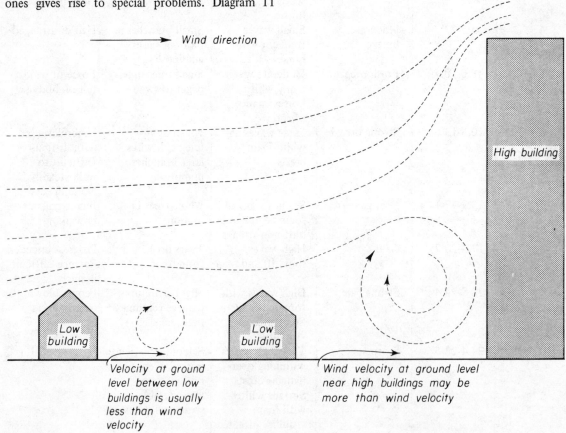

11   The pattern of air movement between low buildings and between low and high buildings

34

used. Town planning authorities are conscious of the problem and many now require promoters of major developments to demonstrate that unacceptable velocities will not occur in pedestrian areas. While general principles of air movement, some of which have been described, are well understood, and can be borne in mind design, there are no simple in drawing board means of predicting velocities. Schemes to be studied will have to be tested in a wind tunnel. Suitable tunnels exist at NPL and some universities. Normal architectural models can often be used. It is important, however, to ensure that the tunnel used can model not only the velocity gradient of the wind but also the pattern of turbulence.

In addition to influencing form and siting, wind must be taken into account in some aspects of the internal planning of buildings. It is clearly desirable to avoid locating doors near corners where velocities will be high. (Dustbins, similarly, should not be located at high velocity positions.) Sometimes entrance areas link doors on different facades, as shown in diagram 9, where pressures are very different. Unless revolving doors are used there may be substantial inconvenience. Exhausts and intakes for ventilation systems must also be considered in relation to pressures round the building. In recent years the velocities of air flow and the total resistance of ventilation systems have increased, thereby reducing the effect of wind but it is still important for proper performance that the exhaust and inlet should not be subject to pressure differentials which might result in excessive, or reduced flow. Positions on opposite faces of the building should be avoided for the reasons shown in diagram 12. One solution is to place both exhaust and intake on the same facade. If this is done it is desirable, particularly in high buildings, to have the intake below the exhaust since otherwise the natural buoyancy of the usually warm exhaust coupled with the updraught on the face of the building may convey exhaust air to the intake. A neat solution is to have the exhaust at roof level since this is normally a low pressure zone. The intake may then be on any facade. Diagram 13 shows these arrangements.

Chimneys carry effluent gases up to a level at which they will be adequately diluted before they reach ground level or other buildings. The chimney itself should therefore project beyond the zone immediately affected by the building. If it is high enough the velocity at which the flue gases leave the chimney may significantly aid proper dispersion. In the case of small buildings past performance of chimneys in similar circumstances is the best possible guide, although differences in topography and surrounding buildings must be taken into account. When siting chimneys in circumstances without precise precedent, and in all cases where large chimneys are involved, wind tunnel tests should be carried out.

The structural loading implications of wind are outside the scope of this book. Vibrations resulting from wind also require specialist consideration. It is, however, worth mentioning briefly that the variations in velocity due to turbulent flow result in pressure variations. If very rapid measurements are taken it is possible to show that the

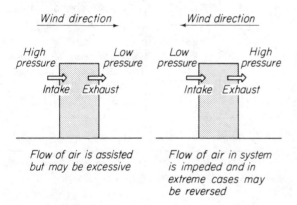

12 Bad position for ventilation intake and exhaust

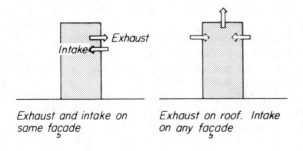

13 Good position for ventilation intake and exhaust

AIR MOVEMENT

power resulting varies with frequency. If a part of the building cladding or structure has a natural frequency which corresponds with a high power from the wind damage may result. Vibration can also be imparted to buildings by the shedding down the air stream of vortices (rotating masses of air) formed at the sides. Vortices are shed alternately from each side and the result is vibration of chimneys and towers and even parts of buildings themselves. Metal chimneys are clearly very much at risk from this phenomenon. The spiral strakes now seen round the top part of chimneys of this type represent a very successful solution developed by the National Physical Laboratory (see diagram 14).

such as Scandinavia, special controlled vents are often incorporated into window frames because operation of the normal opening light would give only crude control. Ventilation of this sort is designed on the basis of experience supplemented by regulations which call for opening lights equivalent

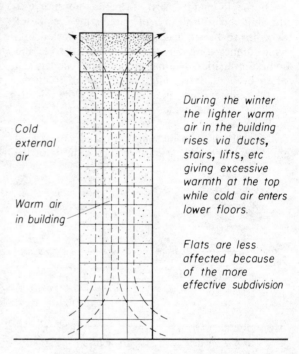

Cold external air

Warm air in building

During the winter the lighter warm air in the building rises via ducts, stairs, lifts, etc giving excessive warmth at the top while cold air enters lower floors.

Flats are less affected because of the more effective subdivision

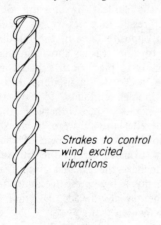

Strakes to control wind excited vibrations

14 Strakes on metal chimney

15 Stack effect in tall building

VENTILATION

Air change and movement within buildings, or ventilation is a vital aspect of design which has governed form throughout history. Acceptable electric light sources and electric fans have now given a situation where natural ventilation is not essential in every case, but the majority of buildings are still naturally ventilated.

Two factors control natural ventilation. One is the pressure variations due to wind. The other is the 'stack' effect which results from warm air in the building rising and being displaced by colder external air. In low buildings with small rooms it is easy to achieve acceptable levels of ventilation, and to control both wind and stack effects by means of windows with controllable opening lights. In countries with more severe climates than England,

to a percentage (in England and Wales 10% was traditional) of the floor area and for adequate space round buildings to allow air movement. BS CP 3 chapter 1 (c) Ventilation gives methods for estimating the ventilation which will take place through openings as a result of wind or stack effects. The results can be regarded as indicative only since it is very difficult to take into account the varying air pressures due to fluctuating wind effects and the topology and surrounding buildings and varying inside and outside temperatures.

In high buildings, above about 10 floors, wind and stack effects become very marked. In the winter the temperature difference between the cold external air and the air inside the building will be markedly different, often by as much as 20 °C. Under these conditions the warm air inside the building tends to rise through stair wells, lift shaft, ducts and any other openings, making the top of the building excessively hot and the bottom unacceptably cold. Flats do not suffer from this problem because of the relatively effective subdivision into individual dwellings but high buildings of other types have to be mechanically ventilated to overcome the problem.

Ventilation criteria

The basic human requirement for fresh air in order to gain an adequate supply of oxygen is well known. In buildings however this never forms a standard for ventilation since there would be acute discomfort long before any danger to life arose. In fact ventilation standards are based on keeping various types of contamination of the air or overheating to acceptable levels. The principal factors which govern ventilation rates in various circumstances and provide criteria for the performance of ventilation systems are:

Air movement

Usually associated with means for ventilation although it could be separated. Some degree of air movement is essential for feelings of freshness and comfort. Desirable speeds vary with temperature and conditions. In domestic buildings and other similar situations a velocity of 0.10-0.33 m/s is considered reasonable.

Body odour (including smoking)

This is the usual criterion where no other special factors apply and standard ventilation rates for most rooms are based on keeping body odour at an unnoticable level.

Fumes, smell, products of combustion

Where these are likely to be offensive, or injurious they should be removed by extraction at source. It is neither satisfactory nor economical to allow the fumes to enter the whole volume of the room. Hoods or extracts should be placed to capture the fumes as soon as they are emitted. High velocities are often required to ensure that the fumes are entrained and this may result in inevitably high ventilation rates in associated rooms unless the source of fumes is enclosed and provided with its own air supply as sometimes occurs in fume cupboards.

Bacteria

Special precautions may be required in hospitals or other places where bacterial concentrations could be critical. The precautions may consist of specially high air change rates and minimisation or avoidance of recirculation. In high risk areas air may be disinfected by bactericidal sprays or ultra violet light. In the case of operating theatres or sterile areas mechanical ventilation may be used to produce special patterns of air movement so that bacteria are swept away from areas where they might do harm.

Excess heat

High rates of ventilation may be used to remove excess heat (1 m^3 can convey 1.3 kJ for each °C of temperature difference between the heated control and the outside air). The method of calculation is described in chapter 7. Radiant heat cannot be controlled in this way. Where high levels of electric lighting are provided it is often necessary to extract air through the lighting fittings so that the heat from them is removed at source and does not escape into the room.

Relative humidity

It is usually considered that relative humidities between 30% and 70% are acceptable from the point of view of comfort and health and little effort is normally made to control humidity except in air conditioned buildings. Levels as high as 70% would, in almost all cases, require substantial vapour inputs within the building and would almost certainly be associated with condensation and mould growth if the vapour inputs cannot be controlled. One of the best ways to reduce high levels of relative humidity is to increase ventilation rates. Very low levels can be achieved during the winter when, on cold dry days, low levels of humidity externally can result in even lower levels when the air is heated. If air at 0 °C and 70% RH is raised to 20 °C the RH will drop to about 18%. Relative humidities of this order can give rise to complaints of dry throats and will cause woodwork to shrink and crack.

In some industrial buildings high levels of relative humidity are caused by or required for special processes. Under the provisions of the Factories Act 1961 the Minister may make regulations covering ventilation rates. In the case of factory rooms where humidification takes place the first schedule of the Act appears to indicate that relative humidities of up to 85-87% are admissible. Clearly in these cases special precautions will be required against condensation.

In modern buildings there is an increasing number of cases where natural ventilation does not give satisfactory conditions and mechanical ventilation must be employed. The main situations which call for the use of mechanical ventilation may be tabulated:

1 Internal rooms.
2 Large closely populated rooms where distribution of natural ventilation would be inadequate (BCSP 3 Chapter I (c) *Ventilation* suggests all rooms occupied by more than 50 people).
3 Rooms where volume per occupant is too low for efficient natural ventilation. (BCSP 3 Chapter I (c) *Ventilation* suggests under 3.5 m^3 per person.)
4 In cases where specially close control of environment is required, particularly in relation to relative humidity and dust particles in the air.
5 Where natural ventilation is impossible as windows cannot be opened because of external atmospheric pollution or noise.
6 In tall buildings where wind and stack effects would render natural ventilation impracticable.
7 Extract ventilation (or an excess of extract over input) may be required to deal with fumes or smells from cooking or other special processes.

Table 6 Recommended minimum rates of fresh-air supply to buildings for human habitation

Types of buildings	Recommended minimum rates of fresh-air supply to buildings
Assembly halls	28 m^3 per hour per person
Canteens	28 m^3 per hour per person
Factories and workshops*	
Work rooms	22·6 m^3 per hour per person
Lavatories and WCs	2 air changes per hour
Hospitals Operating theatres and X-ray rooms	10 air changes per hour
Wards	3 air changes per hour
Houses and flats	
Bathroom and WCs	2 air changes per hour
Halls and passage	1 air change per hour
Kitchens	56 m^3 per hour
Living rooms & bedrooms:	
8.5 m^3 per person	20·5 m^3 per hour per person
11.5 m^3 per person	18·5 m^3 per hour per person
14 m^3 per person	12 m^3 per hour per person
Pantries and larders	2 air changes per hour
Places of entertainment	28 m^3 per hour per person
Restaurants	28 m^3 per hour per person
Schools Occupied rooms (classrooms, laboratories, practical rooms, etc.):	
2.8 m^3 per person	42 m^3 per hour per person
55.6 m^3 per person	28 m^3 per hour per person
8.5 m^3 per person	20·5 m^3 per hour per person
11.2 m^3 per person	18·5 m^3 per hour per person
14 m^3 per person	12 m^3 per hour per person
Cloakrooms	3 air changes per hour
Corridors, lavatories and WCs	2 air changes per hour

*The conditions in factories are regulated by the Factories Acts, and regulations made thereunder From BSCP 3, Chapter I (c) *Ventilation*

4 Daylighting

The sun is important in building design for several reasons. It is the source of natural lighting; the heat from its rays may be utilised to advantage or give rise to severe discomfort through over-heating; and its penetration into buildings, at least in northern latitudes, is a strongly felt need for many people. Until recently the health-giving properties of the sun's rays were thought important but medical opinion has changed considerably on this issue and in addition the germicidal effects of sunshine in interiors are not now considered significant in relation to modern cleaning methods. Another aspect of building design which is strongly affected by the sun is the shading of areas round buildings which can affect both the vegetation and the human enjoyment of adjacent open spaces. The critical effect of these factors can be seen in almost all buildings. The depth, shape and spacing of buildings is largely governed by daylighting considerations. In some types of buildings, particularly those with large glazing areas, acute discomfort due to glare or overheating can occur unless careful design consideration is given to the penetration of the sun's rays.

In past ages when the seasons influenced everyone's life to a very great extent, knowledge of the sun's behaviour was almost universal. Nowadays artificial lighting, central heating, food preservation, roads passable in all weathers and occupations largely indoors make us very much less conscious of the basic pattern of the sun's movement. In view of the importance of this factor in modern building design it is worth devoting careful attention to the way in which the sun's rays behave. There are two cases to consider: light from the overcast sky and direct rays of the sun.

LIGHT FROM THE OVERCAST SKY

In some parts of the world where clear skies are the rule it is the direct rays of the sun which must be taken into account for daylighting analysis but in Britain and many other areas of the world critical daylighting conditions arise when the sky is covered by clouds and it is this condition which must be taken into account in design. The variation in illumination received from the sky varies very greatly according to the movement of clouds and sun. The assumption made as a basis for daylighting analysis is of a heavily overcast sky giving a total unobstructed illumination (on an horizontal plane) at ground level of 5000 lux. This condition is described as the Standard Overcast Sky and measurements show that between 8.00 hours and 17.00 hours the illumination from the sky will be 5000 lux or more for 85 per cent of the time.

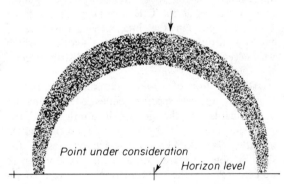

16 *Graphical representation of the variation of brightness in the overcast (CIE) sky*

The luminance of the sky is not uniform all over. It is some three times as bright at the zenith as at the horizon. An equation defining a luminance distribution which has been proved to represent conditions in many parts of the world was proposed by Moon and Spencer and adopted in 1955 by the Commission Internationale de l'Eclairage

DAYLIGHTING

Table 7 Minimum daylighting standards

Room	Daylight factor*	Penetration	Daylight area
Kitchen	2 per cent	Cooker, sink and preparation table should be placed within the daylit area	50 per cent of area with minimum 5 m²
Living room	1 per cent	Half the depth facing the main window	7.5 m²
Bedroom	0.5 per cent	Three-quarters depth facing the window	6 m²

Recommended minimum standards of daylighting for dwellings from BSCP 3 Chapter 1 *Lighting*, Part 1 *Daylight*. BSI, 1964

Offices	1 per cent	4m	not specified
Drawing offices	5 per cent	over whole area of office	

Recommended minimum standards of daylighting for offices from PWBS No 30 *The Lighting of Office Buildings* HMSO 1952
*Measured or working plane

(CIE) as a standard for use in design.* It is sometimes referred to as the Moon and Spencer and sometimes as the CIE sky. Diagram 16 shows a section through the imaginary hemisphere of the sky surrounded by a band, the width of which varies according to the luminance distribution of the CIE sky.

The variation of outdoor illumination makes it unrealistic to use specific values of illumination as standards for the design of interiors and the problem is overcome by the use of a ratio, known as the Daylight Factor, which expresses the illumination received at a point indoors as a percentage of the illumination received simultaneously by an unobstructed point out of doors. The daylight factor is made up of three compon-

CIE sky. Originally in daylighting estimation the sky was assumed to have uniform brightness. A more realistic distribution of brightness which reduces oversizing of rooflights and undersizing of low windows and which has been established as occurring in many parts of the world is defined by the formula

$$B_\theta = B_z \frac{1 + 2 \sin \theta}{3}$$

θ is altitude
where B_θ is the sky brightness at altitude θ
B_z is the sky brightness at the zenith.

ents. The direct light from the sky falling on the point in question (the sky component), the reflected light from external surfaces (the externally reflected component), and light received by reflection from the internal surfaces of the room (the internally reflected component). It is important to bear in mind that in the early stages of daylighting analysis the importance of the reflected components and the variation of sky luminance was not fully appreciated and that the term Daylight Factor will sometimes be found used as synonymous with sky component.

Table 7 shows some recommended minimum standards of daylighting for dwellings and offices.
The daylight factors for dwellings are to be based on the following assumptions of reflection factors.

Walls   40 per cent
Floor   15 per cent
Ceiling 70 per cent

The Illuminating Engineering Society have issued new recommendations for daylighting in 1973 (IES Code). They differ slightly from those quoted above but are less detailed.

Daylight factors are normally measured or

LIGHT FROM OVERCAST SKY

estimated in terms of the light falling on a horizontal plane 0.85 m above ground or at the level of the working plane if this is different from 0.85 m.

The intensity of light falling on a horizontal surface is not necessarily a complete measure of satisfactory lighting. In rooms enjoying daylighting from windows, however, much of the light penetrates almost horizontally thus ensuring that, if the working plane is adequately illuminated, the walls will be well lit, giving a pleasant interior. The general direction of the light also gives a pleasant definition to the form of three dimensional objects. These conditions are not met in top lit rooms where, if the same standard is applied, the walls will be poorly lit and a gloomy environment achieved. This problem is normally overcome by requiring the daylight factor to be very much higher than would be the case with side lit rooms. Top lit factories should have a minimum daylight factor of 5%. The roof glazing required to achieve this may, if not protected from the sun, give rise to acute overheating. The traditional north light roof is a response to this problem.

Calculations of the glare from side lit windows are now possible (see IES Code and Technical Report no. 4) but they are rarely made in practice.

A wide variety of techniques is available for daylighting analysis in buildings. They range from simple tables giving values of daylighting for specific cases through many types of numerical or graphical analysis up to the measurement and often the visual evaluation of lighting conditions in models made specially for daylighting study. For architectural design, while work is on the drawing board, a very widely used method of analysis is the use of the Building Research Station daylight protractors. The following notes set out the principles and a procedure for use of the protractors and the presentation of the resulting information.

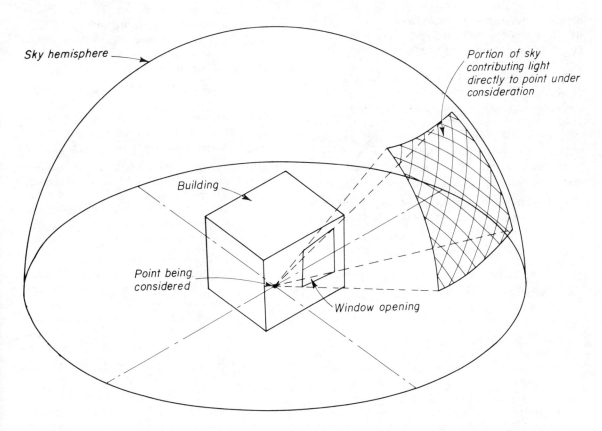

17 *Principles of daylight penetration directly from sky through window*

DAYLIGHTING

Use of BRE sky component protractors

General principles

Source of light
The overcast sky

Minimum design standards are based on an illumination of 5000 lux from the whole unobstructed sky. The brightness of the sky is assumed to vary from zenith to horizon in accordance with the formula adopted by the CIE.

Light reaching any point within a room will come

(i) directly from the sky via windows or other opening (sky component). See diagram 17.

(ii) By reflection from landscape and buildings outside (external reflected component).

(iii) By reflection from other surfaces in the room (internal reflected component).

The measure of daylighting is the daylight factor which gives as a percentage the relationship between the light actually falling on a point within a building and the light which would fall on that point from the unobstructed sky. It is expressed as a percentage.

The daylight factor comprises the internal reflected component plus the external reflected component and the sky component, if any.

The BRS protractors (second series) compute the sky component and enable the external reflected component to be established.

In order to give their answers the sky component protractors

(i) Establish the proportion of the whole sky which contributes light directly to the point under consideration.

(ii) Make allowance for effect of glazing.

(iii) Make allowance for the angle of incidence of light with the working plane.

The protractors are designed to be used on drawings.

Plans and sections of the rooms concerned including details of windows and of any external obstructions and their relationships to the window are required.

Since the behaviour of light is consistent, irrespective of scale, any conveniently sized drawings may be used. Plan and section need not be to the same scale.

There are five protractors for use (available from Government bookshops):

Vertical glazing	protractor 2
Horizontal glazing	protractor 4
30° glazing	protractor 6
60° glazing	protractor 8
Unglazed apertures	protractor 10

In addition, protractors are available based on a sky of uniform brightness.

Methods of analysis

The information required about daylighting will usually be either:

(a) Value of daylighting at one or more particular spots known to be critical

or

(b) The general distribution of daylighting across the working plane in a room. This will involve the establishment of daylighting values at a grid of points covering the workling plane. The following should be borne in mind in laying out the grid.

(i) A rectangular grid is uaually convenient. spacing will depend on the size of the room and closeness of analysis required. Generally 0.75 m to 1.0 m spacing is convenient in rooms up to 200 m^2

(ii) Space adjacent to walls may sometimes be known to be occupied by cupboards or circulation and need not be considered. Where this is not the case it will often be convenient to fix grid points near the wall one half grid space away.

(iii) Work may be saved if grid points are arranged symmetrically about the centre-line of the window (or the main window if there are more than one) and in lines parallel to the window. Provided there are no symmetrical obstructions values need only be computed for half the points and values on section will be identical for each line.

(iv) For easy identification of points grid lines in one direction should be labelled A, B, C, etc, and those in the other direction 1, 2, 3, etc.

USE OF PROTRACTORS

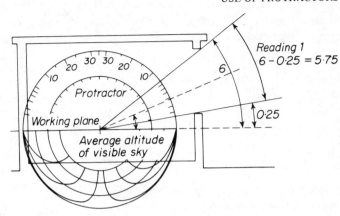

18 *Use of BRS Sky Component Protractor*

Detailed use of protractors

Use of protractor on section (vertical glazing used as example). On a section of the room normal to the window mark the working plane and the point under consideration. Draw lines from the grid point under consideration to the highest and lowest limits at which the sky can be seen from the grid point. Place the base of the sky component for long windows side of the protractor along the working plane and the centre over the grid point. Read off the value as shown in the diagram. The lowest limit may be the sill, or the skyline of the landscape or surrounding buildings if this can be seen above the sill from the point being examined.

The resulting value is the sky component which would be received at the point from an infinitely long window having the section indicated. A correction must be made for the limited length on plan of normal window. Note also, using the normal degree scale protractor the mean altitude of the visible sky.

Use of protractor on plan. On a plan of the room draw lines from the grid point to the edges of the visible sky. Place the correction factor side of the protractor with its base parallel to the window and its centre over the point under consideration. Using the mean altitude found choose the appropriate concentric scale and read the correction factors for each side of the window. Add them to obtain a final correction factor where readings are on opposite sides of the centre zero. Where both readings are on the same side subtract the smaller from the larger.

To establish the sky component of the daylight factor multiply the value for long apertures found on section by the plan correction factor.

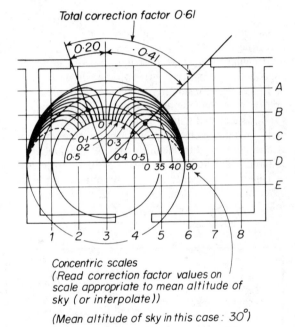

Concentric scales
(Read correction factor values on scale appropriate to mean altitude of sky (or interpolate))

(Mean altitude of sky in this case: 30°)

19 *Use of BRS Sky Component Protractor on plan*

In the example shown in the diagrams

$$5.75 \times 0.61 = 3.51$$

When more than one window contributes light:

(i) In the same wall as the main window. Add the plan correction factors for each window and multiply the section value by this sum. See diagram 20.

(ii) In other walls or roof: compute separate direct components and add.

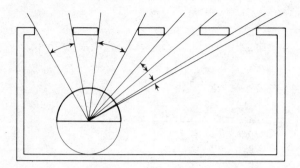

20 BRS Sky Component Protractors: Plan correction for several windows

Glazing bars may be allowed for by making a percentage reduction of light received.

Irregular external obstructions or non-rectangular windows may often be dealt with satisfactorily by considering them as equivalent to one or more rectangular windows. The rectangular windows assumed will be determined visually.

If window cleaning will be infrequent or atmosphere is dust-laden, estimate and make a suitable percentage reduction of light received.

Direct components can be computed for planes other than horizontal. Consideration of the basic geometry in relation to the particular problem will usually reveal a method for solution.

In practice it will usually be found unnecessary to draw the lines delineating the edges of visible sky. Readings can be made by using a set square or straight edge. In school work, however, lines should be shown to demonstrate correctness of method.

Light from the landscape outside (the external reflected component can be computed if its brightness relative to the sky it obstructs is known. In practice it is found that obstructions have an average brightness of about one-tenth of the average sky brightness. A satisfactory allowance will be made if a conversion factor of 0·1 is used to reduce the appropriate sky component. A separate set of protractor readings with appropriate average altitude should be made.

It is usually convenient to carry out the calculations in a tabular form. (See diagram 21)

Presentation of data

Visualisation of distribution. The direct components computed will often be marked on the plan against their appropriate points. Interpretation and understanding of the distribution can

a	b	c *	d *	e *	f *	g	h
Grid reference	Window reference	Mean altitude of visible sky from grid point	Section protractor value	Plan protractor value	Sky component d x e	Internal reflected component (from nomogram)	Daylight factor f + g

* For external reflected components appropriate values should be entered in c, d and e, and the result of d x e multiplied by 0.1 and entered in column f

21 Typical tabular layout of daylight calculation data

usually be assisted, particularly for laymen, by graphical methods.

(a) 'Contour' plans of distribution. Diagram 23 shows a typical contour plan of daylighting distribution which will be self-explanatory. More complex distribution patterns would result from additional windows or rooflights, but no special difficulty would arise from this.

(b) Graphs of intensity. A visual indication of variation in intensity can be given by drawing on section a graph of the varying intensity across the room. The horizontal scale will be that of the room. The extent of the vertical scale can be determined by convenience in including the curve within the section, while its subdivision between the upper and lower limits of daylighting forming the extremes of the scale will be logarithmic to make the graph variation correspond more closely with the sensitivity of the human eye (suitable logarithmic scales can usually be projected from a slide rule). Diagrams 16 and 17 show typical examples.

Internal reflected component

The Building Research Station have developed a series of nomograms which enable the internal reflected component of the daylight factor to be determined quickly. Separate nomograms are required for side-lit and top-lit rooms.

Diagrams 24 and 25 show nomograms to give the average and minimum values of the internal reflected component for side-lit rooms. The scales are self-explanatory.

The average reflection factor can be determined by summing the products of the areas of each different surface and their reflection factors and dividing this sum by the total area. (The reflection factor of windows may be taken as 0.15.)

Use of the nomograms involves joining appropriate points in scales A and B and reading the unobstructed internal reflected component on scale C. If there are external obstructions they can be taken into account by joining the angle of the obstruction on scale D to the unobstructed value already found on scale C and projecting to cut scale E at the required value.

The decision to use average or minimum nomograms depends on the purpose for which the

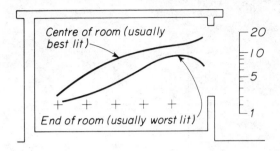

22 Sections showing daylighting distribution

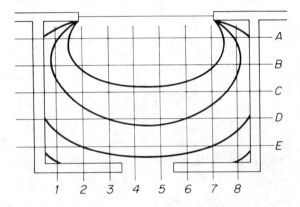

23 Contour plans showing daylight distribution

results are required and the form and window position in the room.

Waldram diagram for daylighting analysis

The Waldram diagram was the main method of daylighting analysis before the development of the BRS protractors and it still has applications where it is difficult to organise the visible sky into rectangular units.

Half of the hemisphere of sky is mapped on to a rectangular chart (diagram 26) so that the area of any portion of the diagram has the same relationship to the whole areas as the light contribution for the part of the sky represented has to half the light from the whole sky.

To make use of the diagram, the outlines of the sky seen through window openings or above complicated roof lines are plotted on the diagram, the area enclosed measured and expressed as a percentage of twice the total area of the diagram

DAYLIGHTING

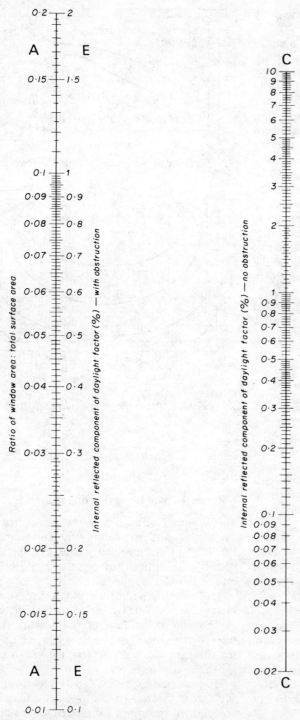
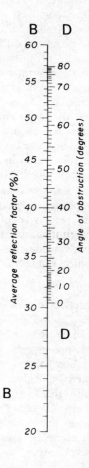

24 *Average Internal Reflected Component (side lit rooms)*
Reproduced from *Architectural Physics: Lighting*

INTERNAL REFLECTED COMPONENT

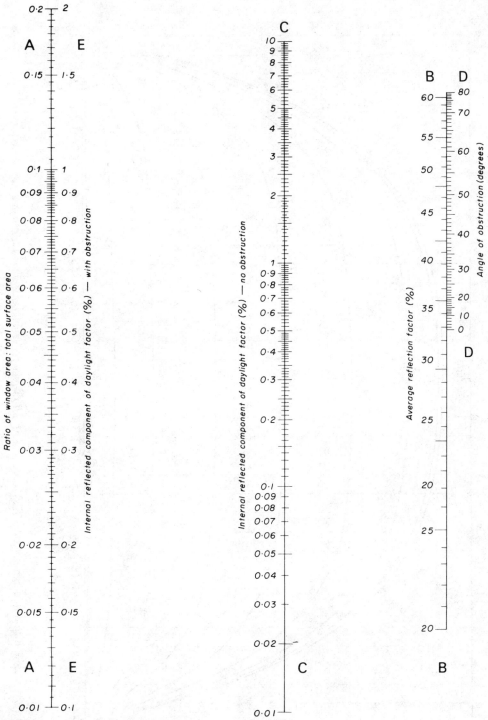

25 *Minimum Internal Reflected Component (side lit rooms)*
Reproduced from *Architectural Physics: Lighting*

DAYLIGHTING

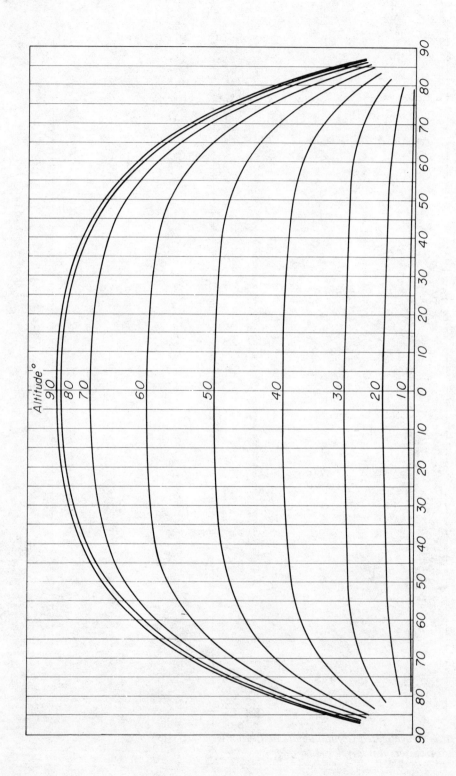

26 Waldram diagram (for CIE sky and making allowance for light transmission through clear window glass)

DROOP LINE DIAGRAM

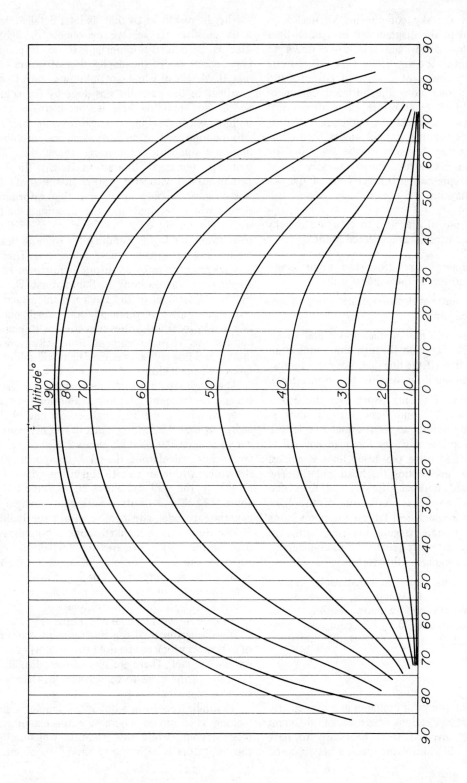

27 *Droop line diagram for use with Waldram diagram*

thereby giving the sky component. Vertical lines remain vertical in the diagram but horizontal lines such as window heads and sills curve down to vanishing points. The laborious task of plotting these curves by means of a series of altitudes and azimuths of points along the horizontal member can be simplified by the use of a droop line diagram (27) which shows the curves representing horizontal lines running at right angles to the line of sight. Measuring the areas can be simplified by the use of a sheet divided into rectangles by the area of each representing 0.1 per cent of the sky component (diagram 28).

Method of use of diagrams to estimate the sky component of a window at a selected point.

1 On plan draw a line from the point being considered normal to the window wall.

2 Measure azimuth angles of vertical edges of sky seen between jambs.

3 Measure the altitude of the upper and lower limits of the sky seen through the window on the normal line (if the line does not cut window, project head and sill heights to normal line).

4 Place a piece of tracing paper over the droop line (diagram 27) and draw vertical lines at appropriate azimuth angles (normal line is 0°). Select appropriate altitudes for head and sill at 0° azimuth and using the droop lines as guides, trace curves representing the head and sill. The area enclosed between the vertical lines and curves represents the proportion of the light from the sky which will fall on the point being considered. (Where irregular obstructs exist the outline maybe transferred to the basic diagram, and altitudes and azimuths plotted.)

5 Transfer tracing paper to calculation diagram 28 and count the number of rectangles enclosed within the window outline. Each rectangle represents 0.1 per cent sky component at 0.01 per cent external reflected component.

Block spacing
When laying out groups of buildings and particularly in town planning situations where sites in different ownership are involved it is necessary to have criteria which will control the spacing between and the height of blocks so that, in later detailed design it is possible to achieve reasonable daylighting. This is particularly important in town planning applications where the daylighting of future buildings, the details of which are unknown, must be safeguarded. In the past this was done by fixing street widths and building heights, or at the rear, by building heights, taking an angle from the first floor cills on one side of an open area and prohibiting buildings opposite from rising above this. While buildings remained limited in height (in Great Britain 80'-0" with two storeys in the roof) there was no great problem apart from the uniformity of streets which resulted and the dreariness of light wells. The advent of desire for more sophisticated town forms and higher buildings led to more flexible methods of ensuirng adequate daylighting. These do not take the form of statutory requirements. The government issues a booklet, Sunlight and Daylight by the Department of the Environment, which sets out authoritative recommendations which may be adopted by local authorities concerned with planning applications. These recommendations will ensure that. in any building layout adequate light will fall on the face of all the buildings.

There are separate recommendations for residential and non-residential buildings. For non-residential cases all sides of the buildings, except end or flank walls less than 15 m long, must have a sky component of at least 0.97% at all points 2 m above ground level. For calculating the sky component only sky visible within an area 45° on either side of a line normal to the wall and above 20° altitude but not above 40°. These limits ensure that daylight can penetrate to adequate depths within the building. Similar tests can be applied along a boundary (or road centre line) to preserve the daylighting for an adjacent site to be developed. In this case it is ensured that an appropriate amount of daylight falls, from the inside of the site being developed on to all points 2 m above the boundary lines.

Table 8 shows the range of criteria.

The Department of the Environment issues a set of indicators which can be used to check whether the criteria are met. There are sets of special indicators for each of the cases to be considered. Diagram 29 shows a typical indicator.

The indicators form a quick and simple means of making a check on daylighting for planning and block layout. While any situation which satisfies the indicators is satisfactory there can be circum-

CALCULATION SHEET

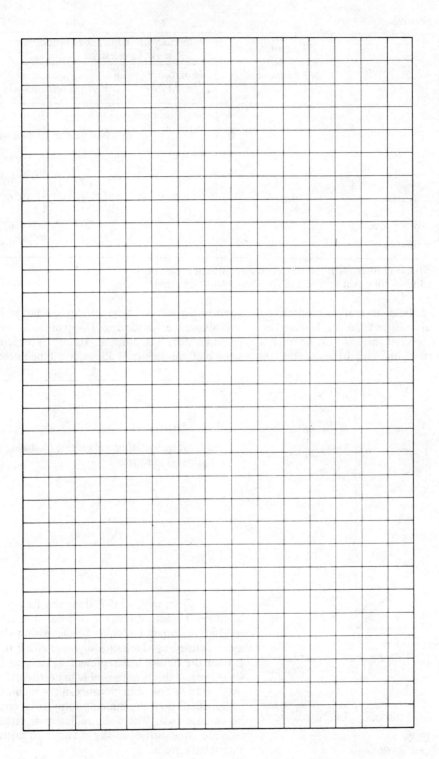

28 *Calculation sheet for use with Waldram diagram. Each rectangle represents 0.1% Sky Component or 0.01% External Reflected Component*

DAYLIGHTING

Table 8 Daylighting criteria

Type of development	Application	Sky component % (2 m above GL)	Limits within which sky component must be achieved		
			Horizontal (azimuth)	Vertical (altitude)	
			angle on both sides of normal line to wall	Not below	Not above
Residential	Block spacing*	0.84	45°	10°	30°
	Boundary	4.3	65°	19°26'	49°6'
Non-residential	Block spacing	0.97	45°	20°	40°
	Boundary	2.9	65°	36°3'	59°13'

* facades of residential buildings facing south (or in any direction east or west of south, including due west) are tested for sunlight not daylight

stances which it is difficult to analyse precisely by means of the indicators. The Waldram diagram can be used to give a precise analysis of whether a particular building configuration satisfies the criteria given in table 8. A diagram is required not including any allowance for glazing. Diagram 30 is specially prepared for this purpose. Its use is exactly the same method described above. It will be noted that,

Diagram

Original scale 1:200
Scale as reproduced 1:400

29 Residential block spacing (building to building) indicator

Note: From any point P along the facade of the building, it must be possible to adjust the indicator so that no other buildings falling within the segment defined by the indicator come closer to point P than the height line representing its own height (less 2 m). If this is not possible indicators D1 or D3 may be tried but, if in no case can the requirements be satisfied, then the building form or layout must be reconsidered. Where the indicator is satisfied it indicates that sufficient sky is visible to satisfy the the stated criteria.

WALDRAM DIAGRAM

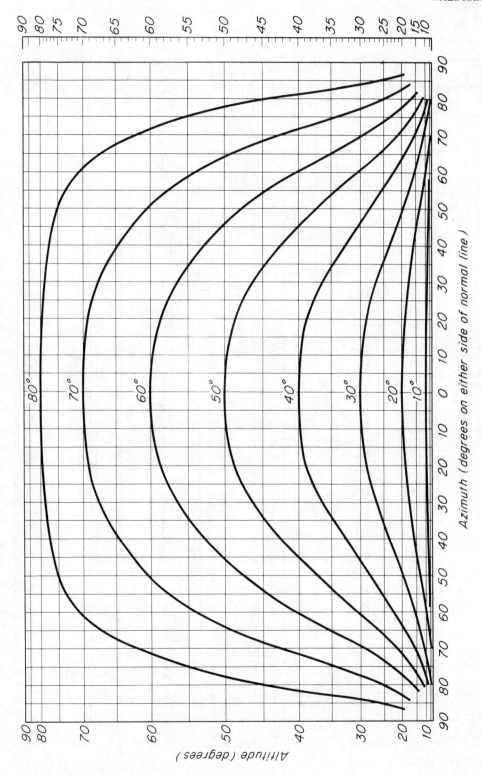

30 Waldram diagram for CIE sky with light falling on horizontal place (with droop lines for plotting horizontal obstructions)

DAYLIGHTING

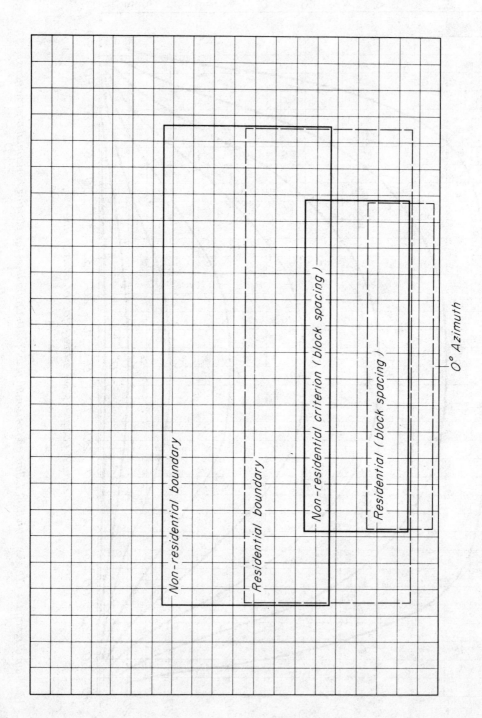

31 Calculation sheet for use with Waldram diagram, marked with limits for the satisfaction of non-residential boundary and block spacing criterion. Each square represent 0.1 per cent sky component. (For satisfaction of daylighting criterion only squares within appropriate rectangles may be counted)

in this case, the droop lines are superimposed on the diagram itself. Diagram 31 is a calculation sheet with the limits of altitude and azimuth for the various applications indicated. Only squares within the outlines defining the altitude and azimuth limits may be counted for satisfaction of the daylighting criteria. Diagrams 30 and 31 may be used for normal analysis of daylight penetration for any situation where there is no glazing. In this case the whole of the calculation sheet may be employed.

MODEL ANALYSIS

It is difficult for graphical methods of daylighting analysis to cope with very complex geometrical situations, particularly if many repetitions are needed. Mathematical and computer methods can overcome the problems of repetition but complex forms, varying colours and complicated glazing bars and mullions still present difficulties if precise results are required. In this circumstances model studies are of considerable value. Light behaves in exactly the same way in a scale model, provided it is accurately made, as in full size and there is the important additional benefit that visual inspection can also be made to assess the effects of colour and appearance. There seems little doubt that, in many cases, where close estimates of daylighting levels are required, visual appraisal of the actual appearance resulting will also be important.

To permit easy visual inspection models of interiors should be reasonably large scale. For small rooms 1:10 would be suitable.

A suitable lightmeter is needed to assess the daylight factor in a model. The BRS Daylight meter or the more elaborate BRS/EEL Daylight photometer are the most widely used instruments. Readings are taken inside the model and compared with readings from the whole unobstructed sky to give the percentage of the total available light actually falling on the point investigated.

The natural sky may be used as a source of light, but a position must be found with an unobstructed view of the sky and, to establish the daylight factor, a uniformly cloudy sky must be used. Both these conditions are difficult to meet and model studies are usually carried out in artificial skies. The most common pattern is the mirror type sky shown in diagram 32. Few design offices are likely to possess artificial skies but many building research organizations and schools of architecture have them and make them available for use by designers.

DIRECT SUNLIGHT

Although it is at present not possible to produce any definite data of human requirements for sun penetration into buildings it is clear that some degree of sun penetration is a strongly felt need on the part of many people. This is recognised in the building codes of a number of countries. Table 9 quotes the recommendations for Britain given in BSCP Chapter I (b) *Sunlight*. More recently, however, the recommendations made in Sunlight and Daylight by the Department of the Environment is that all sides of residential buildings having any southern orientation (ie any orientation from just south of east, through south and including due west) should have at least one hour of sunshine on 1st March (above an altitude of 10°). This is a much less stringent recommendation than that in the code of practice but its intention is to provide a realistic standard capable of achievement which takes into account the external as well as the internal environment (sunshine bearing less than 22.5° from the side of the building is not excluded since it contributes to pleasant external conditions). It is to be hoped that the Department of the Environment intention will be achieved and that residential developments will conform to its new suggestions.

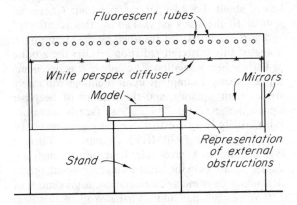

32 Mirror type artificial sky

Note The interflections between the mirrors make the perspex diffuser appear to extend to infinity and can be arranged to give a distribution of light similar to the CIE sky.

DAYLIGHTING

Table 9 Minimum sunlighting standards

Room	Preferable time of day for sun penetration	Minimum period for penetration
Living	Afternoon	At least one hour, at some period of the day, during not less than 10 months of the year from February to November
Bedroom	Morning	
Kitchen	Morning	

Recommended minimum standards for sunlight penetration for dwellings from BSCP Chapter I (b) *Sunlight* (Dwellings and Schools)

It must be remembered that solar radiation has problems as well as advantages and the penetration of the sun's rays into a building can give rise to acute discomfort because of glare. Direct sunshine can give illumination levels on working planes in buildings of 60 000 lux or more which may be compared with 300-500 lux which might represent a good level of artificial lighting or daylighting away from windows. The eye is able to accommodate itself over a very wide range of illumination levels. Moon quotes a range of adaptation of 100 000 to 0.1 lux at 1 000 000 to 1 but the eye can only be adapted to one level at a time and patches of bright sunshine can create a bright and cheerful environment or may cause intolerable glare. Careful consideration of form, fenestration and orientation in relation to sunshine is vital in the design of buildings and may have to be supplemented when necessary by screens, louvres or blinds to control the degree of sun penetration.

Sunshine entry to rooms is not taken into account until the horizontal angle between the sun's rays and the plane of the window is more than 22.5° and more than 5° above the horizon.

Solar heat gain

Overheating due to solar gain has not, in the past, been a serious problem in Britain. Buildings had massive walls with small openings which minimised the direct effect of the sun's rays on the interior. Since heights were limited vegetation and adjacent buildings provided additional screening. Many modern buildings are quite different in their reaction to solar heat. Large proportions of the walls are glazed, the construction and cladding are lightweight and quickly warm up and tall buildings are not shaded by trees and other buildings. These changes, coupled with the property of glass which enables the high temperature radiation of the sun to pass through while preventing the passage of the lower temperature radiation from within the building, mean that acute overheating is now encountered in buildings. The situation is made worse by increased external noise levels which made it difficult to have windows open and so dissipate the heat. In tall buildings ventilation problems may have dictated the use of mechanical ventilation and the windows may not even be openable.

The solar radiation striking the earth's atmosphere brings about 1.5 kW per m^2 per hour (measured normal to the sun's rays). Part of this is reflected and part scattered and absorbed by the atmosphere. Some of this scattered radiation reaches the ground as Diffuse Sky Radiation. The radiation which penetrates directly to impinge upon buildings will rarely strike them squarely and the amount of heat per square metre of the building's surface is therefore substantially less than that originally present in the sun's rays. The 1970 IHVE's *Guide to Current Practice,* Book A, gives solar heat gain through unshaded windows in the United Kingdom which reach, depending on time and orientation, a maximum of 530 W per m^2 per hour. Virtually all of this heat will pass immediately through unshaded single glazed windows and contribute to warming the interior. Solar heat falling on walls will take time, of the order of some hours with most typical constructions, to pass through the wall. During this time external temperatures will have dropped and a substantial

DIRECT SUNLIGHT

part of the heat gain is likely to be lost externally. The small windows and massive walls of some types of tropical buildings show how buildings have been constructed to minimise the effects described.

In England an extension to a school has been constructed in which the heat gain from occupants, lighting and the sun made it possible to dispense with a normal heating installation. It is not possible to do without heating in many buildings but this example does demonstrate the importance of the sun's rays and the need to take them into account in design both to conserve energy requirements for heating and to avoid summer overheating. In section 5 hear a method for estimating solar heat gain is described.

Correct orientation of windows is critical if maximum thermal advantage is to be taken of solar heat gains. The amount of heat which enters a south facing window during the heating season is substantially greater than for any other orientation. If, as is usual in domestic buildings, the south facing windows are curtained at night during the period of maximum heat loss, more heat will enter an unobstructed south facing window during the winter than will be lost through it. This is not an argument for providing very large windows on southern elevations specifically to save energy since mean radiant temperatures internally will be reduced and although energy may be saved overall, larger heating installations will be needed for overcast periods. It is, however, a powerful argument for considering the orientation of buildings more carefully than at present and for planning internally so that where large windows are appropriate they may be concentrated on southern facades. It is important to remember that substantial obstructions will reduce the seasonal heat gain and that only elevations oriented very close to due south enjoy the full benefit.

Overheating in summer may be a risk in buildings with large south facing windows. The admittance method (page 92) may be used to establish the temperatures likely to be reached in summer. South facing windows can have relatively easy control of solar heating. Special window glasses to reject solar heat are not appropriate since they will limit winter gains. Internal blinds, although they prevent the sun's rays from falling directly upon occupants of rooms, allow the sun's heat to pass through the glass, heat the blind and thereby the room itself. External blinds are particularly effective as are blinds in ventilated spaces between glazing. The most convenient aspect, however, of solar control for south facing windows is that a relatively small projecting canopy over the window can give control of sun penetration in summer, while allowing full benefit from the rays in winter. Diagram 38 can aid in the design of this type of canopy.

Geometry of the sun's movement

Control of sun penetration from the point of view of either illumination or warmth depends initially on the form of the building in relation to the sun's movement. A clear appreciation of the geometry of this movement is essential in the early stages of design. In the final stages, very detailed analysis may be required not only of the angles of the sun's rays but also of the heat gain resulting. The basic relation-

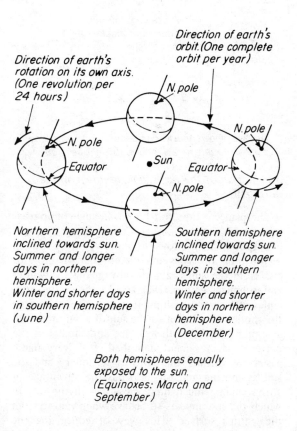

33 *Principles of earth and sun relative positions*

ship between earth and sun is that the earth moves round the sun making a complete orbit once a year. It is in fact the completion of the orbit which defines the year itself. As it moves along its orbit the earth revolves on its own axis exposing its whole surface to the sun's rays during each 24 hours and thus producing night and day. The axis is not normal

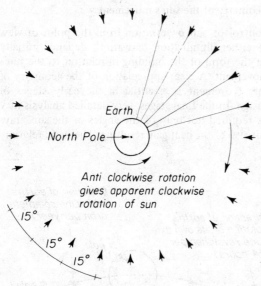

34 *A plan of the earth looking down on the north pole*

to the plane of the orbit but at an angle of approximately 22.5° from normal. Consequently opposing hemispheres incline towards and away from the sun as the orbit proceeds and thereby benefit more or less from the sun's rays thereby giving the winter and summer seasons. Diagram 33 shows the principles of the relationships. It is not easy to imagine the motion in relation to the sun of a building sited at a point on the earth's surface and subject to the patterns of movement described. It is very much easier to consider the earth as stationary and the sun as moving and this also has the advantage of corresponding much more closely with the way in which the sun appears to behave when viewed from the earth's surface. This way of conceiving the relationship of earth and sun is easily expressed geometrically. Diagram 34 shows a plan of the earth looking down on the north pole with arrows indicating the changes in direction of the sun's rays at hourly intervals. Diagram 35 shows a view of the earth from above the equator showing how the direction of the sun's rays resulting from the inclination of the earth at a particular time of year and the hourly movement of the sun throughout the day may both be represented by means of a solar ray cone. A series of such cones would be required in order to represent the apparent movements of the sun through a season because of the constantly

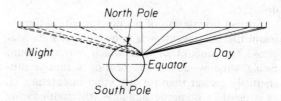

Solar ray cone showing apparent movement of the sun during one day for a position in northern hemisphere in summer

35 *A view of the earth from above the equator*

changing relationship of the earth's inclined axis. The angle, in relation to the equator at which the sun's rays fall on the earth through the year is known as the declination. Diagram 36 shows how the declination varies during the year. These diagrams show the principle of the apparent movement of the sun in relation to the earth. They do not, however, give a comprehensive picture of the apparent movement in relation to specific points on the earth's surface. Diagram 37 shows by means of lines on a hemisphere how the sun moves in relation to a point on the ground at 52° N latitude. The long lines running round the hemisphere represent the path of the sun at various times of year. The highest line representing the path of the sun in mid-June and the lowest and shortest the path for mid-December. The shorter cross-lines represent hours. The diagrams show a perfect symmetry of movement. In fact there are minor alterations. These are important in navigation but of no significance in building and the perfect geometry may be assumed for building design.

DIRECT SUNLIGHT

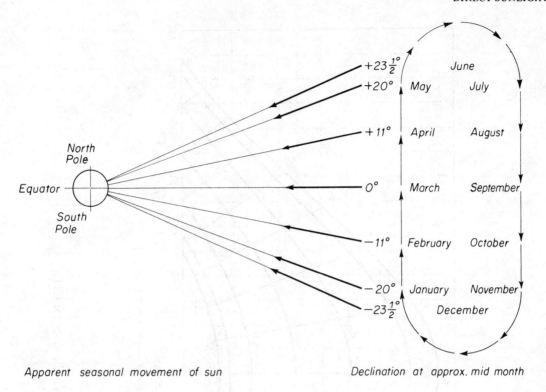

Apparent seasonal movement of sun   *Declination at approx. mid month*

36 *Showing how the declination varies during the year*

Diagram 38 is a sun path diagram for 52° N which may be used to give information about the angles of the sun. Azimuth angles are measured from 0 at due south.

The insolation of a particular point throughout the season may be established by projecting an outline of the window opening on to the diagram in a similar way to that used with the Waldram diagram. Since it is not convenient to have the droop lines marked on the sun path diagram itself a separate diagram is provided (diagram 39). A sheet of tracing paper is used to draw the window outline and then transferred to the sun path diagram. The normal line to the window is positioned on the Azimuth scale at a point corresponding to its own angle from south. Periods of insolation can be identified within the outline of the opening. Diagram 40 shows the principles involved. In addition to window opening it is possible to project the outline of buildings round open spaces to establish insolation there.

The chart is graduated horizontally in degrees measured from south (azimuth) and vertically in

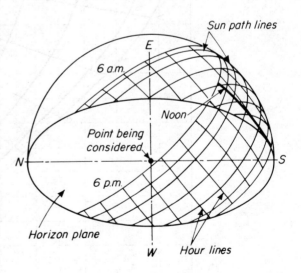

37 *Apparent movement of the sun*

DAYLIGHTING

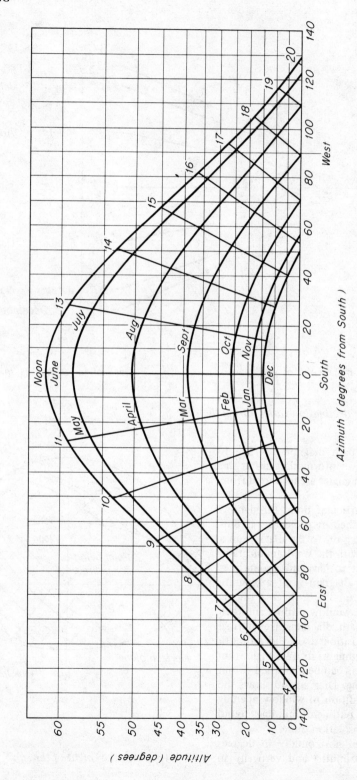

38 Sunpath diagram for 52°N latitude
Note All times are solar times — 12.00 noon due south. (Correction must be made for BST)

DIRECT DAYLIGHT

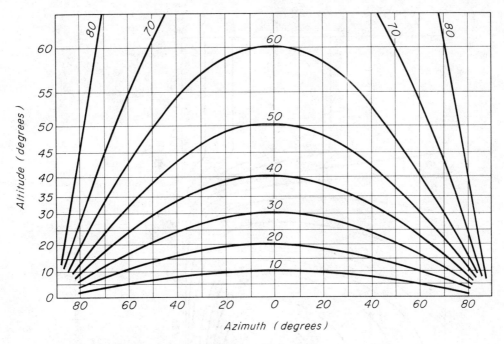

39 Droop line diagram for use with diagram 38

degrees of altitude. The curved lines represent the movement of the sun during one day at the middle of each month. The cross-lines represent hours in solar time or GMT. The slight variations between solar time and GMT are not significant for buildings. The diagram can be used with reasonable accuracy anywhere in England and Wales. (Noon altitude error is less than 1° at all latitudes between Southampton and Nottingham and less than 2° at York.)

Use of diagram

1   Azimuth and altitude of the sun for any time of day and year may be read from the chart. These

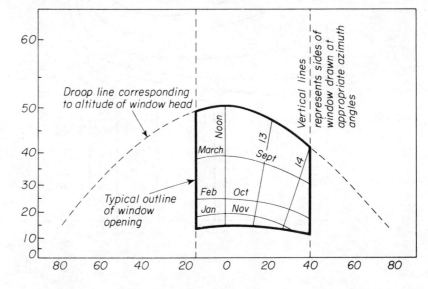

40 Method of use of sun path and droop line diagrams

61

DAYLIGHTING

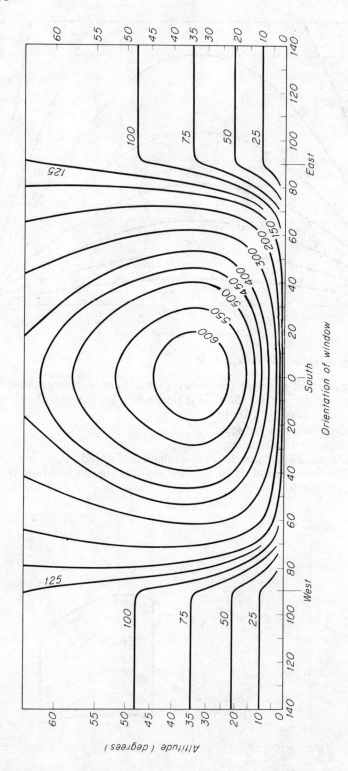

41 Solar gain through 4mm vertical glazing, W/m² (Tube used in conjunction with Diagram 38.)

SUNPATH DIAGRAM

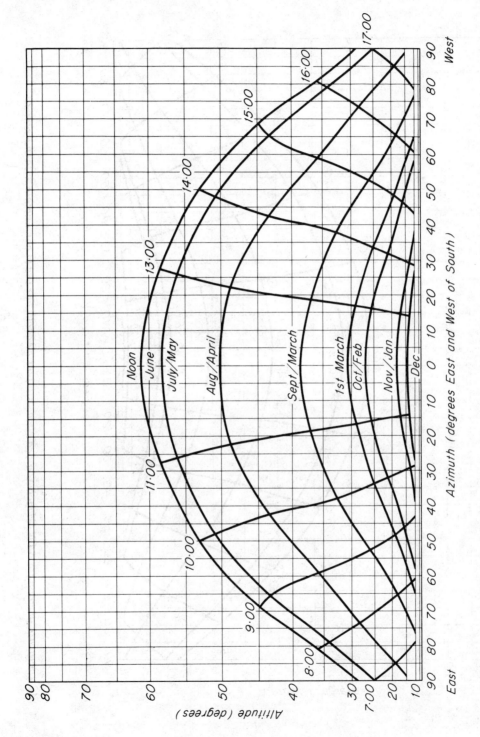

42 Sunpath diagram for 52° N (London, Birmingham and Southern England) on 1 March. For use in conjunction with Waldram diagram 30

DAYLIGHTING

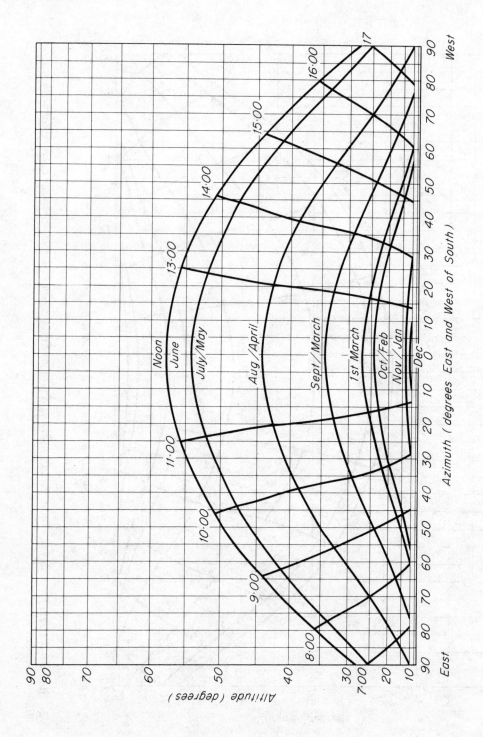

43 Sunpath diagram for 54 N (Manchester and Newcastle) on 1 March. For use in conjunction with diagram 30

SUNPATH DIAGRAM

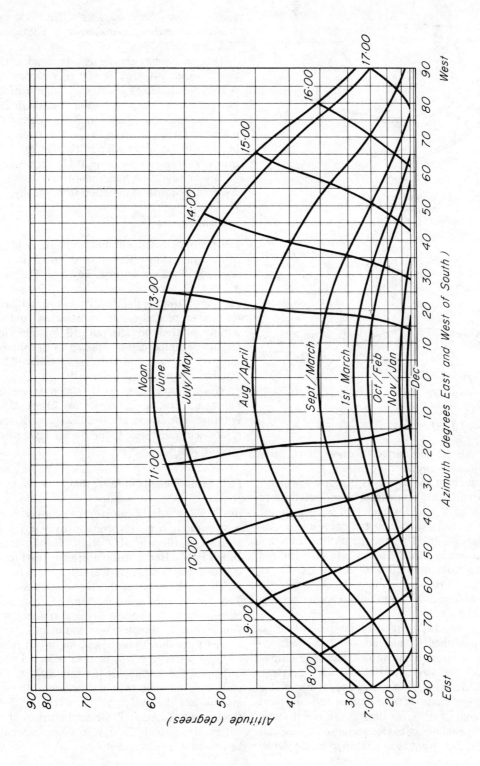

44 *Sunpath diagram for 56 N (Edinburgh and Glasgow) on 1 March. For use in conjunction with diagram 30*

DAYLIGHTING

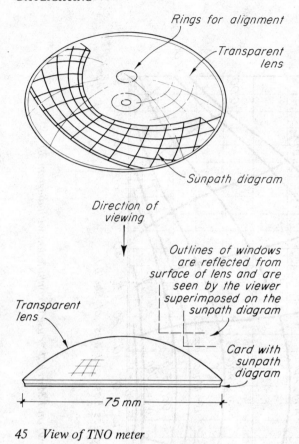

45 View of TNO meter

angles may be transferred to drawings to carry out graphical analysis of insolation.

2 The outline of the sky seen through a window from a particular point may be projected on to the sun path diagram and the periods during which the sun can fall on the point under consideration observed within the outline. External obstructions should be taken into account and excluded from the outline of the visible sky. In the case of exterior situations the skyline of the surroundings can be projected on to the diagram in a similar way.

Recently another method of insolation analysis has been published in this country. It is based on 'gnomonic' projection and allows the use of normal elevations of windows rather than distorted diagrams. It is of particular interest where appearance of the windows from inside, or the extent of the view must be considered at the same time as sun penetration. To use the method an extensive range of diagrams is necessary. Sets have been published in the *Archi-* *tects Journal Environmental Handbook* and in *Windows and Environment,* published by McCorquodale, 1969.

Diagram 38 may also be used to estimate the heat gain through windows at any time of day or year. A tracing is made of the sun and hour lines and the base of the diagram with the 0 azimuth marked. The tracing is then transferred to diagram 41 and positioned with 0 azimuth on the tracing corresponding to the orientation of the window. The heat gain per m^2 may now be read off against any time of day or year. The horizontal lines on diagram 41 represent the diffuse radiation from the sky.

Sunlight criteria

The methods described above can clearly be applied to the sunlighting criterion of the DoE. To simplify checking the DoE criteria, however, Diagrams 42, 43 and 44 show sunpaths for 52, 54 and 56° N capable of covering England, Wales and Scotland with reasonable precision and constructed so that the outline plotted on Diagram 30 for checking the Daylight criteria may be applied to the appropriate sun path diagram. Periods of insolation will be indicated by unobstructed sunpath lines.

The sunpath line for 1 March is specifically indicated, making it simple to check the DoE criterion. The 0 azimuth from Diagram 30 must be positioned over the appropriate window orientation on diagrams 42, 43 or 44.

Field measurements of insolation

In the past study of insolation in existing buildings was laborious, either involving a survey and graphical analysis or the use of a Robin Hill Camera. The Dutch Public Health Laboratories at Delft have produced a very simple device called the TNO meter which enables very rapid determination to be made. The device, shown in diagram 45, consists of a circular sunpath diagram under a transparent lens. When noon is oriented due south and the meter viewed from directly overhead the bright reflection of windows or sky will be seen reflected from the surface of the lens apparently directly over the sun path diagram. It is possible, therefore, to establish immediately the duration of insolation at that spot throughout the year. There are two sizes of meter, the smallest, about 75 mm in diameter, can easily be carried in a pocket.

5 Heat

THERMAL COMFORT

Heat presents a more complex situation since a wide variety of factors, including most of those already mentioned, must be considered in balance with one another. The appropriate balance is governed by the human body's requirements for thermal comfort. While it is not appropriate here to attempt to cover the physiology of the body's need for and reaction to heat (a detailed treatment and further references can be found in *Basic Principles of Heating and Ventilation* by Bedford), a general description of the factors involved in thermal comfort is essential to any appreciation of the aspects of the natural environment which are concerned. Heat is constantly produced by bodily processes and must be dissipated to keep the body temperature at its correct level. The body normally loses heat by radaition, convection and evaporation, and if human beings are to be thermally comfortable, not only must the appropriate quantity of heat be lost, but a proper balance must be maintained between the various modes of loss. The rate of heat loss in each of these aspects is governed by the surrounding environmental conditions. The net heat loss by radiation is controlled by the mean radiant conditions; the rate of loss by convection is dictated by the air temperature and rate of air movement; and evaporation losses by breathing and sweating depend on air temperature, relative humidity and air movement. The acceptable value of each of these features is not fixed but can vary in conjunction with one or more of the others. It is also possible for the body to vary its own balance of losses (eg increased sweating). There are, however, limits to each of the factors beyond which a satisfactory balance for comfort is not possible.

There have been a number of attempts to produce a unified means of representing thermal comfort. The kata thermometer, now used as a sensitive omni-directional anemometer, was originally used by Sir Leonard Hill as a measure of warmth. Unfortunately it is more sensitive to air movement than the human body and is consequently not a reliable measure of human reaction to warmth. A series of experiments in the laboratories of the American Society of Heating and Ventilating Engineers led to the development of a more sophisticated device called the Scale of Effective Temperature which takes into account air temperature, air movement and relative humidity. No account is taken of radiation. In England an instrument called a eupathoescope reacting to thermal conditions in the same way as the human body was developed by Dufton. The instrument is influenced by air temperature. air movement and radiation but takes no account of relative humidity. The readings of this instrument are in terms of a scale called Equivalent Temperature. Values in this scale can also be established by measuring air temperature, air movement and radiation separately and combining them by a nomogram. Table 11 shows for standard air movement conditions a table of air temperatures and mean radiant temperatures to give equivalent temperatures of 18.5°C and 25°C. The heavy lines on the table indicate conditions beyond which one of the factors will cause discomfort which cannot be balanced by variations in the others. Extensions to the effective temperature and equivalent temperature scale have been made to cover mean radiant temperature and relative humidity respectively and the resulting scales are described as Corrected Effective Temperature and Equivalent Warmth. Table 10 gives a brief chronological summary of Comfort Indices.

These comfort scales do not, however, solve the problem of defining human comfort, since they mask in one unified value directional variations of radiation, temperature gradients in the air or other factors which could cause unsatisfactory conditions and which must be designed separately. In addition, recent research has cast doubt on the validity of specific fixed temperatures as a close guide to thermal comfort. There is evidence that people adapt their clothing and immediate environment to suit prevailing conditions and, with-

Table 10   Selected summary of thermal indices and comfort concepts

Dry Bulb Temperature   Long established.

Wind chill–Sensible cold   Concepts developed in early nineteenth century to measure not temperature, but rapidity of heat loss. One instrument developed was the Heberden thermometer.

Wet Bulb Temperature   Nineteenth century.

General Board of Health   1857 Report   called for:
Walls warmer than air.
Air temperature higher at floor than at head.

Globe Thermometer Temperature   1877,   Aitken (and Vernon 1930) (Combined effect of radiation and air movement.) Resultant Dry Temperature Missenard 1935.

Kata Thermometer Cooling Power   1914 Hill
The cooling time of Kata thermometer gave indication of the cooling power of the environment and consequently of comfort. The thermometer did not cool in the same balance as the human body and the thermometer is now used as an omni-directional anemometer.

Effective Temperature   Houghton and Yagloulou 1923 Single scale of temperature which combined air temperature, relative humidity and air movement. A basic scale dealt with unclothed subjects and a normal scale with fully clothed ones.

Corrected Effective Temperature   Vernon and Warner 1933   Developed effective temperature by taking radiation into account by means of Globe thermometer in place of dry bulb.

Equivalent Temperature   Dufton 1929
Based on Eupathescope. Cooled in the same balance as the human body in relation to air temperature, air movement and mean radiant temperature. Did not take humidity into account.

Resultant Temperature   Misserard 1933
Effective temperature with radiant component.

Equivalent Warmth   Bedford 1936
Based on observations made in actual (not laboratory) conditions and analysed statistically.

Equatorial Comfort Index   Webb, BRS 1961
Similar method to Bedford, based on data related to Singapore.

Calidity, Webb BRS 1967   Term developed to apply to thermal comfort aspect of the environment.

PMV (Predicted Mean Vote) and PPO (Percentage Fanger 1970   Sophisticated method of analysis based upon extensive laboratory and field studies which enables the percentage of dissatisfied occupants of a room with particular thermal characteristics to be predicted.

Note Some studies have appeared to demonstrate a 70 per cent acceptance of comfort by people in rooms with the following values of the indices.

Globe thermometer	16.5–19.5°C
Effective temperature	14 –17°C
Equivalent temperature	14.5–19°C
Air temperature (dry bulb)	15.5–19.5°C (England)

in reasonable limits, are most comfortable when thermal conditions remain fairly stable irrespective of actual values.

Inextricably associated with thermal comfort and with ventilation is another set of comfort factors described as 'freshness'. An environment which is thermally satisfactory may yet give rise to complaints of 'stuffiness'. A variety of phenomena have been investigated in order to discover the basis of this including ionisation of the air and the influence of radiation on the nasal membranes. Bedford concludes, however, that important factors in freshness are the correct relationship between some of the factors involved in thermal comfort (eg mean radiant temperature higher than air temperature) and adequate skin stimulation of the occupants of heated interiors by variation of air velocity. Table 12 sums up requirements for a

Table 11 Air temperatures and mean radiant temperatures to give Equivalent Temperatures of 18.5 and 21°C

Air temperature deg C	Mean radiant temperature deg C to give	
	Equivalent temperature 18.5°C	Equivalent temperature 21°C
10	32	37.5 } 1
12	28	34
15.5	25	31
18.5	21	27 } 2
	18.5	24
24	14.5	20 } 3

For any given equivalent temperature the conditions with higher MRT than AT are normally preferred

1 12°C minimum acceptable air temperature
2 Most satisfactory
3 Noticeably less comfortable when mean radiant temperature is less than air temperature

From 'Designing for Thermal Comfort', by Alexander Hardy, *Architects Journal,* 20 November 1958

Table 12 Requirements for a pleasant environment according to Bedford

1 Rooms should be as cool as is compatible with comfort.
2 The velocity of air movement should be about 10 m per minute in winter (less than 6 m per minute may cause stuffiness). Higher rates are desirable in summer.
3 Air movement should be variable rather than uniform.
4 The relative humidity should not exceed 70 per cent and should preferably be substantially below this figure.
5 The average temperature of internal surfaces should preferably be above or at least equal to the air temperature.
6 The air temperature should not be appreciably higher at head level than near the floor and excessive radiant heat should not fall on the heads of occupants.

pleasant environment quoted by Bedford. The following dry bulb temperatures (°C) represent currently recommended standards of warmth.

Domestic		Factories (according to type of work)	
Living rooms	20–21		
Bedrooms	13–16	Sedentary	18
Kitchens	16	Light	16
		Heavy	13
Offices		General spaces	
General	20	Entrance, stairs,	
Machine rooms	19	etc	16
		Lavatories	18
Shops	19	Cloakrooms	16
Classrooms	17	Hospital wards	19

For most normal conditions relative humidity should be 30–65 per cent.

The standards quoted for thermal comfort and freshness usually give a particular value for general application and make no provision for sex or other differences. There seems to be no basic general variation in comfort requirements of men and women but different conventions of dress usually mean that women call for higher temperatures than men. Old people require higher values than young ones. Different countries have different standards. This may be influenced by the extremes of climate experienced and by the relative economics of providing and running heating installations in the countries and different clothing may help to sustain the differences. The North America and British temperature standards have been significantly different but with the rapid extension of the use of central heating in this country there is evidence that the standards here are tending towards the American levels. Apart from general differences in heating standards individuals vary very greatly in their reaction to thermal environment and these differences are more marked than geographical ones. It is possible for different individuals to feel too hot and too cold respectively in the same thermal conditions. There is no one set of conditions which will satisfy everyone even in one locality The aim in design is therefore to satisfy a majority and to reduce to a minimum the inevitable proportion of dissatisfied occupants.

Humphreys of the Building Research Establishment has demonstrated in several studies (published as BRS Current Papers) that globe thermometer temperatures form a satisfactory measure of thermal

HEAT

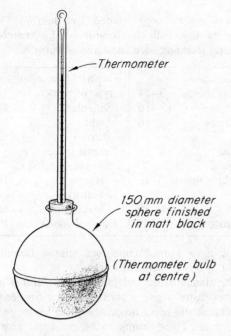

46 Use of TNO meter

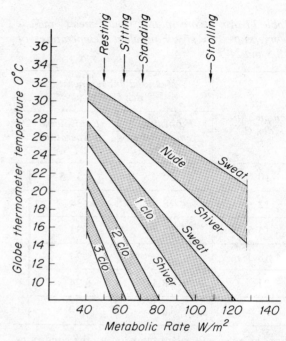

47 The relationship between activity, globe thermometer temperature and comfort

environment in temperate regions. Diagram 46 shows a globe thermometer.

Humphreys has also demonstrated that while occupants of buildings cannot easily respond to fast changes of temperature they can adapt themselves to a wide range of conditions if these are long term. Clothing in particular can be varied and location of desks and seats adapted. In addition control of heating and ventilation can be adjusted to counterbalance long term variations. A scale of 'clo' units has been developed to represent the effects of different clothing levels. Table 13 shows typical combinations of dress and their corresponding clo values together with the appropriate comfort temperatures for sedentary people.

The degree of activity also has a considerable effect upon the comfort temperature. Table 14 shows the metabolic rate for young adult males for various types of activity expressed, as is now conventional, in W/m^2 of body surface area and Diagram 47 shows the relationship between activity,

Table 13 Combinations of dress

Clo value	Type of clothing	Comfort temperature for sedentary subject °C
0	Nude	28.5
0.5	Light slacks and T-shirt	25
1.0	Business suit or slacks and pull-over	22
1.5	Heavy suit and waistcoat woollen socks	18
2.0	Heavy suit and overcoat woollen socks and hat	14.5

Table 14 Metabolic rates for young adult males

	W/m^2 of body surface
Sleeping and digesting	47
Lying quietly and digesting	53
Sitting	59
Standing	71
Strolling (2.5 km/hr)	107
Walking (4.2 km/hr)	154

THERMAL PROPERTIES OF BUILDINGS

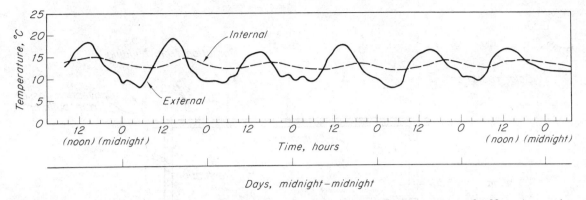

48 Effect of enclosure in modifying amplitude and timing of diurnal temperature cycle. Note: in massive well insulated buildings the effect will be most marked. In lightweight constructions the amplitude and time of the internal temperature cycle will approach more closely to the external conditions

globe thermometer temperature and comfort. It will be observed that active people have a significantly wider range of tolerance than sedentary ones, on the other hand comfort can be maintained through substantial temperature variation by varying clothing.

THERMAL PROPERTIES OF BUILDINGS

These considerations make it clear that the important factors in the natural environment affecting thermal comfort are air temperature, air movement, relative humidity and radiant conditions. Radiant conditions range from exposure to the rays of the sun to those of a clear cold night when the radiation balance is reversed and the body is losing heat by radiation to the very cold sky. Experience tells us, without the need for elaborate analysis, that these conditions rarely combine in nature to give satisfactory thermal conditions and buildings have an important function in providing thermal comfort. The interaction of building and thermal conditions is complex. Apart from the fact that several factors are involved the fabric of the building itself reacts to thermal changes and does so over a period of time. This contrasts with the behaviour of the building in relation to wind or light. In these cases the presence of a wall gives in effect a total and immediate barrier. Heat on the other hand is not suddenly arrested, it is delayed but passes through over a period and it is necessary to take into account the resistance of the materials to the passage of heat, the way in which they warm up or cool down and the time taken to do so. One of the major effects resulting from the delay in heat transfer to and from buildings is a smoothing out of the extremes of temperature. Diagram 48 shows in principle the type of relationship which exists between internal and external temperatures.

The cooling effect on occupants of wind and radiation to the cold night sky is clearly reduced inside buildings and during the winter solar radiation will contribute to a general raising of internal temperatures. The interiors of buildings in temperate zones in the winter therefore, provide, without any heat input, improved thermal conditions. Air movement is controlled, extremes of air temperature and radiate conditions are reduced, a more equable temperature is maintained which may be significantly above the mean external temperature.

In the summer, however, even in temperate zones it is possible for the heat from the rays of the sun

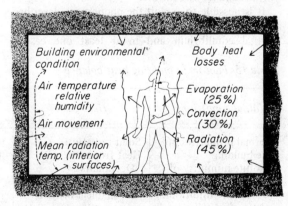

49 Heat balance of human body in relation to its environment

HEAT

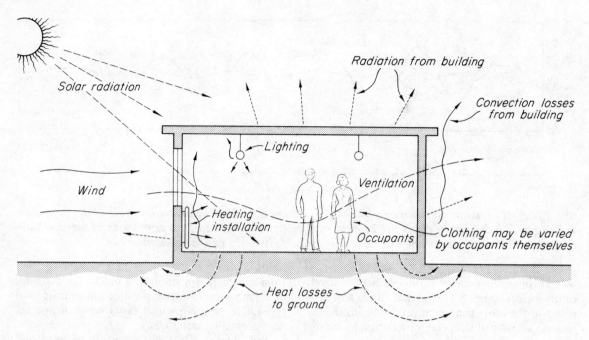

50 Heat balance of building in relation to external and internal influences

to raise inside temperatures to beyond comfort levels. Buildings with large windows may present a major problem particularly if they are of lightweight construction, which implies low thermal capacity and fast warm up. The situation may be made very much worse if external noise makes it difficult to open windows and obtain adequate ventilation. Towns often present environments of this type. It is quite possible to find summer temperatures so high that parts of buildings are uninhabitable and common to find summer discomfort.

It is possible to exploit the phenomena described and to design buildings which remain in good thermal equilibrium and in at least one case a building has been designed which uses solar heat, heat from occupants and lighting and dispenses with any heating installation. Most buildings will, however, require an installation which can provide additional heat when required and in some cases extract excess heat. It is clear, however, that the proper design of the fabric of the building itself is vital to the success of the thermal environment inside and will be a controlling factor in the economy of the initial and running cost of the heating installation.

Diagram 49 shows the balance of the human body with its thermal environment and diagram 50 shows some of the factors of the building's

Table 15 Factors in building heat balance

Initial conditions	+	Heat gains	−	Heat losses	→	Satisfactory comfort conditions

Thermal properties of the building
Absolute humidity of the air
(unless air conditioning with humidity control is provided)
		Solar radiation
		Heat from
		  Occupants
		  Lighting
		  Mechanical installations and equipment
		  Heating installation
				Radiation to sky
				Convection to air outside
				
				Ventilation losses into ground
				Refrigeration plant
						Mean radiant temperature
						Air temperature
						Relative humidity
						(A satisfactory rate of air movement must also be given by ventilation arrangements)

THERMAL PROPERTIES OF BUILDINGS

Table 16 Typical heat gains and heat requirements

Heat Gains (wild heat)			Heat requirements	
Source	Extent of input (sensible heat) gain W)	Actual input governed by	Nature	Governed by
Occupants	100-350 per person	Number of occupants Sex and age of occupants Activity Period of occupation	Warming up	Thermal properties of building and contents Initial and final temperatures
Solar gain	0-530 W per m^2 of surface	Orientation Form and construction Time of day Time of year Situation of building	Ventilation	Rate of air change Outside and inside air temperatures
Lighting	1 watt of lighting contributes 1 watt to heating	Wattage and period of use of lighting	Losses through fabric	External radiant temperature External air temperature Wind velocity Moisture content of construction
Electrical apparatus	1 watt of lights contributes 1 watt to heating	Wattage and use of apparatus	Heating cold materials brought into building	Not normally critical for heating
Gas apparatus	17 000 kJ per m^3 of gas	Rates of consumption and periods of use of apparatus		
Water heating	Depends on nature, extent and use of system. Often a substantial proportion of the heat input is given out as space heating			
Other processes	Depends on circumstances			
Heating installation	Should maintain balance between gain and requirements			

thermal behaviour together with aspects of the natural environment which must remain in balance together if thermal comfort is to be maintained inside. Tables 15 and 16 show the main factors which must be maintained in proper balance. It will be noted that only the contribution of the heating and refrigeration installations and to some degree the ventilation can be varied in operation. The other factors are either not capable of modification such as the heat gain from occupants or must be settled by the basic design such as the thermal properties of the fabric of the building and the form and fenestration in relation to solar heat gain.

Construction and planning in relation to heat balance

In order to achieve efficiency and economy in heating and ventilation, careful thought must be given to construction and planning. The importance of the thermal capacity of buildings has been

Table 17 How planning and siting affect heat loss

Aspect	Bad	Good	Possible effect on total heat loss % (approx.)
Plan shape	Complex or long, narrow shapes increase wall area and consequent heat loss. Length of wall 40 m. Floor area 64 m². The heat loss from solid floors is increased by narrow widths	Square shape gives minimum wall area (spherical building ideal but impracticable). Length of wall 32 m. Floor area 64 m². Square shape best	10 1¼
Flue position	Flues on external walls lose heat (Cold flues are less efficient and encourage condensation)	Internal flues contribute heat to rooms on upper floors	6
Exposure	Exposed sites increase heat losses through the fabric and increase the rate of air change	Sheltered situations reduce heat losses through the fabric and reduce the rate of air change	20
Orientation	Large windows give considerable heat losses and may cause discomfort in winter	Windows facing south and receiving the sun's rays in winter may reduce fuel consumption over the season (Overheating in summer may be avoided by proper screening)	Depends on particular case too closely for generalised assessment

mentioned earlier. The importance of providing a structure with adequate resistance to the loss of heat is apparent in relation to both economy of heating costs and comfort, since walls with low insulation values will have low surface temperatures, and this will affect the radiant conditions within the building.

The arrangement and plan of the building has a considerable effect on heating economy. Compact plans will have substantially smaller heat losses than complex shapes, and orientation and areas of glazing are important. Table 17, shows a number of factors which affect heat loss. This table is based on domestic buildings, but the principles are of general application.

In addition to the form of the building, attention must be given to the detailed losses through various parts of the building. Diagram 51 compares the basic rate of heat losses through typical elements of construction, and then compares the actual losses

THERMAL INSULATION

through that element in an actual building. It will be observed that the elements losing heat at the maximum rate do not necessarily represent the maximum heat loss in the building, and this may influence whatever precautions against heat loss are considered desirable. It must also be remembered that while windows have a considerable rate of heat loss, south facing windows may actually give a heat gain over the whole of the heating season.

THERMAL INSULATION

The conservation of heat in buildings has become more important of recent years because of the great increase in cost of fuels and the improved standards of comfort which people expect. Developments in new systems of construction have tended the other way, ie to lighter, thinner walls of concrete units, roofs of metal decking, etc., which, without added insulation, would lose heat from the building more quickly than traditional constructions. The provision of insulation has been simplified by the development of many new materials in a convenient form for incorporating in most constructional systems.

Before considering the insulation of walls, roofs, etc., it should be emphasised that much heat can be lost by the escape of warm air round the edge of doors and windows, up chimneys and other means of uncontrolled ventilation.

Careful detailing of doors and windows, use of 'weather stripping' and provision of draught lobbies can help to cure much of this loss.

Part of building	Comparative rates of loss through various constructions (Watts/m^2 deg C = U value)	Comparative rates of heat loss through various parts of a building (based on 115 square meter two storey brick house)
ROOF Tiles, felt and plaster board ceiling. 25 mm fibreglass U = 0.91	▓▓	▓▓
WALLS 280 mm unventilated cavity with insulating inner leaf U = 1.31	▓▓	▓▓▓▓
WINDOWS and DOORS Single glazing U = 5.1	▓▓▓▓▓▓	▓▓
GROUND FLOOR Solid U = 0.86	▓▓	▓▓
VENTILATION	—	▓▓▓

51   Comparison of the rates of heat loss through various building elements compared with the actual losses through those elements in a typical building

Until 1974 the Building Regulations required only a modest standard of insulation in walls, floors and imposed no limit on the area of glazing. In 1975 amendments to the regulations have considerably increased the insulation which walls, floors and roofs of dwellings must provide. In addition the average U-values of perimeter walls, including window openings, is controlled. The new U-values require either new constructions, or, if existing conventional constructions are used, substantial additional insulation is necessary. However, one of the most significant changes is the requirement that an average U-value (1.8 W/m²°C) must not be exceeded in the perimeter walling. In many cases, if single glazing is used, the area of windows will have to be curtailed. The following table gives the main U-value requirements from the regulations.

Max U-value (W/m^2 deg C)

External walls	1.0
Wall between a dwelling and a ventilated space (eg permanent vents exceed 30% of wall area)	1.0
Wall between a dwelling and partially ventilated	1.2
Wall between a dwelling and any part of a building to which regulations do not apply	1.7
Note: The overall average U-value of perimeter walling including windows must not exceed 1.8.	
Wall or partition separating room from roof space	1.0
Floor permanently exposed on the underside	1.0
Floor covering partially enclosed space space	1.0
Roof	0.6

BS CP 3 Chapter II Thermal Insulation does not prescribe standards of insulation but sets out the significant factors involved and suggests ways of ensuring satisfactory thermal environment.

At present these regulations only apply to new dwellings, it must be anticipated, however, that similar legislation for other types of building will not be long delayed.

Heat transfer

Before considering heat insulation of buildings it is advisable to consider some of the related principles of the transference of heat. Heat will flow in a solid object, in liquid or in gas, or between them until the temperature of each is equal. Transfer of heat can be by conduction, convection or radiation.

Conduction is the direct transmission of heat through a material. The rate of conduction, ie conductivity, depends partly upon the density of the material. Metals have a high conductivity, wood has low, and gases have even less conductivity. This *conductivity* (k) is the amount of heat in wall that passes through 1 m² of the material of 1 m thickness for 1 deg C difference in temperature of the inner and outer surfaces.*

Convection is the transmission of heat in fluids and gases by circulation. When a liquid or gas is heated it is displaced by the colder more dense liquid or gas round it, and it tends to rise. In doing so it will impart some of its heat to anything in its path. The greater the movement, the greater the speed of transfer. For this reason air movement in insulating cavities must be avoided.

Radiation is the transference of heat from one body of radiant energy through space to another. All bodies emit radiant energy, the temperature of the body defines the wavelength and the rate of emission depends upon the temperature and the nature of the surface. When radiant energy reaches an opaque body, part of it is reflected and the remainder absorbed and converted into heat. Bodies are therefore continuously radiating heat and receiving radiations back at varying wavelengths. Part of the radiation from the sun has a short wavelength and will pass through glass, radiation from an electric fire has long wavelength and much of it will not pass through glass. Short-wave radiation will be mostly reflected by a white surface, and not absorbed; thus a white surface will help to keep a roof cool in summer. Generally, white or bright surfaces have a high reflection value and small absorption and low emissivity. Aluminium foil is an example which is becoming useful in many ways. Dark matt surfaces have a high absorption and high emissivity.

When heat is passed out of a building through

---

*For insulation it is usually more convenient to use the *resistivity* of a material, which is the reciprocal of the conductivity, ie $1/k$

The resistivity multiplied by the thickness of that material gives the *resistance* (R) of that construction.

THERMAL INSULATION

the structure, all three methods of transference are used. Heat is conducted through the solid parts of the wall or floor or roof, it is radiated across cavities and from the outside surface: it is also convected from the outside surface by wind passing across that surface.

The overall rate of transmission is known as the thermal transmittance, the unit being the heat in watts that will be transferred through 1 m^2 of the construction when there is a difference of 1 deg C between the temperature of the air on the inside and that on the outside; this is called the 'U' value or 'air-to-air' heat transmittance co-efficient'. These coefficients are calculated from the conductivity of the materials or k mentioned previously, and the surface resistances.

Resistance of surfaces

A surface resists the net transfer of heat according to its emissivity, its absorptivity and its reflectivity. At normal temperatures the emissivity and the absorptivity of a surface are the same, and the surface reflects what heat it does not absorb, eg the emissivity of aluminium foil may be 0.05, so its absorptivity will be 5 per cent of the heat falling in it (from a body at normal temperature), and it reflects 95 per cent of such heat. The absorptivity of a surface for high temperature, ie solar radiation, may be quite different. This is important when insulating against sun's heat. Surfaces can lose heat by convection, so the resistance of outside surfaces is governed by climate, (temperature and speed of wind). The main effect of these factors on the resistance of a surface are therefore:

1 Cooling wind across an external surface will reduce its resistance.
2 The resistance of a corrugated surface can be about 20 per cent less than a plain surface of the same material because of its larger area.
3 Surfaces of low emissivity, ie bright metallic surfaces, will have a high resistance, but this may be nullified if convection takes place as (1) above.
4 When a surface is radiating to an area of very low temperature, as to a clear sky in very cold, calm weather, the resistance of the surface can be decreased considerably.

5 The resistance of a horizontal surface will depend upon whether the transfer of heat is upwards or downwards, as convection will assist to take the heat away above and to keep it near the surface below.

The values of resistance of non-metallic surfaces have been computed and those normally used in Britain are set out, with much other useful data, in the IHVE *Guide 1970*, published by the Institution of Heating and Ventilating Engineers, from which the following tables are taken:

Table 17   Internal surface resistances m^2 deg C/W

Surface	
Walls	0.123
Floors or ceilings	
for upward heat flow	0.104
for downward heat flow	0.148
Roofs, flat or sloping	0.104

Table 18   External surface resistances, m^2 °C/W (for ordinary building materials)

Surface	Exposure		
	Sheltered	Normal	Severe
Walls	0.08	0.055	0.03
Roofs	0.07	0.045	0.02

Notes
'Sheltered' includes the first three storeys of buildings in the interior of towns.
'Normal' includes the fourth to eighth storeys of buildings in towns and most suburban and courts premises.
'Severe' includes ninth and higher floors of buildings in towns, fifth and higher floors in suburban districts and buildings exposed on coasts and hills.
Ordinary building materials covers the great majority of materials and colours in common use. For corrugated surfaces reduce values by 20%. For bright metallic surfaces see IHVE Guide.

These surface resistances are in the same units as the resistances of specific thicknesses of building materials. It is possible, therefore, to add up the resistances for the materials making up a wall, floor or roof and the external and internal surface resistances to obtain a total resistance. From this the

HEAT

thermal transmittance of the construction may be determined by calculating the reciprocal of the total resistance.

$$\text{Thermal transmittance (U-value) W/m}^2{}^\circ\text{C} = \frac{1}{\text{sum of resistances of sur-surfaces and layers of material. m}^2{}^\circ\text{C/W}}$$

Much of the basic data on thermal properties of building materials is given in terms of their thermal conductivity (k) which is the amount of heat in watts that will pass through one square metre of material one metre thick for a 1°C temperature difference between the faces. The resistance of a given thickness may easily be established:

$$\text{Resistance of a layer of given thickness (m}^2{}^\circ\text{C/W)} = \frac{\text{thickness (m)}}{k\,(\text{W/m}^\circ\text{C})}$$

Resistance of cavities and air spaces

Many forms of building construction result in air spaces between layers of building materials and these

Table 19 Thermal resistances of air spaces

Sealed or ventilated	Space		Thermal resistance $m^2{}^\circ C/W$	
	Width	Surfaces*	Heat flow horizontal or upwards	Heat flow downwards
Sealed	6 mm	obm	0.11	0.11
	6 mm	ref 1	0.18	0.18
	19 mm or more	obm	0.18	0.21
	19 mm or more	ref 1	0.35	1.06
		reflective multiple foil insulation and air space	0.62	1.76
Ventilated	19 mm or more	loft space with asbestos cement pitched roof	0.14	
		as above with reflective foil on upper surface of ceiling	0.25	
		loft space with pitched roof, felt and building paper	0.18	
		air space between tiles and felt or building paper	0.12	

* obm signifies ordinary building materials
  ref 1 signifies one or both sides faced with reflective materials

air spaces contribute significantly to the insulation. With surfaces of traditional building materials, ie brick, stone, concrete etc having high emissivity the normal practice is to take the resistance value of a cavity as 0.18 m^2°C/W, provided that it is at least 19 mm across and unventilated. It has been found that smaller cavities have lower resistance but increasing the width of the cavity does not appreciably affect the resistance. If the cavity is ventilated, the resistance will be reduced depending on the speed of circulation of the air. If the cavity is horizontal and the heat flow is downward (eg through a floor) the resistance is increased.

If one surface of an unventilated cavity is of low emissivity, such as aluminium foil the resistance is increased since radiation is substantially reduced. Multiple foil sheets providing a series of low emissivity surfaces and cavities provide excellent insulation.

Roof spaces for pitched roofs also have an insulation value, although it is difficult to seal them and they must normally be considered as ventilated.

Table 19 gives values for the thermal resistance of cavities and air spaces.

Table 20 Typical thermal properties of building materials

Material	Density kg/m^2	Thermal conductivity (k) W/m deg C	Thermal resistivity (1/k) m deg C/W
Asbestos cement	1600	0.36	2.78
Asphalt	1700	0.50	2.00
Brickwork, common	1700	0.84	1.19
Compressed straw slabs	260	0.09	11.1
Concrete			
ballast	2000-2400	1.0-2.0	0.5-1.0
cellular	450-950	0.10-0.28	3.6-10
clinker	1500-1700	0.33-0.40	2.5-3.0
foamed slag	650-1100	0.14-0.25	4.0-7.2
vermiculite	350-950	0.07-0.28	3.6-14.3
Cork board	140-320	0.05-0.06	16.7-20
Expanded polystyrene	15-30	0.03-0.04	25-33
Fibre board	460	0.05	20.00
Glass	2500	1.02	0.98
Glass wool	50	0.035	28.6
Hair felt	80	0.04	25.00
Hardboard	640	0.10	10.00
Plaster board	960	0.16	6.25
Plaster, dense	1300	0.50	2.00
Plywood	530	0.14	7.14
Roofing felt	1000	0.20	5.00
Slates	2600	1.80	0.57
Stone	2000-2800	1.30-2.80	0.36-0.77
Tiles, roof	1900-2250	0.83-0.94	1.06-1.20
Timber	640	0.14	7.14
Vermiculite	80-100	0.03-0.04	25-33
Wood-wool slabs	470-800	0.08-0.14	7.14-12.5

More extensive tables of thermal properties of building materials and construction may be found in the Institute of Heating and Ventilating Engineers *Guide 1970*.

HEAT

Calculation of U-values

An example of the calculation of the U-value of a 275 mm brick cavity wall plastered internally is set out below:

	Resistances $m^2\,°C/W$
112 brick outer skin	
$\dfrac{0.11 \text{ (thickness m)}}{0.84 \text{ (k)}}$ =	0.13
Unventilated 50 mm air space =	0.18
112 mm brick inner skin =	0.13
15 mm plaster $\dfrac{0.015}{0.5}$ =	0.03
External surface resistance =	0.06
Internal surface resistance =	0.12
Total =	0.65
∴ U value = $\dfrac{1}{0.65}$ =	1.54 $W/m^2\,°C$

U-values for any desired construction may be calculated in this way. (Note that with pitched roofs the area of the roof itself will be the actual area measured in the plane of the roof and will be larger than the equivalent area of ceiling.)

When considering the improvement that might be obtained by using an insulating lining the appropriate U value can be re-expressed as resistance, the resistance of the insulation added, and the reciprocal of this total gives the U value for the whole.

In constructions such as post-and-panel walling, where different parts may have different resistances, the proportion of the areas having different values must be calculated and multiplied by the appropriate resistances, and these must be totalled to get the true mean U value.

Where parts of the construction have a very much lower resistance than others, as with a metal framework exposed to both inside and outside, and having panels of good insulating material between, heat may pass through these metal members quickly and draw heat even from the edges of the panels. This is referred to as the *cold bridge effect*. It is not normally a critical element in heat loss but may because of reduced surface and interstitial temperature, give rise to trouble from condensation. Diagrams 52 and 53 show the effect and typical temperatures.

Tables 21, 22 and 23 and diagram 54 show typical U-values.

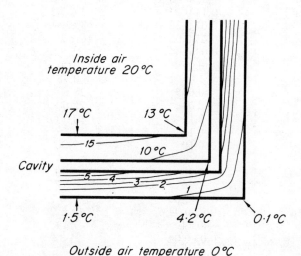

52   Reduction in wall temperatures at corner

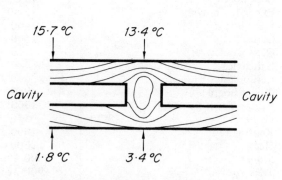

53   Cold bridge effect

U–VALUES

Table 21 Standardised U-values for wall and roofs of normal exposure

A WALLS There are, since the 1975 amendments to the Building Regulations, three different standards for domestic wall u-values (1.0 for external walls, 1.2 for walls to partially ventilated spaces and 1.7 for walls between dwellings and parts of buildings to which the amendments do not apply) In addition, in order to comply with the requirement for an overall u-value for walls and windows of not more than 1.8, it may be necessary to have walls with a low u-value than the statutory maximum of 1.0. This table has, therefore, been set out in considerable detail. Resistances have been shown so that the u-values of variations may be quickly established.

Basic construction	Thickness mm	External finish	Internal finish	Resistance m^2 °C/W	U-value W/m^2 °C		
					over 1.7	1.7–1.0	Below 1.0
SOLID BRICKWORK	105	–	–	0.3	3.3		
	"	–	15 mm hard plaster	0.33	3.0		
	"	–	15 m lightweight plaster	0.39	2.5		
	"	–	25 mm cavity & foil backed 10 mm plaster board	0.67		1.5	
	"	–	25 m exp. polystyrene + 10 mm plasterboard	1.12			0.9
	"	Tile hanging	15 mm lightweight plaster	0.55	1.8		
	"	Tile hanging	25 mm cavity + 10 mm plasterboard	0.7		1.4	
	220	–	–	0.43	2.3		
	"	–	15 mm hard plaster	0.46	2.1		
	"	–	15 mm lightweight plaster	0.52	1.9		
	"	–	25 mm cavity + 10 mm plasterboard	0.67		1.5	
	"	–	25 mm cavity + foil backed 10 mm plasterboard	0.77		1.3	
	"	–	25 mm exp. polystyrene + 10 mm plasterbaord	1.25			0.8
	"	Tile hanging	15 mm lightweight plaster	0.68		1.5	
	"	Tile hanging	25 mm cavity + 10 mm plasterboard	0.83		1.2	
	"	Tile hanging	25 mm cavity + 10 mm foil backed plasterboard	0.95		1.1	
	"	Tile hanging	25 mm exp. polystyrene + 10 mm plasterboard	1.41			0.7
	335	–	–	0.55	1.8		
	"	–	15 mm hard plaster	0.58		1.7	
	"	–	15 m lightweight plaster	0.64		1.6	
	"	–	25 mm cavity + 10 mm plasterboard	0.79		1.3	
	"	–	25 mm cavity + foil backed 10 mm plasterboard	0.91		1.1	
	"	–	25 mm exp. polystyrene + 10 mm plasterboard	1.37			0.7
	"	Tile hanging	25 mm cavity + 10 mm plasterboard	0.95		1.0	
	"	Tile hanging	25 mm cavity + foil backed 10 mm plasterboard	1.07			0.9
	"	Tile hanging	25 mm exp. polystyrene + 10 mm plasterboard	1.53			0.7

HEAT

Table 21 continued

Basic construction	Thickness mm	External finish	Internal finish	Resistance m² °C/W	U-value W/m² °C		
					over 1.7	1.7–1.0	Below 1.0
CAVITY BRICKWORK* 105 mm brick inner and outer leaves, 50 mm cavity.	260	–	15 mm hard plaster	0.64		1.6	(1.0)
	"	–	15 mm light weight plaster	0.7		1.4	(0.9)
	"	–	25 mm cavity + 10 mm plaster board	0.85		1.2	(0.8)
	"	–	25 mm cavity & foil backed 10 mm plasterboard	0.97		1.0	(0.7)
	"	–	25 mm exp. polystyrene + 10 mm plasterboard	1.43			0.7 (0.6)
	"	Tile hanging	15 mm lightweight plaster	0.86		1.2	(0.8)
	"	Tile hanging	25 mm cavity + 10 mm plasterboard	1.01			1.0 (0.7)
	"	Tile hanging	25 mm cavity + foil backed 10 mm plasterboard	1.13			0.9 (0.7)
	"	Tile hanging	25 mm exp. polystyrene + 10 mm plasterboard	1.59			0.6 (0.5)
CAVITY WALL 105 mm brick onto leaf, 50 mm cavity, 100 mm aerated concrete inner leaf.	225	–	15 mm hard plaster	0.96		1.0	(0.7)
	"	–	15 mm lightweight plaster	1.05			1.0 (0.7)
	"	–	25 mm cavity + 10 mm plasterboard	1.17			0.9 (0.6)
	"	–	25 mm cavity + foil backed plasterboard	1.29			0.8 (0.6)
	"	–	25 mm expanded polystyrene + 10 mm plasterboard	1.75			0.6 (0.5)
CAVITY WALL 100 mm aerated concrete inner & outer leaves, 50 mm cavity	250	Tile hanging	15 mm hard plaster	1.45			0.7 (0.5)
	"	Tile hanging	15 mm lightweight plaster	1.51			0.7 (0.5)
	"	Tile hanging	25 mm cavity +10 mm plasterboard	1.66			0.6 (0.5)
	"	Tile hanging	25 mm expanded polystyrene + 10 mm plasterboard	2.24			0.5 (0.4)

*Note U-values in brackets take into account 25 mm fibreglass slab built into cavity

Basic construction	Thickness mm	External finish	Internal finish	Resistance m² °C/W	U-value W/m² °C		
					over 1.7	1.7–1.0	Below 1.0
SOLID AERATED CONCRETE WALL	100	–	15 mm hard plaster	0.66		1.5	
	"	–	15 mm lightweight plaster	0.72		1.4	
	"	15 mm light weight	15 mm lightweight plaster	0.81		1.2	
	"	Tile hanging	15 mm hard plaster	0.82		1.2	
	"	Tile hanging	15 mm lightweight plaster	0.88		1.1	
	"	Tile hanging	25 mm cavity + 10 mm plasterboard	1.03			1.0
	"	Tile hanging	25 mm cavity + foil backed 10 mm plasterboard	1.15			0.9
	"	Tile hanging	25 mm exp. polystyrene + 10 mm plasterboard	1.61			0.6

U-VALUES

Table 21A Note 1   Properties of materials used

Material	Conductivity W/m °C	Resistance m² °C/W
Brickwork 105 mm	0.84	0.125
220 mm	"	0.262
335 mm	"	0.399
Plaster 15 mm hard	0.5	0.03
15 mm lightweight	0.16	0.09
10 mm plaster board		0.06
Cavity, unventilated	–	0.18
with foil face		0.3
behind tile hanging		0.12
Tile hanging	0.84	0.038
25 mm expanded polystyrene	0.033	0.76
13 mm expanded polystyrene	"	0.39
Glass fibre 50 mm	0.035	1.43
100 mm	"	2.86
Aerated Concrete 100 mm	0.22	0.45
150 mm	"	0.68
Softwood 100 mm	0.13	0.77
Weatherboarding 20 mm	0.14	0.14
External Surface		0.055
Internal Surface		0.123

Table 21A Note 2   Calculation of variations from constructions tabulated

$$\text{U-value} = \frac{1}{\text{Resistance}}$$

$$\text{Resistance} = \frac{\text{Thickness (m)}}{\text{Conductivity (W/M °C)}}$$

To establish the effect of additional layers calculate the resistance. (See table 20 for basic thermal properties).
Add value found to the resistance for the type of construction given in table 21A.
Divide 1 by the value found to establish new U-value.

Table 21B   Roofs

Only one U-value for roofs, 0.6 W/m² °C, is given in the 1974 Amendments to the Building Regulations. While, at the time of writing this applies only to domestic buildings there can be little doubt that new, similar, standards will be introduced for non-domestic buildings.

	U-value W/m² °C
Pitched roof: Slates or tiles or battens with felt sacking, and plaster board ceiling	1.9
as above but with 50 mm fibreglass insulation between joists	0.54
Flat roof: 20 mm asphalt or 150 mm concrete	3.3
as above with ventilated cavity + 50 mm fibreglass or 10 mm plaster board	0.53
20 mm asplate or screed on 50 mm woodwool, with ventilated cavity and 10 mm plaster board ceiling	1.16
as above with 50 mm fibreglass or plaster board	0.44
Bituminous felt or boarding with ventilated cavity and 10 mm plaster board ceiling	1.85
as above with 50 mm fibreglass or plaster board	0.51

Note: It will be observed that few traditional uninsulated constructions have adequate insulation but that 50 mm of fibreglass gives satisfactory performance in all the cases quoted.

HEAT

Table 22 U-values for floors

Floors, solid (from BRS Digest 145)

Dimension m	U-values W/m² °C	
	Four exposed edges	Two exposed edges at right angles
30 x 30	0.26	0.15
30 x 15	0.36	0.21
30 x 7.5	0.55	0.32
15 x 15	0.45	0.26
15 x 7.5	0.62	0.36
7.5 x 7.5	0.76	0.45
3 x 3	1.47	1.07

Floors, timber suspended (from BRS Digest 145)

Dimension m	U-values W/m² °C		
	Bare or with lino, plastics or rubber finish	Carpet or cork finish	Any surface finish and 25 mm quilt over joists
30 x 15	0.39	0.38	0.30
30 x 7.5	0.57	0.55	0.39
15 x 15	0.45	0.44	0.33
15 x 7.5	0.61	0.59	0.40
7.5 x 7.5	0.68	0.65	0.43
3 x 3	1.05	0.99	0.56

Floors, intermediate (from IHVE guide)

Construction	U-value W/m² °C	
	Heat flow down	Heat flow up
Timber: 20 mm on joists with 20 mm pasteboard ceiling	1.4	1.6
Concrete: 150 mm with 50 mm screed	2.2	2.7
as above with 20 mm wood floor	1.7	2.0
200 mm hollow tile floor	1.6	1.9
as above with 20 mm wood floor	1.3	1.5

U VALUES

Table 23 U values for windows and roof glazing

Windows (normal exposure) (from IHVE guide)

Type	U-value, $W/m^2\ °C$	
	wood frame	metal frame
Single glazed	4.3	5.6
Double glazed	2.5	3.2*

*with thermal break in frame

Glazing (normal exposure)

Type	U-value
Roof glazing	6.6 $W/m^2\ °C$
Horizontal daylight with skylight over (ventilated space)	3.8 $W/m^2\ °C$

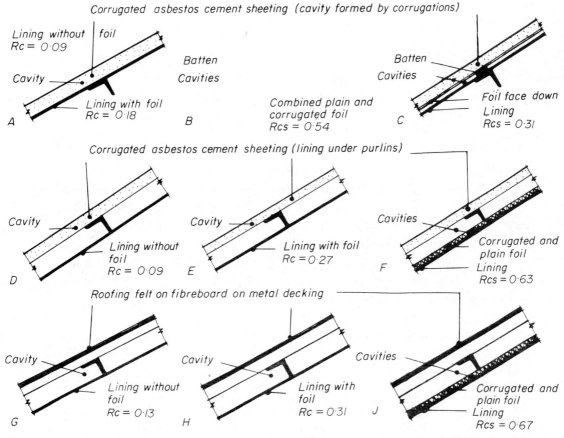

Rc(s) indicates thermal resistance of cavity (cavities) m^2 deg C/W   "foil" = single sided aluminium foil

54 Showing the resistance values appropriate to the cavities in a variety of pitched roof constructions

HEAT

HEAT LOSS CALCULATIONS

When the thermal transmittance (U-value) of the various elements of construction and the ventilation rates can be established (see pages 76 and 38) it is possible to estimate the overall heat losses from whole buildings. There are two important applications of this type of calculation. The first is the determination of the maximum hourly rate of heat loss. This figure is needed in order to select appropriate heaters for rooms and to size boilers and other parts of the heating installations (Diagrams 119, 120 and 121 enable sizes for boiler rooms and flues to be estimated approximately on the basis of maximum hourly rate of heat loss.) The second is the estimation of heat requirements over a whole heating season. This enables the cost of fuel to be estimated which may be an important factor in the selection of a heating system.

Maximum hourly rate of heat losses

This value is composed of heat losses through the fabric of the building plus the heat required to warm the air which is ventilating the building. The fabric loss for any space in the building is found by taking the areas of different types of construction multiplied by the thermal transmittance (U-value) multiplied by the temperature difference which it is desired to maintain between the two sides of the construction.

The normal design temperature difference assumed between the interior of habitable rooms and the exterior is 20 deg C (outside temperature 0°C, inside temperature 20°C). Where rooms inside are maintained at different temperatures the heat passing through internal partitions and floors must be included in the calculation.

$$QF = \Sigma \quad A \times U \times TD$$

Heat Flow through fabric	The sum of all the cases of the following	Area Thermal Transmittance	Temperature difference
		For each different type of construction	
(W)		(m²) (W/m² deg C)	(deg C)

$$QV = \text{Volume of space} \times \frac{\text{Number of air changes per hour}}{3600} \times \text{Volumetric specific heat} \times \text{Temperature difference}$$

(W) (m³)   (Number of air changes per second)   (1300 J/m³ deg C)   (deg C)

or Population × Air change rate per person × Volumetric specific heat × Temperature difference

The total hourly rate of loss is the sum of QF and QV.

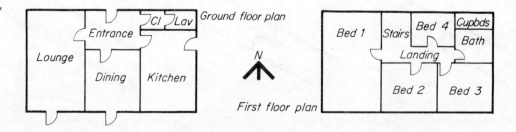

55 *Plan of building used as subject for heat loss calculation*

HEAT LOSS CALCULATIONS

The ventilation losses in a room are based on the volume of air passing through the room requiring to be warmed multiplied by the volumetric specific heat of air multiplied by the temperature difference. The volume of air to be warmed is usually determined by multiplying the appropriate number of air changes per hour (see Table 6) by the volume of the room. In some cases, as the table shows, a specific volume of air per person is used instead of a general rate of air change. Rates of air change are almost invariably quoted in terms of hours. For convenience in calculation this must be reduced to seconds by dividing by 3600. The amount of heat required to raise a cubic metre of air 1°C (volumetric specific heat) is 1300 joules.

For practical calculations involving numbers of rooms each with windows and different types of construction it is essential for clarity to set out the calculation in tabular form. The following notes and diagram 55 describe a simple building and Table 24 shows a typical calculation.

The individual room totals can be used to size the room heaters and the grand total to size the boilers. Allowances are added to the values calculated to allow for rapid heating up, for firing losses and for losses from distribution pipework.

Where a quick but approximate value is needed, for example, in the early stages of architectural design in order to estimate a boiler room size, the calculation can be simplified by taking into account only the external skin of the building and the total ventilation loss. (Table 27, Seasonal Heat Losses, shows a suitable form.)

Example of conventional heat loss calculation in SI units

Gives the maximum rate of heat loss in extreme conditions assuming a steady state. Gives data for sizing heating installation.

The building analysed in the example is a brick built two-storey house of about 140 m² in floor area.

Constructional details

Ground floor   solid.
First floor   timber joists, boards, plaster ceiling.
Roof   tiles on felt, 25 mm fibreglass insulation on joists, plaster ceiling.
External Walls   Cavity wall unventilated, 105 mm brick, 50 mm cavity, 150 mm insulating block, 12 mm plaster.
Windows   single glazing in wood surrounds.
Doors   fully glazed (treat as windows).
Rate of air change per hour   2 changes.
The heating installation will not provide hot water.
Desired temperatures   Ground floor   20°C.
                       First floor   16°C.
External temperature for design   0°C.
Exposure   Normal.

The calculation is accomplished by establishing the U-value of the various elements of construction and for each room in succession multiplying the area of each particular type of construction by the U-value and by the temperature difference between the air in the room and the air on the other side of the construction. The individual results for each type of construction are summed to give the total heat loss for the room which can be used to select appropriate heating appliances. The total heat loss from all the rooms plus an appropriate allowance for firing can be used to select a central boiler. Separate circuits would be designed in terms of the sum of the heat requirements in rooms served by each circuit.

Some form of tabulation of the calculation is desirable to assist working and to record the results. Table 24 shows a useful layout and gives the example calculation.

The result of the calculation is the nett heat requirement for the building. In selecting a boiler an appropriate allowance must be made for the efficiency of the system (often taken as 70% for gas or oil installations) together with losses through exposed pipes and an allowance for water heating if this is needed.

Seasonal heat loss

In addition to the maximum rate of heat loss it is often desirable to make an estimate of seasonal heat requirements. The Building Research Station, in BRS Digest 94, November 1956, has put forward a simple method for small houses, based on the concept of an average temperature difference between inside and outside over the heating season and average seasonal thermal transmittance through the various types of construction.

In this method it is assumed, as a result of

HEAT LOSS CALCULATIONS

Table 24 Example heat loss calculation

Construction	U value	Kitchen			Dining			GROUND FLOOR Lounge			Entrance		
		Area m^2	Temp diff °C	Rate of heat loss through element W	Area m^2	Temp diff °C	Rate of heat loss W	Area m^2	Temp diff °C	Rate of heat loss W	Area m^2	Temp diff °C	Rate of heat loss through element W
Ground floor (thermoplastic tile)	1.00	17.5	20	350	15.6	20	312	22.0	20	440	15	20	300
Intermediate floor (heat flow up)	1.4	17.5	4	98	15.6	4	87	22.0	4	123	15	4	84
Roof	0.6	—			—			—			—		
Walls*	1.0	17.0	20	340	7.6	20	152	28.4	20	568	13.3	20	266
Windows	5.7	5.0	20	570	3.0	20	342	6.0	20	684	3.5	20	399
Ventilation	1300† x 2 / 3600**	Vol 40 m^3	20	580	Vol 35 m^3	20	510	Vol 50 m^3	20	725	Vol 34 m^3	20	490
Room total				1938			1403			2540			1539

Total carried forward 7420

* Where more than one wall material occurs in a room additional lines should be used
† Volumetric specific heat of air J/m^3 deg C
** $\frac{2}{3600}$ represents 2 air changes per hour as a rate per second, necessary if watts are to be used

Note: Total wall area = 154 m^2, Total window area = 31.1 m^2. Therefore overall U-value = 1.8 which is acceptable.

HEAT LOSS CALCULATIONS

Table 25

Brought forward 7420 w

Construction	U value	GROUND FLOOR Cloaks			FIRST FLOOR* Bed 3			Bed 2			Bed 1		
		Area m²	Temp diff °C	Rate of heat loss W	Area m²	Temp diff °C	Rate of heat loss W	Area m²	Temp diff °C	Rate of heat loss W	Area m²	Temp diff °C	Rate of heat loss W
Ground floor	1.00	4	20	80	—			—			—		
Intermediate floor (heat flow up)	1.4	4	4	22	—			—			—		
Roof	0.6	—			12	16	115	14.5	16	139	22	16	211
Walls	1.0	11.0	20	220	16.5	16	264	7.6	20	152	30.4	16	486
Windows	5.7	1.0	20	114	2	16	182	2	16	182	4.0	16	365
Ventilation	$\frac{1300 \times 2}{3600}$	Vol 10 m³	20	145	Vol 27 m³	16	310	Vol 33 m³	16	380	Vol 50 m³	16	570
Room total				581			871			853			1632

Carried forward 11,357

* Heat gains from floor below have been ignored

HEAT

HEAT LOSS CALCULATIONS
Table 26

Construction	U value	Landing			Bed 4			Bath			Brought forward 11,357 w
		Area m^2	Temp diff $^\circ C$	Rate of heat loss W	Area m^2	Temp diff $^\circ C$	Rate of heat loss W	Area m^2	Temp diff $^\circ C$	Rate of heat loss W	
Roof	0.6	10.2	16	98	8	16	77	4	16	38	
Walls	1.0	8.0	16	128	8.4	16	134	5.6	16	90	
Windows	5.7	1.6	16	146	2	16	182	1.2	16	109	
Ventilation	$\frac{1300 \times 2}{3600}$	Vol 23 m^3	16	265	Vol 18 m^3	16	210	Vol 9 m^3	16	105	
Room total				637			603			342	

FIRST FLOOR

Total (maximum rate of heat loss) = 12,939 w

∴ Net rate of heat input required = 13 kW

Table 27 Seasonal heat loss calculation

The following calculation based on the BRS method gives an idea of seasonal heat requirement for the example building:

Heat losses

	Area	×	Seasonal transmittance coefficient W/m² °C	×	Average temperature difference deg C	=	Rate of heat loss through elements Watts
Roof	70 m²		0.85		10		600
Windows	32.5 m²		3.4		10		1100
External walls	160 m²		1.71		10		2700
Ground floor	70 m²		1.14		10		800
Ventilation	350 m³ × 2/3600 changes/sec × 1300				10		2500
					Rate of loss		__7700__ W

Total seasonal heat loss = 7700 W × 20 × 10⁶   (20 × 10⁶ is length of heating
 = __154 GJ__   season in seconds)

Heat gains

Windows:	South facing 12.2 m² at 680 MJ/m²	=	8.5 GJ
	North facing 20.3 m² at 250 MJ/m²	=	5.1 GJ
	Total window gains		13.6 GJ
Body heat (1 GJ per person)			4.0 GJ
Electricity (on estimated consumption)			3.3 GJ
Cooking (gas)			6.0 GJ
Water heating			2.0 GJ
	Total seasonal gains		28.9 GJ

Net heat requirement for season = 154 − 29 GJ
 = __125 GJ__

extensive field studies, that mean temperature differences between inside and outside can be established to represent the actual varying temperature differences over the heating period. The recommended temperature differences values were 10°C for full central heating and 6.7°C when heating was mainly confined to the ground floor. It may be thought desirable, in view of the tendency towards higher temperatures to make some upward adjustment of these figures to take account of changes in desired temperatures.

U-values for this method of calculation remain the same as for maximum rates of heat loss with two exceptions. Windows, the account for the effect of curtains, should be taken at 3.4 W/m² deg C instead of 5.7 (single glazing). The values for roofs were also abated but the figures are no longer relevant since higher insulation standards are now demanded. A reduction of the U-value for pitched roofs to 75 per cent would be consistent with the original BRS figures. Ventilation rates over over the season were found to vary from one change per hour to three changes per hour depending on degree of exposure and amount of window opening. For normal circumstances a rate of two changes per hour (2/3600 changes per second) is reasonable. The length of the heating season was taken at 33 weeks which is approximately 20×10^6 seconds.

A particular feature of the method is the inclusion of allowances for heat gains in order to reach a realistic estimate of heat requirements. Windows gain significant amounts of heat. The heat gains from unobstructed windows were given in the digest as

South facing	680 MJ/m²
East and West facing	410 MJ/m²
North facing	250 MJ/m²

and the value for cooking, electricity, water heating and heat gains from people are shown in Table 27.

Gross heat requirement will include an allowance for the efficiency of the system. A very useful chart of system efficiencies is given in BRS Digest (First Series) 133. They range from 40 per cent for open fires to 70 per cent for central heating and 100 per cent for electric fires.

$$\frac{\text{Gross heat}}{\text{requirement}} = \frac{\text{Net heat requirement} \times 100}{\text{\% efficiency of system}}$$

Table 47 gives the calorific values for a range of fuels and the value of fuel required can easily be calculated at current costs.

For large installations engineers use different systems for assessing seasonal losses. One is the 'degree day' method where for various areas of the country values have been established giving the temperature differences below a base temperature accumulated over the heating season. No account is taken of heat gains. A simpler method based on actual performance of installations is to take into account the number of hours of full capacity firing which occur during the heating season. To use this method data must be available for a building similar construction and use. Heat gains and losses are, however, automatically taken into account. A typical figure for an office or commercial building with full central heating and 40–50 per cent of wall area as window is 2500 hours of full capacity boiler output. The length of the heating season is approximately 33 weeks (5500 hours). It will be appreciated that this method is essentially similar to the BRS method for houses described previously and the calculation layout (heat loss section only) is the same except that instead of reducing the design temperature difference (reduced from 20 deg C to 10 and 6.7 deg C in BRS system) the length of the heating season is reduced. 2500 hours firing in 5500 hours of heating season is similar to a mean temperature difference over the heating season of 11 deg C instead of the maximum design temperature of 20 deg C.

Internal surface temperatures

In problems both of thermal comfort and condensation it is necessary to estimate surface temperatures. In steady static conditions, which are assumed for normal heating calculation, the temperature difference between the air and any part of the construction bears the same relationship to the total temperature difference from air (internal) to air (external) as does the thermal resistance of the section of construction giving rise to the difference to the total thermal resistance. In the case of interior surfaces.

$$\frac{\text{Temp diff} \begin{pmatrix} \text{room air to} \\ \text{wall surface} \end{pmatrix}}{\text{Temp diff} \begin{pmatrix} \text{room air to} \\ \text{external air} \end{pmatrix}} = \frac{\text{Int surface resistance}}{\text{Total thermal resistance of construction}}$$

The effects of corners and cold bridges must be considered in relation to condensation (see diagrams 52 and 53).

When intermittent heating is used or other marked fluctuations from steady static conditions are anticipated, then effects should be taken into account (eg heavy construction will be slow to warm up and surface temperatures lower than those estimated by the above method may be present for some time and give rise to discomfort at condensation.

ADMITTANCE METHOD

In recent years many buildings have suffered from overheating in the summer. Large windows, lightweight construction and facades exposed to the unobstructed sky have become increasingly common features of building design and can all contribute to overheating, as can the limitation of window opening often imposed by external noise levels in urban areas. Common sense in design can avoid

ADMITTANCE METHOD

overheating in many circumstances but it is necessary to be able to make estimates of the degree of overheating so that appropriate window areas, materials and levels of ventilation may be employed.

There is no legislation governing this aspect of building design but some government departments are calling for control of summer overheating in new buildings for which they are responsible.

As yet prediction methods are limited but one developed by the Building Research Establishment after extensive analogue studies is relevant and useful. It is called the Admittance method and can establish peak temperatures likely to be reached in rooms in the summer. The method is appropriate for rooms with only one external wall (ie surrounded by similar rooms on both sides, above, below and at the rear). This is not uncommon in office buildings, where overheating has been a particular problem, and may give a useful indication in many other cases.

The method first establishes the mean internal temperature over a 24 hour period and then establishes the maximum variation from the mean. These two values enable the peak temperature to be determined.

While many of the same concepts and values are used as in heat loss calculations, some new terms are employed. The most important are:

Admittance factor Measure of the heat entering a surface due to a 24 hour cycle of fluctuation of temperature. It takes into account the thermal capacity of the constructional element. Dense materials can absorb more heat and have higher admittances than lightweight ones.

Environmental temperature A single value which takes into account both air temperature and radiant conditions. Its use is essential in the admittance procedure. It is also being used in very accurate heat loss calculations particularly where radiant heatersaare employed.

Environmental temperature =
$\frac{2}{3}$ mean radiant temperature + $\frac{1}{3}$ air temperature

Solar gain factor Proportion of heat in the sun's rays transmitted into room taking into account the angle of the incident radiation and losses incurred in the glazing and any associated blinds.

Alternating solar gain factor Measure of the solar radiation gain taking place through glazing and blinds modified to take into account the effects of the thermal capacity of the structure.

There are three stages in the estimation
1 Establish the mean conditions taking into account gains and losses.
2 Determine the magnitude of the variation from the mean.
3 From 1 and 2 establish the peak environmental temperature.

The following step by step procedure with the arithmetic processes set out in simple boxes, together with the accompanying tables of data provide a quick and simple way of carrying out the calculation. At each stage the labels of the boxes should be read, the appropriate values entered and the arithmetic processes carried out.

STAGE 1   **Mean conditions**

A   GAINS

(i) Solar

Select daily mean solar gain from col c table 28		Solar gain factor for type of glazing from table 29		Window area (m^2)		Mean solar gain (W)	
	x		x		=		
	x		x		=		1

(ii) Other casual gains

Source	Rate (W)	x	Duration (hours)	÷ 24		Daily mean gain (W)	
Lighting (installed wattage)		x		÷ 24	=		
Occupants (see table 35) No. X rate per occupant		x		÷ 24	=		
Processes		x		÷ 24	=		
		x		÷ 24	=		
		x		÷ 24	=		
			TOTAL				2

HEAT

TOTAL GAINS

Mean solar gain W from (1)	+ Mean daily gain (W) from (2)	= Total mean gain (W)
	+	=	3

B LOSSES

(i) Fabric

External elements	Area (m²)	×	U-value (W/m² °C) From tables 21-23	=	Rate of loss (W/°C)
Window		×		=	
Wall		×		=	
		×		=	
			Total			4

(ii) Ventilation (see Table 30 for air change rates)

Ventilation rate (air changes/hr)	×	Room volume (m³)	× 0.33	=	Mean rate of ventilation loss (W/°C)
	×		× 0.33	=		5

This result is adequate for air change rates up to and including 2. For higher rates complete:

1	÷	Value from box (5)	=	Subtotal a
1	÷		=	

0.21	÷	Total area of all internal surfaces m²	=	Subtotal b
0.21	÷		=	

Total	a	+ b	=	Subtotal c
		+	=	

1	÷	Subtotal c	=	Mean rate of ventilation loss (W/°C)
	÷		=		6

ADMITTANCE METHOD

C MEAN ENVIRONMENTAL TEMPERATURE

Fabric loss (W/°C) from (4)	+	Ventilation loss (W/°C) from (5) or (6)	=	Total rate of loss (W/°C)
	+		=	

Total heat gain (W/°C) from (3)	÷	Total loss (W/°C) from (7)	=	sub total	+	Mean outdoor temp from table 31 (°C)	=	Mean internal environmental temperature (°C)	
	÷		=		+		=		10

STAGE 2 Variation from mean

Decide the hour of peak indoor temperature. In cases of doubt several probable times should be checked. Solar gains usually predominate in fixing the peak times and inspection of tables 28 (solar intensity) and 31 (typical temperatures) should show likely peak times. With lightweight construction (see table 32 for definition of light and heavy weight) all factors operate simultaneously. For heavy weight construction the solar gain for 2 hours prior to the time under consideration should be used to allow for heat absorption by the fabric.

A SOLAR VARIATION

Peak intensity of solar rad (W/m²) from Table 28 col d	−	Daily mean solar intensity from (W/m²) Table 28 col c	=	Effective peak input (W/m²)	×	Area of glass m²	=	Sub total	×	Alternating solar gain factor for type of glazing from Table 33	=	Swing in effective heat gain (W)	
	−		=		×		=		×		=		11

(Structural variations may normally be ignored. If in doubt consult IHVE *Guide 1970*.)

HEAT

B   CASUAL HEAT GAIN VARIATION

(i) Peak

Casual heat sources contributing at peak hour	Rate (W)
Lighting	
Occupants	
Processes	
Total	12

(ii) Variation

Peak from (12) (W)	Mean − from (2) (W)	= Casual gain variation (W)
	−	= 13

C   AIR TEMPERATURE VARIATIONS

Area of glazing (m^2)	U-value for type of glazing see Table 23 $(W/m^2\ ^\circ C)$	Heat transmitted $(W/^\circ C)$	Vent losses from (5) or (6)	Fabric & vent losses	Ext air temp swing from table 31 $(^\circ C)$	Air temp variation of effective input (W)
	×	=	+	=	×	= 14

D   TOTAL VARIATION IN EFFECTIVE HEAT INPUT

Solar variation from (11) (W)	Casual variation from (13) (W)	Air variation from (14) (W)	Total variation (W)
	+	+	= 15

ADMITTANCE METHOD

E   INTERNAL ENVIRONMENTAL TEMPERATURE SWING

(i) Area x admittance

Internal surface	Area (m^2)	Admittance x factor from table 34	Area x = admittance (W/^{o}C)
Window		x	=
Outside walls		x	=
Partitions		x	=
Floor		x	=
Ceiling		x	=
	Total		

16

(ii) Swing

Area x admittance from (16)	+ Vent gains from (5) or (6)	= Subtotal
	+	=

17

Total swing in effective heat input from (15)	÷ Area x admittances + vent gain subtotal from (17)	= Swing in internal environmental temperature (^{o}C)
	÷	=

18

STAGE 3   **Peak environmental temperature**

Mean environmental temperature from (10) (^{o}C)	+ Swing to environmental temperature from (18)	= Peak environmental temperature (^{o}C)
	+	=

HEAT

Example

As an example of the method consider a room 4 m × 4 m × 3 m high with a cavity brick external wall including a single glazed window of 5 m² facing south. The floor is woodblock or concrete the internal partitions plastered half brick walls, the ceiling plastered concrete. Two people work in the room for eight hours a day and 40 W/m² of lighting is installed but is not likely to be used in June when an estimate of peak temperature is desired. The windows will be opened during the day but closed at night. The resulting calculation is:

STAGE 1   **Mean conditions**

A   GAINS

(i) Solar

Select daily mean solar gain from col c table 28	×	Solar gain factor for type of glazing from table 29	×	Window area (m²)	=	Mean solar gain (W)	
155	×	0.76	×	5	=	589	(1)

(ii) Other casual gains

Source	Rate (W)	×	Duration (hours)	÷ 24	=	Daily mean gain (W)	
Lighting (installed wattage)		×		÷ 24	=		
Occupants (see table 35) No. × rate per occupant	280	×	8	÷ 24	=	93	
Processes		×		÷ 24	=		
		×		÷ 24	=		
		×		÷ 24	=		
			Total			93	(2)

ADMITTANCE METHOD

TOTAL GAINS

Mean solar gain (W) from (1)		Mean daily gain (W) from (2)		Total mean gain (W)	
589	+	93	=	682	(3)

B LOSSES

(i) Fabric

External elements	Area (m^2)	×	U-value ($W/m^2\ °C$) from tables 21-23	=	Rate of loss ($W/°C$)
Window	5	×	5.7	=	28.5
Wall	7	×	1.7	=	11.9
		×		=	
		Total			40.4

(4)

(ii) Ventilation (see Table 30 for air change rates)

Ventilation rate (air changes/hr)	×	Room volume (m^3)	× 0.33	=	Mean rate of ventilation loss ($W/°C$)
3	×	48	× 0.33	=	47.5

(5)

This result is adequate for air change rates up to and including 2. For higher rates complete:

1	÷	Value from box (5)	=	Subtotal a		0.21	÷	Total area of all internal surfaces m^2	=	Subtotal b
1	÷	48	=	0.021		0.21	÷	80	=	0.003

Total	a	+ b	=	Subtotal c		1	÷	Subtotal c	=	Mean rate of ventilation loss ($W/°C$)
	0.021	+ 0.003	=	0.024		1	÷	0.024	=	42

(6)

HEAT

C MEAN ENVIRONMENTAL TEMPERATURE

Fabric loss (W/°C) from (4)	+	Ventilation loss (W/°C) from (5) or (6)	=	Total rate of loss (W/°C)	
40	+	42	=	82	(7)

Total heat gain (W/°C) from (3)	÷	Total loss (W/°C) from (7)	=	sub total	+	Mean outdoor temp from table 31 (°C)	=	Mean internal environmental temperature (°C)	
682	÷	82	=	8.3	+	16.5	=	24.8	(10)

STAGE 2 Variation from mean

A SOLAR VARIATION

Peak intensity of solar rad (W/m²) from table 28 col d	−	Daily mean solar intensity table 28 col c	=	Effective peak input (W/m²)	×	Area of glass (m²)	=	sub total	×	Alternating solar gain factor for type of glazing from table 33	=	Swing in effective heat gain (W)	
540	−	155	=	385	×	5	=	1925	×	0.42	=	809	(11)

B CASUAL HEAT GAIN VARIATION

(i) Peak

Casual heat sources contributing at peak hour		Rate (W)	
Lighting	−		
Occupants	2 @ 140	280	
Processes	−		
	Total	280	(12)

100

ADMITTANCE METHOD

(ii) Variation

Peak from (12) (W)	Mean from (2) (W)	Casual gain variation (W)	
280	− 93	= 187	(13)

C AIR TEMPERATURE VARIATIONS

Area of glazing (m^2)	U-value for type of glazing see table 23 $(W/m^2\,°C)$	Heat transmitted $(W/°C)$	Vent losses from (5) or (6)	Fabric & vent losses	Ext air temp swing from table 31 $(°C)$	Air temp variation of effective input (W)	
5 ×	5.7 =	28.5 +	42 =	70.5 ×	7.5 =	529	(14)

D TOTAL VARIATION IN EFFECTIVE HEAT INPUT

Solar variation from (11) (W)	Casual variation from (13) (W)	Air variation from (14) (W)	Total variation (W)	
809 +	187 +	529 =	1525	(15)

E INTERNAL ENVIRONMENTAL TEMPERATURE SWING

(i) Area x admittance

Internal surface	Area (m^2)	Admittance factor from table 34	Area x admittance $(W/°C)$	
Window	5	× 5.6	= 28	
Outside walls	7	× 5	= 35	
Partitions	36	× 3	= 108	
Floor	16	× 3	= 48	
Ceiling	16	× 6	= 96	
		Total	315	(16)

HEAT

(ii) Swing

Area × admittance from (16)	+	Vent gains from (5) or (6)	=	Subtotal
315	+	42	=	357

(17)

Total swing in effective heat input from (15)	÷	Area × admittances + vent gain subtotal from (17)	=	Swing in internal environmental temperature (°C)
1525	÷	357	=	4.3

(18)

STAGE 3   Peak environmental temperature

Mean environmental temperature from (10) (°C)	+	Swing in environmental temperature from (18) (°C)	=	Peak environmental temperature (°C)
24.8	+	4.3	=	29.1

This room, even with windows open, will be very uncomfortable indeed and some modification is desirable. Window blinds or heat reflecting glass should be considered. Had the window opening been restricted by external noise, as is often the case in towns, conditions inside would have been substantially worse.

In the early stages of design generalised solutions presented graphically can be helpful in saving unnecessary calculations. Such a generalised solution is given in diagram 56.

ADMITTANCE METHOD

Table 28 Solar intensities of vertical surfaces at 52°N (W/m^2) calculated using sunpath diagrams and overlays for solar heat gain calculations by Peter Petherbridge, published by HMSO

Orientation	Month (Values are for about 21st day)	Hour of day, Solar Time (S at Noon)												24 hour Mean
		6	7	8	9	10	11	12	13	14	15	16	17	18	
EAST	June	595	705	715	660	525	330	130	130	125	110	95	70	55	195
	May and July	550	680	710	660	520	330	130	130	120	105	90	70	50	187
	April and August	400	620	690	650	510	320	120	120	110	105	80	55	30	158
	March and Sept		410	555	555	465	295	95	95	85	70	55	25		112
	Feb and Oct			330	430	375	250								70
SOUTH EAST	June	335	475	595	660	635	540	435	270	125	110	95	75	55	190
	May and July	330	485	630	665	650	580	450	330	120	115	90	70	50	197
	April and August	280	470	610	700	690	620	520	320	110	105	80	55	30	191
	March and Sept		350	575	680	715	670	560	390	195	70	55	30		179
	Feb and Oct			380	550	650	720	525	380	225	50	30			146
SOUTH	June	55	75	195	340	445	530	585	540	445	340	195	75	55	161
	May and July	50	70	230	395	500	580	610	580	500	395	230	70	50	177
	April and August	30	95	280	405	530	620	650	620	530	405	280	95	30	190
	March and Sept		110	285	450	585	695	715	695	585	450	285	110		206
	Feb and Oct			230	400	560	670	695	670	560	400	230			184
SOUTH WEST	June	55	75	95	110	125	270	435	540	635	660	595	475	335	190
	May and July	50	70	90	115	120	330	450	580	650	665	630	485	330	197
	April and August	30	55	80	105	110	320	520	620	690	700	610	470	280	191
	March and Sept		30	55	70	195	390	560	670	715	680	575	350		179
	Feb and Oct			30	50	225	380	525	720	650	550	380			146
WEST	June	55	70	95	110	125	130	130	330	525	660	715	705	595	195
	May and July	50	70	90	105	120	130	130	330	520	660	710	680	550	187
	April and August	30	55	80	105	110	120	120	320	510	650	690	620	400	158
	March and Sept		25	55	70	85	95	95	295	465	555	555	410		112
	Feb and Oct							250	375	430	330				70

HEAT

Table 29 Solar gain factors and alternating solar gain factors for various types of glazing and sun control

Window Construction	Table 29 Solar gain factors	Table 33 alternating solar gain factors	
		light weight building	heavy weight building
Single 4 mm clear glass	0.77	0.55	0.43
Double 6 mm clear glass	0.61	0.47	0.39
Single 4 mm clear glass with internal venetian blind	0.46	0.46	0.43
Double 6 mm clear glass with white venetian blind between glass	0.28	0.25	0.23
Single 4 mm clear glass with external canvas roller blind	0.11	0.09	0.07

These values have been selected from BRS Current Paper 47/68 (Crown Copyright 1968), *Summertime Temperatures in Buildings* by A.G. London, which gives more comprehensive tables and a basic explanation of the principles of the 'Admittance Method'.

Table 30 Ventilation rates for naturally ventilated buildings on sunny days

Position of opening windows	Usage of windows		Effective mean ventilation rate	
	Day	Night	Air changes (h^{-1})	Ventilation allowance $(W/m^3\ ^\circ C)$
One side only	Closed Open Open	Closed Closed Open	1 3 10	0.3 1.0 3.3
More than one side	Closed Open Open	Closed Closed Open	2 10 30	0.6 3.3 10.0

From IHVE *Guide 1970*

ADMITTANCE METHOD

Table 31a Mean and peak temperatures for typical sunny spells

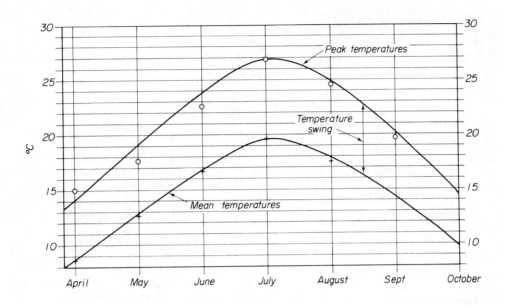

Table 31b Typical hourly temperatures for sunny spells (based on studies for sunniest fine day periods at Kew and Cardington)

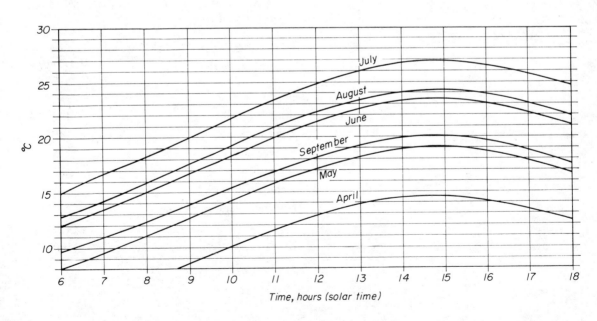

105

Table 32 Building classification by weight

Building classification	Construction	Average surface factor F (table 35)
Heavyweight	Solid internal walls and partitions, solid floors and solid ceilings.	0.5
Lightweight	Lightweight demountable partitions with suspended ceilings. Floors *either* solid with carpet or wood block finish *or* suspended type	0.8

From IHVE *Guide 1970*

Table 35 Heat output by young males for various degrees of activity

Activity	Heat output (W)
Seated, at rest	115
Light work, office	140
Seated, eating	145
Walking	160
Light bench work	235
Moderate work, or dancing	265
Heavy work	440
Exceptional effort	1500

From IHVE *Guide 1970*

Table 34 Admittance factors

Construction	Admittance (Y, W/m² °C)	
EXTERNAL WALLS		
Brick, Solid : 105 mm brick 16 mm dense plaster	4.1	
220 mm brick 16 mm dense plaster	4.4	
105 mm brick 16 mm light weight plaster	3.1	
220 mm brick 16 mm light weight plaster	3.4	
Brick, Cavity : 105 mm brick / 50 mm cavity / 105 mm brick / 16 mm dense plaster	4.3	
as above but light weight plaster	3.3	
as above but light weight concrete block inner leaf	2.9	
INTERNAL WALLS		
105 mm brick, 15 mm dense plaster each side	4.5	
75 mm light weight concrete block 15 mm dense plaster both sides	2.6	
Two fibre board sheets with cavity between	0.3	
ROOFS		
150 mm concrete with 19 mm asphalte on 75 mm screed with 15 mm dense plaster ceiling	5.1	
50 mm woodwool with 19 mm asphalte on 13 mm screed with cavity and 10 mm plaster board ceiling	1.5	
WINDOWS Use U-values	See Table 23	
FLOORS & CEILINGS	Floor	Ceiling
Timber 10 mm timber / cavity / 16 mm plaster board ceiling	0.1	0.3
Concrete 50 mm screed / 150 mm concrete	5.6	5.6
as above but with wood block as carpet floor finish	3.1	5.8

For comprehensive tables of Admittance Values and other thermal data for a wide range of constructions see Building Research Establishment Current Paper 61/74 (Crown Copyright 1974) Thermal Response and Admittance Procedure by N.O. Millbank and J. Harrington-Lynn from which the values have been taken.

HEAT

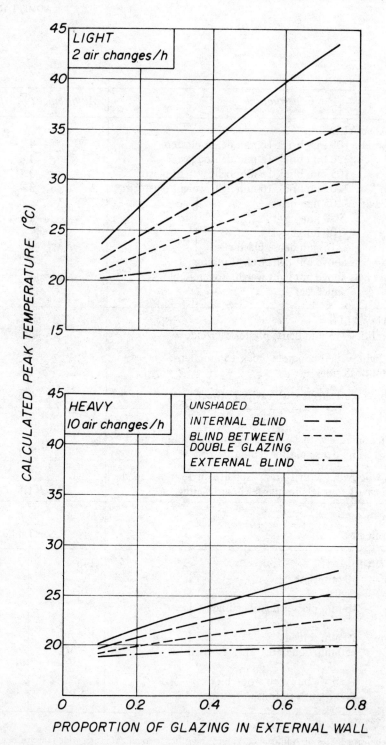

56 Examples of temperatures reached in typical building situations (based on drawings from IHVE Guide 1970)

6 Sound

GENERAL

Very great increases of noise resulting from increased motor and air traffic have occurred at the same time as developments in building materials and techniques giving lightweight components erected with dry and sometimes not well-sealed joints. These factors either separately, or in conjunction, frequently cause discomfort and dissatisfaction in modern buildings. Noise transmitted from one room to another or from one part of the building to other parts can often cause annoyance. In addition to noise penetration and transmission it is necessary in many rooms to consider the way in which sound behaves within the space itself in order to ensure that speech or music can be heard intelligibly.

Some appreciation of the nature of sound and the units of measurement and terminology is essential to more detailed consideration of the practical applications of sound insulation and room acoustics. Sound itself is the sensation produced by a certain range of rapid fluctuations of air pressure affecting the ear mechanism. Vibrations from a source of noise excite a similar movement in the molecules of air, which results in a series of pulses of increased pressure moving outwards from the original source. These vibrations can be transmitted, not only through air, but through any elastic medium including the materials making up the fabric of buildings, and from one medium to another.

The ear is able to react to frequencies of vibration of from 20 to 20 000 Hz (hertz or cycles per second) although with increasing age the ability to distinguish the higher frequencies diminishes. In normal usage the frequency is spoken of as pitch, low frequency sounds corresponding to low pitch and high frequency corresponding to high pitch. Sources of sound producing vibrations at only one frequency, known as pure tone, are rare in practice and normally sounds which are heard are made up of vibrations at many frequencies.

From a freestanding source sound moves outwards in all directions forming a spherical wavefront. The energy is therefore progressively spread as the sound waves move outward from the source and its intensity diminishes in proportion to the square of the distance from the source (diagram 57). When sound waves impinge on surfaces much of the energy is reflected. The reflection is similar to that of a ray of light falling on a mirror, the angle at which the sound leaves the surface being the same as the angle of arrival

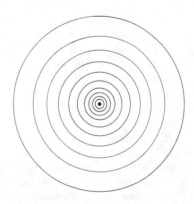

57 Sound propagated outwards from source diminishing in intensity with distance from the source. Multiple sources, although individually the same, will reinforce each other and the effect described will appear to be considerably reduced

(diagram 58). Convex surfaces will give dispersed reflections and concave surfaces concentrated reflections. While sound is propagated in straight lines obstructions to the waves do not cast sharp acoustic shadows. Particularly with the lower frequency, longer wavelength sounds diffraction round the object takes place (diagram 59).

Measurements of sound are normally made in terms of sound pressure which is the increase in air pressure created by the sound. Sound energy is the power per unit area. The ear is capable of

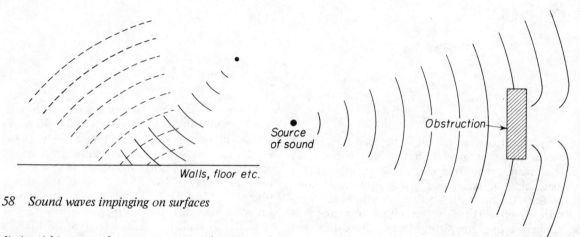

58 Sound waves impinging on surfaces

59 Sound waves diffracting round an obstacle

distinguishing sounds over a very wide range of sound pressures. In order to simplify the expression of values, to give better correspondence with the subjective appreciation of sound and to simplify insulation calculations a logarithmic scale giving the ration of sound pressures to a base level chosen at the threshold of hearing. This is described as a decibel scale (symbol dB). Values on the scale range from zero (threshold of hearing) to 130 (painful sound). An increase of 6dB at any point on the scale represents a doubling of pressure, while a decrease of 6dB represents halving. Doubling of sound energy is represented by 3dB. Decibels cannot be summed by simple addition. A calculation is needed to establish the overall value resulting from the simultaneous occurrence of two or more sounds of known decibel value.

In addition to defining sound levels the decibel scale is used to specify the ratios of sound reduction achieved by various elements of building construction. In the same way the overall insulation value of a wall containing more than one type of construction cannot be obtained by taking a weighted average of dB values in proportion to their areas. In fact the overall value is likely to be little more than the value of the least effective part of the wall unless the area of poor insulation is very small. British Standard Code of Practice 3, Chapter III, 1960 *Sound Insulation and Noise Reduction,* contains a more detailed treatment of the theoretical aspects of sound and the range of units employed than is appropriate here and contains a functional scale which enables combined noise levels to be established, a chart to enable the insulation value of non-uniform partitions to be established and a table of the insulation values of partitions with doors.

For the majority of purposes dB values representing the total sound energy or pressure over the whole frequency range are used. It is recognised, however, that sounds at different frequencies, although having the same pressure or energy, can appear to have different loudness when judged subjectively. Several scales have been developed which divide the sound into octave bands of different frequencies thereby enabling the importance of sounds at particular frequencies to be indicated. Noise Criteria (NC), Speech Interference Level (SIL) and dBA are examples of scales of this sort. The dBA scale is considered to correlate well with subjective judgments of loudness. Most data on typical noises, recommended standards and insulation performance of various constructions are given in normal dB values averaged over 100-3200 hertz and this is the scale that has been generally adopted here.

Some of the sound falling on surfaces is absorbed. The ratio of sound absorbed to the sound falling on the surface is described as the absorption coefficient. The coefficient varies at different frequencies. A value of 1 represents total absorption such as would be given by an open window.

Sound transmission through walls and floors has no direct connection with absorption. Transmission occurs as a result of the constructional members being set in vibration by incident sound and giving rise to vibrations in the air on the far side. The efficiency of walls and floors in preventing sound transmission depends upon their mass. The heavier the construction the less easily it is set into vibration and the higher its insulation value. Transmission straight through a partition or floor is known as direct transmission. It is clearly possible, however, for sound to be transmitted along structural members which link but do not divide rooms. Diagram 60 shows some possible paths of this type of transmission which is described as 'indirect' or flanking. Each particular type of construction will reduce the amount of sound transmitted by a fixed proportion

supporting the 'floating' floor finish. Floating floors should be isolated from structural walls and should not be continuous from room to room. Diagram 62 A, B and E show floors of this type.

Impacts on floors or other constructional members can result in sound transmission not only into adjacent spaces but also, particularly if the building is of framed construction, into other

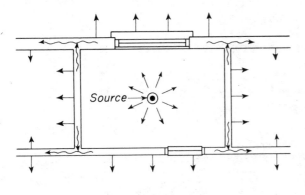

Plan

60 *Possible path for sound transmission through the structure of a building*

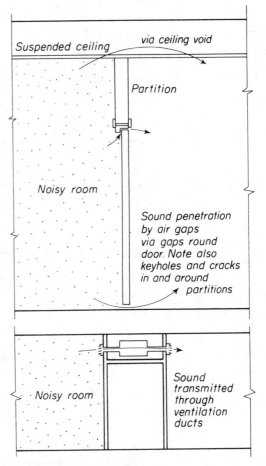

irrespective of the actual noise level. The insulation values can therefore be expressed in decibels. The reduction in noise levels is found by subtracting the insulation value in dB from the noise level on the other side of the construction. In practice this value may be modified by indirect transmission. Any air passages such as keyholes, gaps round doors (diagram 61) or gaps between wall and ceiling can allow sound to pass and seriously diminish the insulation value.

Sound insulation may be improved by *discontinuous construction*. To achieve this a wall is divided into two skins which are separate and consequently vibrations are not easily transmitted from one to the other. It is less easy to make floors discontinuous. A resilient quilt is usually employed laid on a structural floor and itself

61 *Possible path for sound transmission through gaps in building construction*

parts of the building. This problem can be met in two ways. Floating floors (described above and in diagram 62) will limit the transmission of vibrations into the main structure and soft finishes such as cork tiles or carpet will reduce the generation of vibrations. It should be noted that while floating floors provide improved insulation against

SOUND

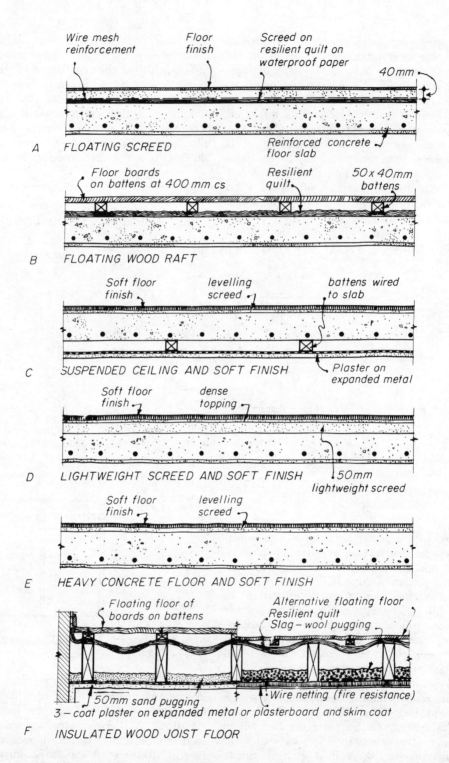

62 *Floors designed to reduce sound transmission*

both airborne and impact noise the soft finishes on solid floors do not improve insulation for airborne sounds nor would increased mass significantly reduce the impact noise transmission.

In addition to external noises and internal noises due to the occupants, considerable noise can be generated by services in buildings. In domestic buildings plumbing can be a source of considerable annoyance as can boilers and pumps and power-operated appliances such as dishwashers, waste-grinders and washing machines. In larger buildings lifts, boiler and refrigeration plants and other mechanical plant may give rise to noise.

NOISE CONTROL

In the practical design of buildings consideration of sound is likely to have three aspects: planning to keep noise sources as far as possible from quiet areas; structural precautions to reduce noise penetration and, in cases where it is appropriate, the internal acoustics of rooms.

Planning in relation to noise

It is clearly desirable to position buildings on sites as far as possible from sources of noise. In addition hard paving should be kept to a minimum and grass and planting used as much as possible. Hard paved re-entrant areas should be avoided. In some cases it may be thought desirable to interpose solid screens between noise sources and windows. For such screens to be effective it is important that they should be close either to the source or to the window. Their effect in intermediate positions may be negated by the diffraction of sound. High buildings do not usually give any useful protection from noise since while upper floors receive less sound from areas immediately adjacent, they receive more sound from distant sources. It is important to recognise not only noise sources external to the site but also site activities which may themselves give rise to annoyance. Children's play areas, refuse collection, deliveries or garage areas may all give rise to annoyance.

Internal planning arrangements should attempt to concentrate quiet rooms on facades remote from external noise sources and should also attempt to group noisy and quiet rooms separately and in isolation from each other.

Constructional precautions to reduce noise

The Building Regulations 1965 call for 'an adequate resistance to the transmission of airborne sound' for walls and for airborne and impact sound in the case of floors. No specific standard of sound reduction is specified although examples of types of construction which will be 'deemed to satisfy' the regulations are given. The Building Standards (Scotland) Regulations 1963 on the other hand specify particular levels of sound insulation which are the same as the grade 1 curves for flats and the noise standard curves for house party walls shown in diagram 41. Recommendations for many other circumstances are given in standard texts.

BRS Digests 102 and 103 (second series) set out recommended standards for insulation values for walls and floors and give examples of various constructions in relation to the standards. Diagram 64 shows curves defining the standards for both airborne and impact sound. It will be noted that the insulation value differs at different frequencies. It will be noted also that the house standard for party walls is more stringent than that for flats. This is a recognition both of the increased difficulty of achieving high standards in flat construction and of the acceptance, in this type of dwelling, of greater noise penetration. The grade II standard for flats does not conform with the present Building Regulations but is retained by BRS in relation to the improvement of existing dwellings or to circumstances to which the Building Regulations do not apply. Grade I insulation is considered to be the highest insulation practicable at present vertically between flats and in constructions reaching this standard noise from neighbours is said to be only a minor disturbance. Grade II insulation leaves at least half the tenants undisturbed but gives acute dissatisfaction to many.

Table 36, *Gradings of party walls in traditional dwellings,* is taken from BRS Digest 102 and shows how various constructions comply with the recommended standards. The table assumes sound construction without air gaps or easier paths of sound through flues or fireplaces. The classification of cavity walls is the same as solid

SOUND

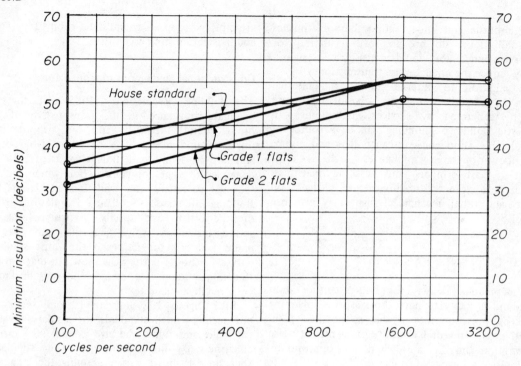

63 Grade curves for impact sound insulation

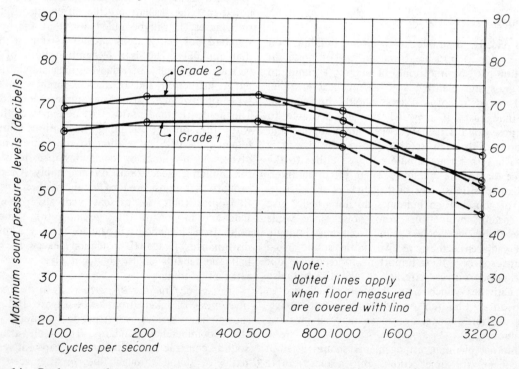

64 Grade curves for airborne sound insulation

NOISE CONTROL

Table 36 Gradings of party walls in traditional dwellings

Construction			Weight incl. any plaster	Grade
			kg/m^2	
Solid walls				
(a)	One-brick wall	pbs	415	P-w
(b)	In situ concrete or concrete panels with joints solidly grouted	plaster optional	415	P-w
(c)	175 mm concrete at 2320 kg/m^3	without plaster	–	P-w
(d)	175 mm concrete at 2080 kg/m^3	pbs	–	P-w
(e)	Lightweight concrete or other material	pbs	415	P-w
(f)	300 mm lightweight concrete at 1200 kg/m^3	pbs	–	P-w
(g)	225 mm no-fines concrete at 1600 kg/m^3	pbs	–	P-w
(h)	200 mm, as above	pbs	365	I
(i)	Lightweight concrete, or other material	pbs	220	II
(j)	Half-brick wall	pbs	220	II
Cavity walls, with wire ties of butterfly pattern and plastered on both sides		*Cavity width, mm*		
Two leaves each consisting of:				
(k)	100 mm brick, block or dense concrete	50	415	P-w
(l)	Lightweight aggregate concrete—sound absorbent surfaces to cavity (see text)	50	300	P-w
(m)	As above	75	250	P-w
(n)	As above	50	250	I
(o)	50 mm lightweight concrete at 1280 kg/m^3	25	–	II
(p)	100 mm hollow concrete blocks	50	–	II

Notes

    pbs = plastered on both sides, in dense two-coat work not less than 12 mm thick and weighing not less than 24 kg/m^2
    P-w = Party-wall grade (house standard in diagram 41)

    All weights, thicknesses and widths are minima

walls since evidence now shows that cavity walls do not have any real advantage over solid walls of the same weight. If lightweight concrete blocks are used with sound-absorbent sides presented to the cavity the weight of the wall can be reduced to 300 kg/m^2 or to 250 kg/m^2 if the width of the cavity is made 75 mm instead of the normal 50 mm.

Table 37, based on BRS Digest 103, gives the insulation grading of the typical floor construction in concrete shown in diagram 40. Some of the constructions are not satisfactory for impact sound, but are included for comparison. These are usually brought up to grade II if a resilient finish is applied, such as 4 mm-6 mm cork tiles, or 6 mm linoleum or soft rubber; or up to grade I if finished with fitted carpet, or rubber flooring on a sponge rubber underlay, or cork tiles 6–12 mm thick.

With floor (B), in diagram 62, the battens must not be fixed to the slab but merely rest on the quilt. Boards should be 21 mm tongued and grooved, battens should be 50 mm wide and minimum 30 mm deep. The resilient layer is best

SOUND

Table 38 Grading of wood-joist floors in flats

Construction	Sound insulating grading	
	Airborne	*Impact*
(i) Plain joist/floor with plaster-board and single-coat plaster ceiling (no pugging) Thin walls As above Thick walls	8 dB worse than Grade II 4 dB worse than Grade II	8 dB worse than Grade II 5 dB worse than Grade II
(ii) Plain joist floor with plaster-board and single-coat plaster ceiling and 15 kg/m² pugging on ceiling Thin walls As above Thick walls	4 dB worse than Grade II Possibly Grade II*	6 dB worse than Grade II Possibly Grade II*
(iii) Plain joist floor with heavy lath-and-plaster ceiling (no pugging) Thin walls As above Thick walls	Probably 4 dB worse than Grade II* Grade II	Probably 6 dB worse than Grade II* Grade II
(iv) Plain joist floor with lath-and-plaster ceiling and 85 kg/m² pugging on ceiling Thin walls As above Thick walls	Grade II Grade II or possibly Grade I*	Grade II Grade II
(v) Floating floor and plaster-board and single-coat plaster ceiling (no pugging) Thin walls As above Thick walls	4 dB worse than Grade II Possibly Grade II*	3 dB worse than Grade II Possibly Grade II*
(vi) Floating floor with single-coat plaster plaster-board ceiling and 15 kg/m² pugging on ceiling Thin walls As above Thick walls	2 dB worse than Grade II Grade II or possibly Grade I*	2 dB worse than Grade II Grade II or possibly Grade I*
(vii) Floating floor with heavy lath-and-plaster ceiling (no pugging) Thin walls As above Thick walls	2 dB worse than Grade II Grade II or I†	Grade II Grade I
(viii) Floating floor with lath-and-plaster ceiling and 15 kg/m² pugging on ceiling Thin walls As above Thick walls	Possibly Grade II* Grade II or I†	Grade II* Grade I
(ix) Floating floor with lath-and-plaster ceiling and 85 kg/m² pugging on ceiling Thin walls As above Thick walls	Probably Grade I Grade I	Probably Grade I Grade I

* Assumed from other measurements
† May give Grade I with very thick walls

Table 37 Grading of concrete floors in flats

Construction	Sound insulation grading	
	Airborne	Impact
(i) Concrete floor (reinforced concrete slab or concrete and hollow pot slab) weighing not less than 220 kg/m², with hard floor finish	Grade II	4 db worse than Grade II
(ii) Concrete floor with floor finish of wood boards or 6 mm thick linoleum or cork tiles	Grade II	Grade II
(iii) Concrete floor with floor finish of thick cork tiles or of rubber on sponge rubber underlay	Grade II	Probably Grade I
(iv) Concrete floor with floating concrete screed and any surface finish	Grade I	Grade I A*
(v) Concrete floor with floating wood raft	Grade I	Grade I B*
(vi) Concrete floor with suspended ceiling and hard floor finish	Probably Grade I	2 dB worse than Grade II
(vii) Concrete floor with suspended ceiling and wood board floor finish	Probably Grade I	Grade II
(viii) Concrete floor with suspended ceiling and floor finish of thick cork tiles or rubber on sponge-rubber underlay	Probably Grade I	Grade I C*
(ix) Concrete floor with 50 mm lightweight concrete screed and hard floor finish	Probably Grade	4 dB worse than Grade II
(x) Concrete floor with 50 mm lightweight concrete screed and floor finish of thick cork tiles or of rubber on sponge-rubber underlay	Probably Grade I	Probably Grade I D*
(xi) Concrete floor weighing not less than 365 kg/m² (reinforced concrete slab 150-175 mm thick) with hard floor finish	Grade I	4 dB worse than Grade II
(xii) Concrete floor weighing not less than 365 kg/m² with floor finish of thick cork tiles or of rubber on sponge-rubber underlay	Grade I	Grade I E*

* See diagram 62 for details of construction

in glass-wool or mineral-wool, 25 mm thick at a density of 8–9 kg/m³. This compresses to 9 mm under the battens. The quilt should be turned up against the walls and the skirting should be fixed to the walls and not to the floor.

In floor (A) the floating floor is a 1 : 2 : 4 concrete slab, 30 mm thick (50 mm thick is better), resting on a similar quilt, which should be covered by an extra layer of paper, at least over the joints in the quilt. Wire mesh is advisable in the floating slab as reinforcement. This slab should not be greater than 15–20 m² in area, or 5–7 m

SOUND

either way, as the edges tend to curl up in drying out, and this is noticeable in large areas.

Partitions should not be built off floating floors. Pipes and conduits should not be allowed to cause direct contact between the floating floor and the sub-floor. They can be on the sub-floor but should be haunched up in mortar so that the quilt passes over them and is continuously supported.

Lightweight screeds are not satisfactory for *impact* if too light. Minimum density should be 112 kg/m^3, and a dense topping is necessary to seal off all air passages.

Suspended ceilings are a benefit against airborne sound. It is important that the ceiling should be moderately heavy, not less than 25 kg/m^2, should not be too rigid, should be completely airtight and that points of suspension should be few and flexible.

Concrete floors are good for insulation because they brace and load the walls which they adjoin, so that flanking paths are not so liable to transmit noise. With wood floors it is possible to add insulation, as shown in diagram 62, to give better insulation but sound will still travel round by the supporting walls if they are thin, eg 112 mm brick or less.

Table 38 shows the grading of wood-joisted floors in flats, with the two conditions of thick walls and thin walls for each type of floor. To fulfil the thick wall condition two at least of the supporting walls require to be 225 mm thick: the walls above need not be so thick. Many systems shown do not in fact come up to the standard but are included for comparison, as they are common types of construction used in houses and are only tolerated because the family has to work out its own noise control as between its own members!

The ceiling of three-coat plaster on expanded metal lath is an important feature of several satisfactory constructions. It is important that the pugging, where it is used, should be directly on the ceiling.

Doors can provide easy passage of sound through otherwise well-insulated walls. Diagram 61 shows how sound can penetrate through the cracks round doors and through keyholes. Modern doors are often of hollow construction and do not in themselves provide much insulation. A hollow hardboard door with normal cracks gives about 15 dB reduction. This may not be critical in most cases where doors open to corridors rather than directly into other rooms. Where insulation must be maintained sealing of cracks by rubber strips and the use of heavier doors is desirable. If a cill is not acceptable automatic draught excluders may be used to close the crack at the foot of the door. Sometimes to improve insulation, doors are used on both faces of a partition. If the air sealing is effective shutting is likely to be difficult and a very much more effective system is to create a sound lock with two doors opening on to a small lobby finished with absorbent materials.

Suspended ceilings in office buildings where, for reasons fo flexibility, partitions do not extend up to the structural floor, can provide an easy path for sound penetration between rooms. Not only can sound pass through the gaps in the construction but if the ceiling is light and rigid it can itself transmit sound from one room to another. It should be remembered that the usefulness of ceilings for sound absorption is in no way affected by sound transmission. To achieve satisfactory insulation using an airtight ceiling as previously described, the space above should be as great as possible and in critical cases additional absorbent may be used above the ceiling.

External walls of traditional construction are normally heavy enough to give adequate sound insulation (275 mm cavity walls give 50 dB average reduction or more). Windows in external walls, however, present a critical problem. An open window will only give 5–10 dB reduction depending on whether it is fully or only slightly open. A closed single glazed window with 3 mm glass will give only about 20 dB. Sealing improves the effect up to about 25 dB and the use of heavy glass can improve things slightly more. To gain a significant further increase it is necessary to use double glazing. Unlike double glazing for thermal resistance which has only a narrow space between the sheets of glass, double glazing for sound insulation is best spaced to give about 200 mm air space between the two sheets of glass. The performance is significantly improved if the reveals are lined with absorbent. Diagram 65 shows a typical construction. One problem associated with this type of window is the prevention of condensation within the air space. It is not possible, in view of the very wide gap, to seal the two sheets of glass hermetically. The diagram

NOISE CONTROL

shows one method for keeping the air space dry. A tube of silica gel is used. When it has absorbed all the water that it can it is removed and heated, after which it can be replaced and will function again. Double glazing with 6 mm plate glass, and 200 mm air space lined with absorbent, can give an insulation value of 43 dB. For cleaning purposes it would be convenient if one leaf of the double window could open. It is difficult, however, at reasonable expense, to obtain a leaf that will open and also remain reasonably tightly sealed when closed.

From the values given above it is clear that for most circumstances where external noise is a problem windows will have to be sealed if satisfactory internal conditions are to be achieved. If this is to be the case air conditioning will be required to give ventilation and avoid overheating.

Many of the new constructional methods coming into use are based on lightweight panels with dry joints. With this type of construction it is particularly difficult to achieve satisfactory sound insulation. BRS Digest 96 *Sound Insulation* and new forms of construction, discusses the problem. The recommendation of the digest for party walls in lightweight construction is of two separate and separately supported skins with a wide (225 mm minimum) space between.

Digest 143, Sound Insulation Basic principles, describes the physical principles involved to complement the other Digests. Digests 128 and 129 deal specifically with insulation against external noise.

External noise indices

Some of the major external sources of noise such as motorways and air corridors are variable in the intensity, duration and frequency of the noises that they generate, and if standards are to be established they must take these factors into account. The problem is particularly important since householders can receive grants for precautions or compensation if they are exposed to some types of noise and these payments have to be rationally based. A number of special indices have been developed. The Traffic Noise Index (TNI) and the Noise Pollution Level (LNP) take into account the factors described but are complex in application and the '10 per cent level' (L_{10}) is thought to be a reasonable and practicable unit. The value is based upon the noise level in dBA which is exceeded for 10% of the time between 6 am and 12 midnight measured one metre away from the facade of the building.

The Noise Advisory Council recommends that for houses of normal construction, without special noise provisions, the L_{10} index should not exceed 70 dBA.

Aircraft noise presents a special problem. The loudness at the various frequencies differs from that of traffic noise and a special scale has been developed which weights the contributions of various frequencies so that the scale values correspond with subjective experience. The scale is called the

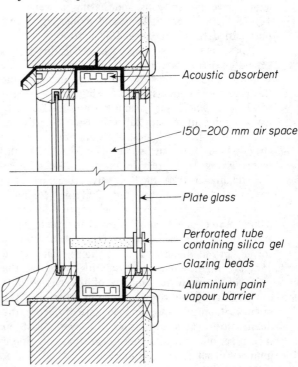

65 *Window designed to reduce sound transmission*

Perceived noise decibels (PNdB) scale. With aircraft in particular it is necessary to take into account the balance of noise and frequency and the noise and number index (NNI) has been developed. This index takes the loudness of the noise (L PNdB) and combines this with the number of aircraft by means of a formula.

Service installation

Where pipes or ducts cannot be sited remotely from quiet rooms it is possible that noise from the pipe or duct will cause annoyance. Flows in

SOUND

soil pipes resulting from WC or bath discharges can present a problem. Air flow in ventilating ducts also gives rise to noise and it is desirable to ensure that where such ducts are sited in rooms or above suspended ceilings the velocity of the air passing through is kept down to a level that does not cause a noise problem. This remedy is not possible for drainage pipework. The Building Research Station recommend a pipe casing, supported separately from the pipes, constructed so as to have no air gaps and having a weight of 10–25 kg/m² for kitchens and a heavier casing, unspecified, for bedrooms and living rooms.

Pipes and ducts may also transmit noises generated elsewhere. Firm fixing to heavy walls will minimise transmission to rooms and isolating the pipework or ducts from machinery by lengths of flexible pipe (usually canvas for ducts) usually has a significant effect. Ball valves are a particular source of noise in domestic installations, particularly where they are supplied direct from the mains giving high pressure on the valve. Use of the diaphragm type of ball valve (diagram 181) specially developed by the Building Research Station, will normally give substantial improvement. Water hammer in pipes can result from vibration of the ball valve arm, which should be solved by a diaphragm type valve. The loose jumper in the statutory isolating valve at the point of entry to the building of the service pipe can also give rise to water hammer as can the same feature of taps supplied directly from the main. Most water authorities, however, insist on this type of fitting in the open that the loose jumper will act as a non-return valve.

Mechanical equipment can also give rise to serious noise. The isolation of pipes and ducts to vibration from pumps, fans and motors has been described. The mechanical equipment should also be isolated from the structure of the building so that vibrations are not transmitted. In the case of very noisy apparatus, such as compressors for refrigeration plant, a separate foundation independent of the structure of the main building is best. In other cases where mounting on the main structure is unavoidable the mechanical equipment should be supported on anti-vibration mountings, which themselves are supported on an inertia block of concrete, sized to have a natural frequency different from that of the machinery, and itself resting on a resilient pad to minimise noise transmission to the structure.

It is important in domestic buildings to recognise that bathrooms and boiler installations are noise sources and that precautions against sound penetration will usually be needed. This is particularly true for bathrooms which cannot usually be planned remote from bedrooms. The Building Research Station recommend partition walls for bathrooms of not less than 75 mm of clinker concrete plastered on both sides. Wcs and cisterns should not be fixed to the dividing wall to a bedroom and solid core doors should be used.

There are no statutory requirements in this country for limitation of noise from plumbing and high levels of noise are not uncommon. 90–95 dB may be achieved. In some other countries there are statutory limits to the noise which is permissible.

ROOM ACOUSTICS

While in domestic circumstances the internal acoustics of rooms is not usually critical, in larger spaces such as lecture rooms, conference and music rooms or in assembly halls and auditoria the satisfactory use of space depends on satisfactory acoustic conditions. This is a problem of considerable complexity in auditoria and expert consultants will normally be employed to advise on the problem. There are a number of very detailed texts dealing with the subject. In the case of the smaller rooms it is possible to distinguish two principal aspects, apart from proper insulation, which govern satisfactory hearing. They are the planning and shape of the room to give good paths for sound from speakers or performers to the audience and the rate at which a sound dies away. This is called the reverberation time.

Direct sound

Ideally all the occupants of a space should be able to hear by means of a direct path of sound between the speaker or performers and each individual occupant. In conference, council or board rooms where any occupant may be called upon to speak, this is achieved by layout of seating so that all the occupants can see one another clearly. A long narrow board table will be much less satisfactory than a circular one. In

ROOM ACOUSTICS

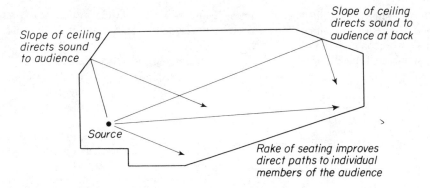

66 Shaping of auditorium to improve the paths of sound from source to audience

rooms where there is a large passive audience, it is possible to improve the direct paths of sound by raising the speaker and also by raking the seating so that each member of the audience is less obstructed by those in front. Sound reflected once from a surface (first reflection) can, if it is directed in the right direction, usefully augment the direct sound. Diagram 66 shows a notional section through a small auditorium where the speaker has been raised, the seating raked, and parts of the ceiling arranged to reflect sound to specific audience areas. It is important that the extra distance which first reflections have to travel over the direct path should not be too great. If the additional distance is over 20 m the reflected sound will arrive at the listener noticeably later than the direct sound and give rise to an echo. The Building Research Station in Digest 82 (second series), *Improving Room Acoustics,* recommend that difference in length of the two paths should not exceed 10 m and preferably be less. Care is therefore needed in disposing reflecting surfaces to ensure intelligible reinforcement of sound without echo. In general reflecting surfaces should be flat, curved surfaces may give rise to sharp concentrations of sound in some places and lack of reinforcement elsewhere.

In the same way that properly disposed reflecting surfaces can assist in distributing sound, inappropriate surfaces will have an undesirable effect. Cross-beams projecting below a general ceiling level are very undesirable since they prevent part of the ceiling from reflecting sound to the audience. Reflecting surfaces do not necessarily have to be part of the main fabric of the building. In very large halls it is not unusual to hang a reflector over the speaker's position. The Building Research Station recommend as a minimum thickness and weight a surface of 25 mm plasterboard, painted on the side used for reflection.

Reverberation time

Sound emitted from a source will be reflected not once but several times within a space. Diagram 67 shows the early stages of propagation. If the inter-reflections continue for too long the echo effect described above comes into effect for a multiplicity of reflected paths and intelligibility suffers. Standards to govern this effect are based on the time taken for the sound level in the room to

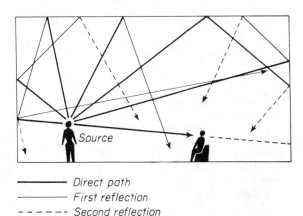

———— Direct path
———— First reflection
- - - - - Second reflection

67 The early stages of propagation of reflected sound

SOUND

decay by 60 dB. This is described as the reverberation time. The BRS recommend the reverberation times in seconds for various sizes and uses of rooms shown in Table 39.

Table 39

Use	Small rooms 750 m^3	Medium rooms 750-7500 m^3	Large rooms over 7500 m^3
For speech	0.75	0.75-1	1
Multi-purpose eg school halls	1	1-1.25	1-1.5
For music	1.5	1.5-2	2 or more

These reverberation times apply over the whole range of frequencies except that a longer time at low frequencies may be desirable for music.

Reverberation time depends upon the volume of the room and the amount of sound absorption given by the surfaces. Sound absorbents perform differently at different frequencies and it is necessary when estimating reverberation times to make separate calculations at different frequencies. For complex analysis a full octave band analysis may be undertaken but for most situations checking at 125, 500 and 2000 hertz is considered appropriate. Several formulae exist for establishing reverberation time, each with special relevance to particular conditions. The original formula and one which is widely used at present is that developed by Sabine and is, in SI terms

$$\text{Reverberation time (seconds)} = \frac{0.16 \times \text{volume in m}^3}{\text{total absorption (sum of areas in m}^2 \text{ multiplied by their absorption coefficients)}}$$

If satisfactory values are not found in a particular case it will normally be necessary to vary the absorption by varying the surface finishes. Theoretically the volume could be varied but this is rarely a possibility in practice. A number of different types of absorbent may be required to achieve a proper balance since the performance of absorbents varies greatly with frequency. Fibre-boards and similar soft materials commonly used for absorbents are very efficient at high frequencies but not at low ones. A thin panel concealing a space lined with absorbent is efficient for low frequencies. Some proprietary acoustic panels have a combination of materials which give a more balanced performance over the range of frequencies. When positioning absorbents, surfaces giving important reflections should clearly be avoided. In general back walls will be a first choice and ceilings a last choice of position.

The absorption coefficients of some common surfaces are given in table 40. See *Materials* for a more comprehensive list.

Table 40 Sound absorption coefficients

Item	Unit	Absorptive coefficient		
		125 Hz	500 Hz	2000 Hz
Air	m^3	0	0	0.007
Audience (padded seats)	person	0.17	0.43	0.47
Seats (padded)	seat	0.08	0.16	0.19
Boarding or battens on solid wall	m^2	0.3	0.1	0.1
Brickwork	m^2	0.02	0.02	0.04
Woodblock, cork, lino, rubber floor	m^2	0.02	0.05	0.1
Floor tiles (hard)	m^2	0.03	0.03	0.05
Plaster	m^2	0.02	0.02	0.04
Window (5 mm)	m^2	0.2	0.1	0.05
Curtains (heavy)	m^2	0.1	0.4	0.5
Fibreboard with space behind	m^2	0.3	0.3	0.3
Ply panel over air space with absorbent	m^2	0.4	0.15	0.1
Suspended plaster-board ceiling	m^2	0.2	0.1	0.04

Table 41 Example of reverberation time calculation

Case considered: Small multi-purpose room 6 m x 10 m x 3.5 m
Desired reverberation time 1 second at all frequencies
Assumed population 10 persons

A first estimation, no absorbents

Surface	Finish	Area m^2	Absorption coefficients and units					
			Low freq (125 Hz)		Med freq (500 Hz)		High freq (2000 Hz)	
			Absorb co-eff	Absorb units	Absorb co-eff	Absorb units	Absorb co-eff	Absorb units
Ceiling	Plaster (or concrete)	60	0.03	1.8	0.02	1.2	0.04	2.4
Walls	Plaster	92	0.02	1.8	0.02	1.8	0.04	3.6
Window	Glass (4 mm)	20	0.2	4.0	0.1	2.0	0.05	1.0
Floor	Wood block	60	0.02	1.2	0.05	3.0	0.1	6.0
Occupants		No 10	0.17	1.7	0.43	4.3	0.47	4.7
Air		210 m^3	–	–	–	–	0.007	1.5
	Total absorption units			10.5		12.3		19.3
	Reverberation time (secs) $\left(= \dfrac{0.16 \times 210}{\text{Total absorption units}}\right)$			3.2		2.7		1.7

This is not a satisfactory pattern. More absorption is required particularly at the lower frequencies. (A total of 33.6 absorption units are needed at each frequency to give the desired reverberation time.) A suspended plaster board ceiling gives very much better absorption at low frequencies than plaster direct on concrete. Plywood panels backed with absorbent have similar characteristics as do windows. It is reasonable therefore to attempt to correct the reverberation time by using a suspended ceiling, increasing the window size and introducing an area of absorbent backed ply. Since a substantial reduction is required at all frequencies an area of carpet, which is a very efficient absorbent, may be included with advantage. The result of these modifications is shown in table 42.

Standing waves

In rooms with parallel walls 10 m or less apart it is possible for particular wavelengths of sound to appear intensified at the expense of others. This effect is emphasised if all the dimensions are the same (in effect a cube) or if the dimensions are related by simple ratios (eg 1 : 2 or 1 : 3). Carefully selected relationships of dimensions, opposing walls out of parallel and absorbents to operate at the critical frequencies are means of avoiding difficulties from this phenomenon. A brief theoretical treatment of this problem is to be found in *Acoustics and Noise in Buildings*, by Parkin and Humphries.

SOUND

Table 42 Example of reverberation time calculation—continued

B Second estimation, using absorbents described

Surface	Finish	Area m²	Absorbent coefficient and units					
			Low freq (125 Hz)		Med freq (500 Hz)		High freq (2000 Hz)	
			Absorb co-eff	Absorb units	Absorb co-eff	Absorb units	Absorb co-eff	Absorb units
Ceiling	Suspended plaster-board	60	0.02	12.0	0.1	6.0	0.04	2.4
Walls	Plaster	62	0.02	1.2	0.02	1.2	0.04	2.5
	Ply panel with absorbent backing	30	0.3	9.0	0.15	4.5	0.1	3.0
Window	Glass	30	0.2	6.0	0.1	3.0	0.05	1.5
Floor	Wood blocks	30	0.02	0.6	0.05	1.5	0.1	3.0
	Carpet	30	0.1	3.0	0.3	9.0	0.5	15.0
Occupants	No. 10		0.17	1.7	0.43	4.3	0.47	4.7
Air		210 m²	—	—	—	—	0.007	1.5
	Total absorption units			33.5		29.5		34.6
	Reverberation time $\left(\dfrac{0.16 \times 210}{\text{Total abs units}}\right)$			1.0		1.1		0.97

The balance of absorbents selected appears to have made a very satisfactory correction

ENVIRONMENTAL SERVICES

7 Thermal installations

Most buildings require an installation to maintain a satisfactory thermal balance. In some parts of the world this will involve only the extraction of heat, in others heating and cooling seasons are involved. In Britain, although an increasing number of buildings, particularly those with high levels of artificial lighting or high ones with large proportions of glazing, must be provided with facilities for cooling, the majority of buildings are thought to give adequate environmental control in summer by their form and fabric alone and it is only in the winter that additional heat is required. To make a satisfactory provision for this it is necessary to decide on the source of energy which is to be used and the means of distributing the heat derived from that energy into the building. The main criteria involved in this decision are amenity and economy although in most cases there will be constraints limiting the range of decision. These may include the preferences of the building owner; the non-availability or unsatisfactory supply record of a particular fuel; the Clean Air Act; impracticable or uneconomic flue requirements; limited site area precluding storage or many other factors arising from a particular situation. Consideration of amenity must include the factors given in Table 43.

Some of these factors, such as the cost of fuel over the life of the installation, are speculative and it is difficult to arrive at a solution by averaging the costs over a period of years except in the case of temporary buildings whose life is defined. In such cases the most economical answer is often to reduce the initial cost as much as possible. A method of approach to the problem which may be helpful in reaching a decision is shown in diagram 68. It shows the cumulative costs of two different installations, of different capital and fuel costs expressed in relation to time. If the two lines did not meet the solution would be clear and simple. It will be more usual, if the systems being considered are reasonable choices, for the lines to cross after a period of years. If this occurs very soon then it will clearly be more economical to employ the system giving lower cost in all but the first few years. Where the changeover is many

Table 43 Amenity factors to be considered in selecting a heating system

1 *Thermal comfort*
   taking into account not only the appropriate type of installation but also the degree of control required to maintain good conditions (also related to economy).

2 *Appearance*
   Internally: form and positions of radiators, exposed pipe runs, grilles, thermostats, etc. Externally: form and position of boiler rooms, fuel stores flue, etc.

3 *Planning*
   Generally: access for fuel delivery and positions of apparatus compatible with the general planning arrangements.
   Individual interiors: satisfactory arrangement of heat emitters to allow freedom in disposing furniture and resolution of conflicting claim for space (eg radiators and windows may compete).

4 *Maintenance*
   both routine attention and servicing and replacement should be possible without gross interference with the use of the building.

The economic factors to be considered must include:

1 *Initial cost*
   (a) Cost of the installation.
   (b) Cost of the accommodation (eg boiler room, flue ducts etc) and access (eg service road for fuel delivery) required for the installation.

2 *Running cost*
   (a) Cost of fuel (which will certainly vary during the life of the building). Appropriate control of output may be important for fuel economy particularly in buildings with rapid reaction to external changes and intermittently used buildings.
   (b) Repayment of loan charges or allowance for replacement.
   (c) Inspection, maintenance and insurance.
   (d) Differences in decorating costs which result from different types of installation.

THERMAL INSTALLATIONS

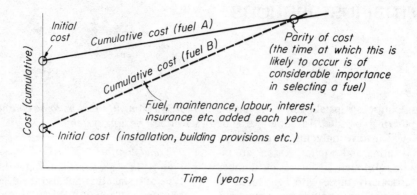

68 *Graph comparing the cumulative costs of alternative heating systems*

years ahead the decision is more difficult and may involve speculative assumptions about future fuel costs. In cases like this where the economic decision is not clear it may be best to assume that economically the possibilities are the same and to base selection on considerations of amenity.

While the above method will be useful in simple cases of comparison it has the disadvantage of giving equal importance to expenditure now and expenditure in the future. It is possible to use the principles of compound interest to evaluate the present investment that is the equivalent of a payment in the future. If all future payments are assessed in this way it is possible to arrive at a present value equivalent to all the expenditure involved and several schemes having different balances of initial running cost can be evaluated and compared. The method is frequently used for business management and is called 'Discounted Cash Flow'.

An example of the use of this method is given in Table 44. One or more heating installations are to be compared. The table shows the costs of one installation, an oil-fired low pressure hot water space heating system using radiators. The installation is to be considered as having a life of 15 years with no residual value at the end. The initial and running costs (including builders' work) are set out in the table, in the pattern in which they occur, together with the conversion factors enabling the expenditure each year to be converted to equivalent present value. These present value factors are selected from tables (see references below). The present value can then be calculated as shown. The various different schemes may then be assessed and a decision reached.

Use of this method involves a decision about the percentage of interest to be taken into account in selecting the present value conversion factors. This should be taken at the average rate of return of money which could be invested by the building owner. Higher rates of interest will reduce the present value of future expenditure while lower ones will increase its importance. When there is doubt as to the appropriate interest rate, more than one may be calculated and the results compared to see whether the effect is significant. In assessing running costs inflation will not normally be taken into account since if it affects all prices equally, comparison is valid ignoring inflation.

In more complex examples than the one given a variety of other factors could have to be taken into account including, in the case of business organisations, investment allowances and taxation rates.

In some cases in buildings, particularly those which are to be rented, it is important for the owner to be able to express the cost in terms of annual expenditure rather than present value. A conversion from present value to equivalent annual expenditure can be made also using tables based on compound interest.

The conversion factor giving annual equivalent values of an initial payment, taken over a period of 15 years at an interest rate of 7 per cent, is 0.11. In the case under consideration therefore, the annual value is £1886 x 0.11 = £207 per annum.

ENERGY SOURCES

Table 44 Discounted cash flow method of estimating equivalent present value

Year	Installation	Fuel	Mortgage interest (less tax)	Annual maintenance	Periodic overhaul	Decoration	Total cash flow	Conversion factor for present value at 7% interest	Present value
Base	550	45	12			60	667	1	667
+1		90	25	5			120	0.93	110
+2		90	25	5			120	0.87	104
+3		90	25	5	20		140	0.82	114
+4		90	25	5		45	165	0.76	126
+5		90	25	5			120	0.71	85
+6		90	25	5	20		140	0.67	94
+7		90	25	5			120	0.62	74
+8		90	25	5		45	165	0.58	96
+9		90	25	5	20		140	0.54	76
+10		90	25	5			120	0.51	61
+11		90	25	5			120	0.48	57
+12		90	25	5	20	45	185	0.44	82
+13		90	25	5			120	0.42	50
+14		90	25	5			120	0.39	47
+15		90	25	5			120	0.36	43

Total present value £1,886

A full description of these techniques and tables and graphs of the conversion factors required can be found in *Building Design Evaluation: Costs-in-Use*, by Peter Stone, published by Spon.

ENERGY SOURCES, HEAT TRANSFER MECHANISMS

The energy for heating is commonly derived from the various types of solid fuels and oil or from gas and electricity. It is also possible to employ the sun's energy or to extract heat from the air or from a river by means of the heat pump and use this for heating. In Britain the balance of cost for some twenty years has been heavily weighted in favour of the very cheap imported oil which has progressively superseded the use of indigenous coal and limited other developments. Because of the unbalance between day and night demand, electricity has been offered at low prices during off-peak hours to encourage use during these periods. Both these situations have now changed. Oil prices have increased sharply and day and night consumption of electricity is becoming more balanced, making substantial inducements for off-peak use self defeating. Purely economic decisions about fuel selection are difficult, partly because of uncertainty about fuel

THERMAL INSTALLATIONS

Table 45 Use of various fuels for heating buildings and the efficiency of their production and distribution

Fuel	% of energy used in domestic heating	% efficiency in terms of primary energy required
Electricity	20	27
Manufactured fuel	24	71
Coal	30	98
Natural gas	16	94
Oil	10	93

prices and partly because problems of conservation, supply and ecology must now be borne in mind. It is clear, however, that in the future, whatever the problems are, much closer attention will have to be given in the design of buildings to economy of fuel consumption.

It is important to remember that maintaining appropriate thermal conditions in buildings is thought to absorb almost 40% of total energy production and that in Britain the percentage is almost certainly higher. The balance of use between the main fuels is shown in table 45 together with the approximate efficiency of the production and distribution processes. (1973 data.)

Each of the usual fuels must be converted to heat in an appropriate piece of apparatus. Traditionally fuel could be moved about buildings but the controlled distribution of heat was difficult, consequently the fuel was burnt in the space where heating was required and the heat emitted directly to the

Table 46 Type of energy to heat transfer mechanism

Fuel or energy source	Direct	Central	
	Heat emitted directly to environment	Heat transferred to a distributing medium for transmission to required situation	
		Apparatus	Usual distribution medium
Solid	Fires Stoves	Boiler Warm air is sometimes produced from solid fuel appliances for ducted distribution in houses, but is uncommon in larger buildings	Water
Oil	Oil heaters	Boiler Oil/air furnace	Water Air
Gas	Convectors Radiant heaters	Boiler Gas/air furnace	Water Air

Electricity	Convectors
	Thermal storage:
	  Floors
	  Block heaters
	Radiant sources:
	  High, medium and low temperature
		Boiler: Electrode immersion heater
		
		Duct air heater
			Water
			
			Air

Solar radiation	Governed by orientation and fenestration	Collector mechanisms	Various (often air or water)
General environment		Heat pump	Refrigerant to air or water

ENERGY SOURCES

Table 47 Fuel data

Fuel		Calorific value		Weight and volume	Storage		Delivery	Plant
		MJ/kg	GJ/m^3	kg/m^3	Domestic	Non-Domestic		
Solid fuel	Anthracite	35	26	75	Floor areas for fuel storage: House with two types of fuel—2m^2 House with one type of fuel—1.2m^2	Allow storage for 4-6 weeks	Domestic: sacks Non-domestic: Direct to fuel stores by lorry (access road required)	Central boiler room. Flue
	Bituminous Coal*	30	22	72				
	Coke	28	11	38				
Oil†	Domestic (35 sec)	45	37	83	Tank Min. bulk delivery 0.5m^3 Economic bulk delivery 2m^3 (2.5m^3 tank is often used)	Tank for 3 weeks (Heating of stored fuel may be necessary)	Hose from road tanker to storage tanks. Storage remote from road may be served by site pipe-line	Boiler room. Flue (Note also oil/air furnaces)
	Light (200 sec)							
Town Gas			0.02				Large meters may require lorry access	Boiler room. Flue Meter
Electricity		3.6 MJ per kwh					Transformers may require lorry access	Switch and distribution gear Transformer station in some cases

* Store not more than 2.5m deep and in batches of not more than 280m^3 to eliminate risk of spontaneous combustion.
† Storage above ground is recommended with catchpit in case of leakage. (Consider access to sludge drain cock)

Useful publications on fuels, fuel delivery and storage can be obtained from: Coal Utilisation Council; Gas Council; British Electrical Development Association; Oil Companies

room. Control was achieved by varying the rate of combustion. Over the last 100 years effective methods of transmitting heat round buildings have been developed. They enable fuel to be converted to heat at a central point, thereby reducing the number of flues required to one, eliminating the need to distribute fuel and simplifying maintenance and control. In new buildings, including houses, this type of central installation is almost universal. The significant exception is where electricity is used as the fuel. The electricity is more easily and economically distributed than heat would be and has no special requirements such as flues. Electric underfloor heating in particular is often installed in new buildings. Table 46 shows the fuels together with the main types of apparatus for heat transfer either by direct emission or by central distribution. Except for electrical appliances and the decorative use of

THERMAL INSTALLATIONS

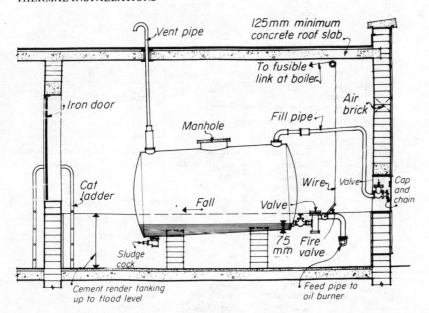

69 Typical store for oil fuel

solid fuel, direct heaters are rarely used except in low cost housing.

The selection of a particular type of fuel, involves the provision of storage and delivery facilities. Even in the case of gas and electricity, which are not themselves stored, meters and switchgear have to be delivered and accommodated. In large buildings these items may be of considerable size, requiring access for large lorries. This is particularly the case where transformer stations are provided. Table 47 gives details of the calorific values, weights and volumes, storage provision and delivery and plant requirements for the main fuels and suggests further sources of more detailed information.

Diagram 69 shows a typical oil fuel store. The sump to receive the oil should leakage occur, the fire valve to shut off the supply in case of fire in the boiler room, the sludge drain, and the fire protection of the store will be noted. As an additional precaution against overflowing it is possible to have an audible alarm fitted to the vent. It is not necessary for the oil fuel store to have direct road access for delivery vehicles. A length of hose is carried by each vehicle and where this will not be adequate a fixed pipeline can be laid across the building site itself.

Diagrams 70 and 71 show domestic and large scale storage for solid fuel. Delivery to the domestic store is usually by porter and sack while in the case of the larger store with access for large lorries, tipping directly into the store is needed.

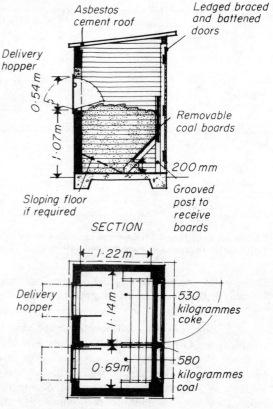

70 Typical store for domestic solid fuel

ENERGY SOURCES

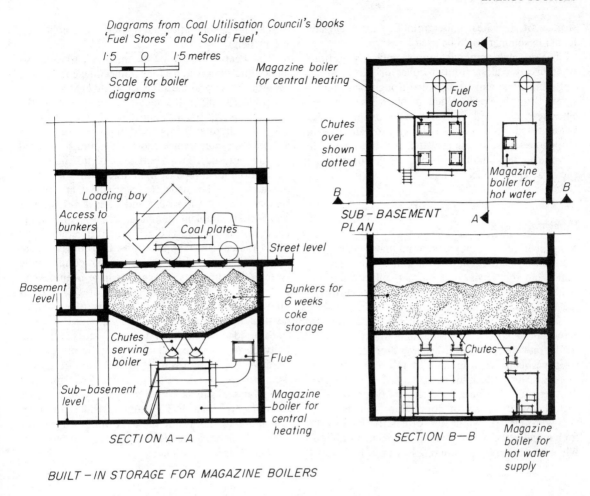

71 *Typical store for solid fuel for a large heating installation*

The importance of the sun in thermal comfort has been emphasised (page 57). In some countries which have a substantial percentage of sunshine it is possible to have solar heating installations which can collect and store the heat and deliver it to the building as required. In Britain the dull periods without sunshine are too long for economic storage of the heat and consequently although the sun when shining can exercise a decisive influence on heating and give reduced fuel bills if the heating system is able to react and reduce output solar heating installations have not been practicable. Some experimental installations giving hot water in summer have been built and plans published for long term

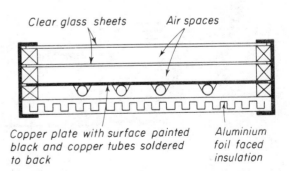

72 *Solar heat collector*

THERMAL INSTALLATIONS

storage of solar heat underground so that heat collected in summer could be used in winter.

The principle employed in most solar installations is to expose glass fronted collector boxes at an optimum angle and to pass air through or to pump water through pipes welded to black metal plates. The heat collected is stored perhaps in a large vessel filled with gravel in the case of air or in a storage vessel in the case of water. It is then available for use in the building when needed. Diagram 72 shows a typical collector box

Heat pump

A number of heat pump systems have been used for heating in this country including one drawing heat from the Thames to warm the Festival Hall. This has been dismantled for many years. However the costs of basic fuels have greatly increased and it is likely that the capability of the heat pump to deliver more energy into a building than is required to operate the pump will become more important. It is particularly significant that heat pumps are most usually driven by electricity. Since the production and distribution of electricity operates at such a low efficiency of conversion from primary energy to electricity the use of electricity directly for heating in the future seems likely to be inhibited. When used to drive a heat pump, however, the overall efficiency of operation compares favourably with other fuels.

The heat pump is a fundamental part of most air conditioning systems for summertime cooling and essential to several methods of waste heat recovery. Diagram 73 shows the cycle of operation of the heat pump. Its functioning depends on the use of a liquid with suitable vaporisation properties (known as the refrigerant, since most heat pumps are used for this purpose). The latent heat of this liquid, by circulation in a pipe system under low and high pressures, can be made to absorb heat at ambient temperature by vaporising and to give out heat at more than ambient temperature by condensing. Ammonia is a cheap and efficient refrigerant often used for commercial applications. For domestic heat pumps such as refrigerators and air conditioning systems less noxious refrigerants have been developed. One widely used refrigerant of this type is known as Freon. If the heat pump is to be used for heating it is essential that the amount of heat transferred through the system represents a substantially greater fuel cost than the electric power used to drive the compressor. If this is not the case it will not be worth installing the expensive heat pump and a simple boiler heating system would be more economical. The same consideration does not apply to the use of the heat pump for cooling since for most air conditioning applications it is the only feasible system.

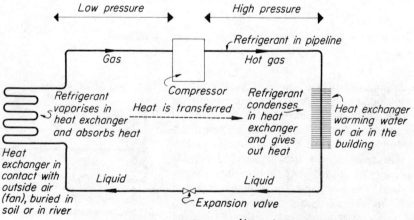

73 *Principles of operation of the heat pump*

ENERGY SOURCES

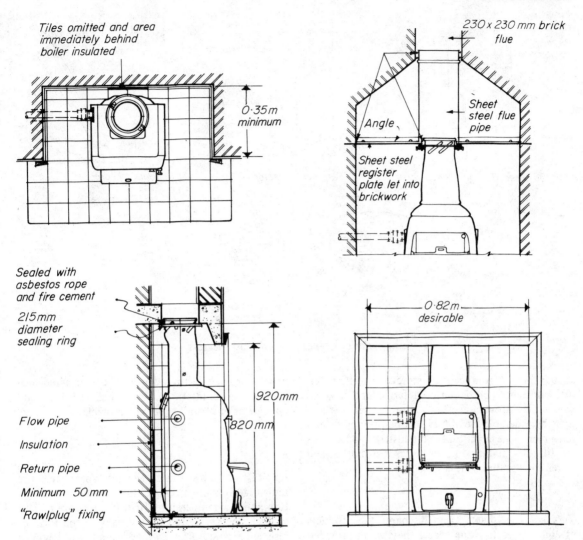

74 *Solid fuel operable stove with back boiler (gives space heating in one room and water heating. Some small radiators can also be served)*

Direct heaters

Solid fuel

The traditional type of open fire is now rarely installed except for decorative purposes. They were troublesome, inefficient and could not burn the types of fuel required in Clean Air Areas. A range of specially designed fires and stoves is now available, however, which will burn smokeless fuels, stay alight overnight and operate with relatively little trouble. They can be fitted with back boilers to provide hot water supply and perhaps one or two small radiators in other rooms. This type of device is widely used in low cost housing. Diagram 74 shows an installation.

Oil

The portable flueless type of oil heater shown in diagram 75 is likely to be used in existing buildings not provided with central heating rather than as part of a system selected and provided with the building. Fixed oil heaters with flues and perhaps with oil

THERMAL INSTALLATIONS

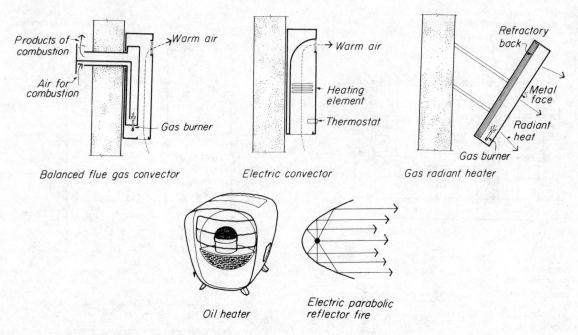

75 *Various types of local space heaters*

supply piped from central storage are, however, often installed particularly in large factory areas where they provide heating at low capital cost and with considerable flexibility of operation and rearrangement. The larger types of industrial type heaters are fitted with fans to give better distribution and faster circulation of the warmed air.

Gas

Very small gas heaters can discharge the products of combustion into the spaces they warm provided that they are well ventilated. Heaters capable of giving full heating in rooms of even modest size will require a flue. The flue requirements for gas appliances are, however, relatively simple. The appliance does not depend for its operation on a draught from the flue. The products of combustion consist mainly of water vapour and the flue does not require cleaning. It is possible to use very small flues and precast concrete flue blocks are available which will bond into brick and block partitions and walls. If the appliance can be fixed on an external wall the air for combustion and its products can be drawn and discharged through the wall immediately behind the appliance. Diagram 75 shows such a balanced flue gas convector. Other types of gas heater provide both radiation and convected heat output for domestic use and high level radiant appliances of various types are available.

Electric

Radiant There are three general types of electric radiant heater: high temperature, medium temperature and low temperature. High temperature radiant sources operate at red heat usually in the form of resistance wire spirals wound round a rod with a parabolic reflector to concentrate the direction of radiation. Diagram 75 shows a section of such a heater. This type of heater is normally purchased for use in existing houses although there are industrial versions employing a more robust heating element set in a reflecting trough which are used for high level radiant heating in industrial buildings. Medium temperature radiant panels do not reach red heat but are too hot to be touched with safety. They are often fixed at high level inclined towards the area being warmed. Electric underfloor heating is one type of low temperature radiant source (see page 138). Another form employs an element mounted on building paper which can be applied to ceilings and other surfaces.

ENERGY SOURCES

Convectors A natural draught electric convector is shown in diagram 75. This type of heater is economical to install and can be thermostatically controlled. The heating element, although still at black heat, is considerably hotter than the heating element of a water operated convector. The air that passes through the heater is therefore made relatively hot which causes complaints from some people and can give rise to noticeable temperature gradients. Fan convectors where the air is drawn through the heating element by an electric fan overcome some of these disadvantages and provide very effective heat output. They are available in a very wide range of sizes, types and noise levels, including models for purchase by occupants and operation from a socket outlet and models for permanent installations. Fixed heaters of this type are often installed at high level.

Electric radiators Pressed steel radiators filled with oil and fitted with an immersion heater are available in a wide range of shapes and sizes. Their heat output is similar to that of a radiator mainly by convecton but having a significant proportion of radiation. They react very much more slowly but are thought by some people to give a better standard of thermal comfort than convectors.

Tubular heaters Tubes of 35 to 50 mm diameter containing a resistance wire coil with a loading of 20 or 30 watts per metre have a balance of radiant and convected output similar to that of radiators They are most often installed in existing buildings and frequently fixed to skirtings below windows or below high level windows or skylights to counteract down-draughts. Multiple mountings support several tubes one above another since relatively long lengths are required for an output sufficient to keep a room at comfort temperature.

Thermal storage Electricity supply undertakings have offered a substantially reduced tariff for current taken only during off-peak hours. Electric thermal storage heaters were designed to take advantage of this. A heating element is enclosed in a block of refractory material of high thermal capacity surrounded by an insulated casing. The block is heated during the off-peak period but gives off its heat slowly, thus providing warmth during the following day. This is not a very controllable system since the output cannot be varied to suit the needs for heat during the day and the building may tend to be very warm at first, slowly cooling as the day goes on. This is not a very suitable characteristic for domestic warming and the thermal storage heaters may have to be supplemented in cold weather. A solution to the problem is provided by heaters which have a fan that can be switched on when desired to draw air through the block thereby giving continuous background heating supplemented when desired by warm air input. All the heaters of this sort are heavy and care must be exercised in positioning them. On timber upper floors they should be placed against the walls on which the joists bear

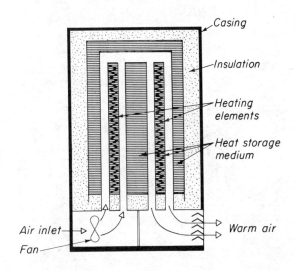

76 Block type electric thermal storage space heater with fan. While fan is not in operation background warmth is provided. Accelerated and controlled heat output is given by the fan which can be time switched and thermostatically controlled

and never in mid-span. The use of block type thermal storage will normally be confined to existing buildings. Their successful operation depends to some degree on the heat storage of the building itself. High thermal capacity, particularly in internal partitions and walls, assists towards successful heating. In buildings with low thermal capacity it is important, if this type of heating is to be used, that the rate of ventilation should be controlled and the standard of insulation should be high. In new premises the same effects can be achieved by under-floor heating which will be very much neater in appearance. Diagram 76 shows a section through a

THERMAL INSTALLATIONS

block type electric thermal storage space heater with a fan for accelerated and controlled output.

Electric underfloor heating

In new buildings with concrete floors, or in existing buildings where concrete floors are to be installed. it is possible to lay a thicker screed than normal (50 to 75 mm thick according to the thermal capacity required) in which a grid of cables is embedded. The screed is then heated during the off-peak period and will give off its heat during the following day.

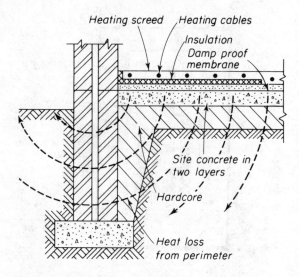

77 *Heat loss from a solid ground floor and the appropriate position of edge insulation for electric underfloor heating*

As in the case of the block heaters the success of the system depends on storage of heat also taking place in the structure. Special attention must be given to the construction and insulation of the floors. Considerable heat loss would take place round the perimeter of solid ground floors and edge insulation should always be provided. Diagram 77 shows how the heat loss takes place and one method of providing insulation. If the ground is wet, heat would be lost rapidly and the insulation should be carried completely across the floor. Where overall carpeting is used above underfloor heating the temperature of the screed will have to be higher in order to contribute the same amount of heat to the room. This would increase the heat loss downwards and in this case also overall insulation should be considered, particularly in the case of rooms of modest size having a relatively large perimeter in relation to their area. The overall insulation prevents the thermal storage of the ground beneath the floor from contributing to the heating and consequently the thickness of the screed will normally have to be increased to compensate for this. The heat loss downwards from upper floors represents a useful gain to the space below but makes the control of heating difficult. It is therefore usually more satisfactory to provide overall insulation. In the case of flats with individually controlled heating overall insulation is essential to prevent large-scale heat loss into adjacent unheated dwellings.

For upper floors the insulation may be laid on top of the slab but it is also possible to apply it to the underside either directly to the slab or suspended from it. This enables the storage capacity of the slab itself to be used and will enable a thinner screed to be employed. Edge insulation is theoretically desirable where heated slabs are in contact with external walls but difficult to achieve in practice. The screeds for upper floors are sometimes laid on fibreglass quilts or other material for sound insulation (floating floors). The insulation will normally be too highly compressed for effective heat insulation.

The laying of screeds for electric underfloor heating is within the scope of normal good building practice. The mix must be as dry as possible and the floor should be laid in bays to minimise shrinkage effects and dried out very carefully. (See *Components and Finishes.*) The layout of the bays for screeding should be planned in conjunction with the layout of the heating cables so that crossing of construction joints is avoided. A rough guide to screed thickness is given:

50 mm thickness

(a) ground floor slabs with edge insulation only and midday boost.
(b) upper floors with insulation under the slab and midday boost.

62 mm thickness

(a) ground floor slabs with edge insulation but no midday boost.

(b) upper floors with insulation under the screed and midday boost.

75 mm thickness

(a) ground and upper floors with insulation under screed and no midday boost.

The great majority of normal floor finishes including cork and carpet can be used in conjunction with underfloor heating. Detailed treatment of heat losses through the fabric of the building can be found on page 87. When installing underfloor heating care must be taken about other wires and pipes which may have to be run across the heated areas. The problem may be acute on upper floors where it is conventional to lay the conduit for lighting in the screed above. Particular care in the wiring layout will be needed and it may be desirable to use MICC cable for the lighting wiring.

The design of electric underfloor heating must take into account not only adequate thermal capacity but also the limitation of the surface temperature to an acceptable value for comfort. 25-27°C respectively for standing or sitting occupants are often used as standards for design. This limits the heat output possible from the floor and in the case of light, not heavily insulated buildings or those with large areas of glazing or high ventilation rates it may not be possible to obtain sufficient heat output. The additional requirement, which will only be called upon at times when the outside temperature reaches its design minimum, must be provided by other means. Block type heaters are a possibility and will be convenient in that the control system for the floor can also be used for them.

Electric underfloor heating requires a grid of cables laid usually at between 75 mm and 225 mm centres. Two types of wiring systems are available. Conduit systems employing asbestos or glass fibre insulated wiring which can be renewed or plastic insulated cables which are permanently embedded in the screed. There is as yet no clear consensus of opinion favouring one system as opposed to the other. With suitable instruments the position of faults in cables embedded in screed can be determined to within a few inches thereby minimising the inconvenience and expense of repair.

The main control of this type of heating is by time-switch which limits the period of electrical input to the off-peak periods. These and the tariffs vary in different parts of the country and more than one tariff may be available in one area, usually related to the length of time during which the off-peak current is taken and particularly to whether a midday boost is included. In addition to the time switch thermostatic control is necessary. For small buildings a simple room thermostat may suffice while for larger installations controls that take account of weather will be desirable and it may also be necessary to divide the building into zones for control purposes.

In general this form of heating provides an extremely neat, clean and trouble-free installation giving a high standard of comfort. Close control is, however, not easy since it can only be achieved during the previous night. If the building overheats during the day windows will be opened to maintain comfort while if more heat is needed it must be provided from other sources. These conditions are often turned to advantage by using the underfloor heating as a background, which at some times of year will be all that is required and providing fast reacting supplementary heating.

Central heating

Heat transfer media

In central heating installations fuel is converted to heat in a central plant and the heat is then distributed round the building to heat-emitting devices by a heat transfer medium. Fluid media for heat distribution have obvious advantages in terms of the ease with which they can be distributed through pipes and ducts. It is clearly necessary, however to select materials for heat distribution which will convey the maximum amount of heat for the minimum amount of material, and for the least cost in terms of material to be purchased. Specific heat gives an indication of the heat-carrying capacity of possible materials. In a building it will be space rather than weight which governs the choice, and the heat transfer efficiency in these terms is given by the volumetric specific heat. The following table shows the specific heat and volumetric specific heat of a number of substances which might be considered for the purpose.

It will be observed that water is the most efficient medium from both points of view, and since this is also the cheapest material on the list, except for air, it is an obvious choice for the purpose, and is

THERMAL INSTALLATIONS

in fact the material most widely used. There are some circumstances when water cannot be used, most particularly in the case of pipes which may be subject to freezing (heating coils in roads or greenhouses are examples, although electric cables are now more likely to be used) or where the medium will be operating in any case below freezing-point (heat pumps, air-conditioning cooling plant, etc). Other materials must then be used, and brine and various oils are often employed. The same problem does not arise with temperatures over boiling-point, since high-pressure systems can be used which raise the boiling point of water, or steam itself may be employed. High-pressure hot water and steam, although they require more complex plant, have advantages in that a given quantity of water can carry more heat, and consequently smaller pipes can be employed. Since the specific heat of water is 4.2 kJ/kg deg C, it follows that a given quantity of water will absorb and convey more heat as its temperature of flow is increased. The average flow temperature of low pressure hot water heating installations is usually about 77°C and cannot be very much increased before boiling takes place. If, however, the boiling point of water is increased by placing the whole installation under pressure the flow temperature can be increased. 175 to 230°C are quite usual temperatures for this type of installation. It is clear, therefore, that for a given quantity of heat smaller pipes can be employed and considerable economy achieved in the transfer of heat over long distances. On the other hand the boiler, heat-emitting appliances and pipework have to be able to resist the high pressures involved (1000 kN/m² perhaps) special pumps and valves are required, stringent safety precautions are necessary and special consideration must be given to pipe expansion. These factors result in substantial installation costs, and high-pressure hot water systems have in the past been limited to large buildings, or district heating systems. The system can be used to supply high buildings, or points above the level of the boiler plant and feed cistern, but there are limits to the possible height since the pressure must not be reduced to the point at which steam will be produced. The flow temperature can of course be modulated to any value below the maximum if this is desirable, as it often is in space heating installations. To maintain economy of main distribution coupled with local modulation it is possible to use high pressure in the mains and provide low pressure hot water for heating by means of calorifiers.

The use of steam appears to give further advantages for heat transfer since the latent heat of vaporisation (2.25 MJ/kg) is also available for heat transfer and the steam will pass rapidly along comparatively small pipes to condense in the heat emitters, giving up its latent heat in the process, providing its own motive power and in very tall buildings not imposing great pressure at the foot of the building. For space heating, however, steam is not often employed for several reasons. The steam-raising plant is more complex and requires more attention, the heat emitters must be protected from human contact since scalding would result, modulation of the heat output is more difficult since the steam temperature is not so easily varied as is water temperature. In addition the condensation of the steam in the pipes and emitters often gives rise to noise. Where steam is available, as is the case in many industrial and hospital buildings, it is often employed for the main distribution of heat for space heating serving local calorifiers which supply conventional low pressure systems. Steam has some advantages for district heating schemes and metering the condensate provides a fairly easy method of estimating the heat used which is often difficult with other systems.

Air, as can be seen from Table 48, is not a very efficient medium for heat transfer. It is, however,

Table 48 Efficiency of various materials as heat transfer media

Material	Specific heat kJ/kg deg C	Volumetric specific heat kJ/m³ deg C
Water	4.2	415
Mercury	0.13	188
Alcohol	2.93	262
Petroleum	2.10	174
Iron	0.46	355
Glycol	3.24	355
Air	1.01	1.2 (Air is compressible but this does not affect the general principle demonstrated)

HEAT TRANSFER SYSTEMS

convenient if mechanical ventilation is to be provided in any case, and it has many advantages for cooling in an air-conditioned building. The cool air has a satisfactory thermal relationship to the warmer structure, and also avoids the condensation difficulties which might arise if panels or radiators were used to extract heat from the building.* The quantity of air which must be circulated for heat transfer will be several times that required for ventilation. This results in ducts which are large enough to present a problem in planning the building and substantial installation costs. Some present developments in air conditioning systems are based on attempts to provide adequate heating or cooling while limiting the amount of air to be moved to that required for ventilation.

Table 49 shows comparative pipe and duct sizes for a range of heat outputs for low pressure hot water, high pressure hot water, steam and air.

Distribution systems

Distribution systems consist basically of a boiler, furnace or mechanism for the production of heat from fuel, a system of pipes or ducts holding a heating medium, leading to heat emitters in the various rooms or parts of the building, and subsequently returning to the boiler. Some motive power must be provided to force the heating medium round the circuit. Early distribution systems both for water and air employed the thermo-syphon or 'gravity' system for their motive power. This meant that the force available to move the water or air was limited to the difference in weight between the hot expanded column rising from the boiler or furnace and the cooler, denser return column. This system imposed severe limitations on planning. The boiler or furnace had to be at the bottom of the installation. The length of run was strictly limited; pipes were large and heat transfer was slow and difficult to control. Steam overcame many of these difficulties but had its own problems. Fans for air distribution systems and pumps for water have changed the situation completely. Quick, efficient distribution is possible through small diameter pipes and ducts with horizontal and downward distributions in patterns that would not previously have been possible. The main factor which governs size both of pipes and ducts is now the maximum acceptable velocity which will not give rise to trouble from noise or corrosion. Diagram 78 shows the principles of operation of a basic heat distributing system.

When low pressure hot water is used the pipe circuits distributing the heat throughout the building are supplied with water from a feed and expan-

Table 49 Comparative sizes of pipes and ducts used for space heating. The sizes are based on typical temperatures, velocities and pressure losses. They give an indication of comparative size and might be used to access approximate space requirements at early design stages

Heating load kw	Pipes (diameter mm)						Air ducts (area m²)		
	Low pressure hot water		High pressure hot water		Steam		Input		Extract
	Flow	Return	Flow	Return	Steam main	Condensation return	Low velocity	High velocity	
10	18	18	12	12	18	12	0.05	0.013	0.037
100	62	62	25	25	37	18	0.5	0.13	0.37
1000	150	150	75	75	100	50	5.0	1.3	3.7

* The ducts carrying air will, however, require careful insulation to avoid wasteful heat gain and also to prevent condensation. A vapour seal (in many cases paint is used) is necessary to prevent condensation on the inside of the insulation (see page 23 for a description of the principles involved).

THERMAL INSTALLATIONS

sion cistern at high level. Diagrams 79 to 82 show some of the patterns of distribution which are possible with some notes on their planning implications. The diagrams show conventionalised two-dimensional arrangements. In practice more than one of these arrangements may be combined in the same installation, various plan arrangements may be employed (eg the ladder system may run completely

Products of combustion

Distribution circuits

Fuel — *Boiler plant* — *Heat emitter*

Motive power for circulation (pump or fan)

78 *General principles of operation of a heat distributing system. (Note that control of time and temperature can be introduced at boiler, pump or emitter)*

round one floor, the rise and drop being sited adjacent to each other in the same duct) or several separate units employed to serve different zones (eg separate ladders may be employed on different facades to make separate control possible). With very simple systems a single-pipe distribution arrangement may be employed for economy. In this arrangement one flow pipe serves radiators progressively,

the water temperature in the pipe being steadily reduced. Radiators at the end will receive water at a lower temperature than those at the beginning, and will consequently have to be of a larger size for the same heat output. There may also be difficulties in the control of heat distribution. The system is, however, neat and economical in installation cost, and can give very satisfactory service in small installations. In general, only radiators can be used in connection with single-pipe systems. Convectors will not operate efficiently at the reduced flow temperatures which arise with low pressure single-pipe hot water systems. The disadvantages of the single-pipe system can be overcome by using separate pipes for the flow and the return, so that all heat emitters receive water at approximately the same temperature. Diagram 83 shows the radiator connections for each method. The pipework of the system does cost more, and if it has to be exposed may be unsightly, but heat distribution is good, radiator sizes may be standardised, or other emitters employed.

It is possible to obtain some of the benefits of each system by using a two-pipe arrangement for main distribution, but serving groups of radiators from a single pipe.

High pressure hot water systems can follow the same basic pattern of distribution. The pressure was often maintained by steam trapped at the top of the boiler but nitrogen pressurisation is now more usual; the feed cistern will be at low level in the boiler room and the water required forced into the system by means of a feed pump (provided in duplicate for maintenance and breakdown). Safety valves and air vents are required instead of vent

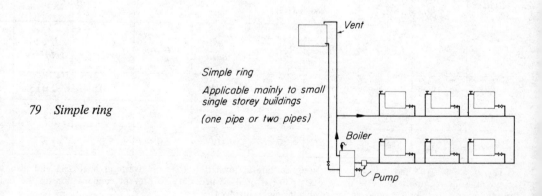

79 *Simple ring*

Simple ring
Applicable mainly to small single storey buildings
(one pipe or two pipes)

HEAT TRANSFER SYSTEMS

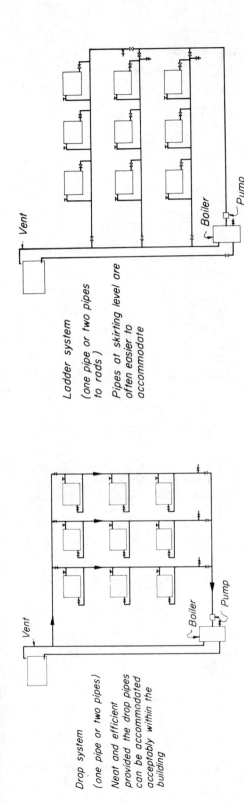
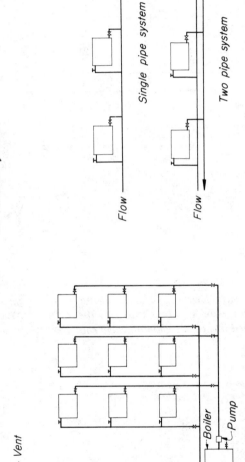

80 Drop system

81 Ladder system

82 Up-feed system

83 Single pipe and two-pipe systems

Ladder system
(one pipe or two pipes to rads)
Pipes at skirting level are often easier to accommodate

Drop system
(one pipe or two pipes)
Neat and efficient provided the drop pipes can be accommodated acceptably within the building

U.P. feed system
(two pipes)
Very convenient for buildings with blocks of various heights or for groups of buildings
Automatic air venting may be desirable at the tops of the radiator towers
System normally used for embedded panel heating

79–83 Low pressure hot water space heating distribution systems

143

THERMAL INSTALLATIONS

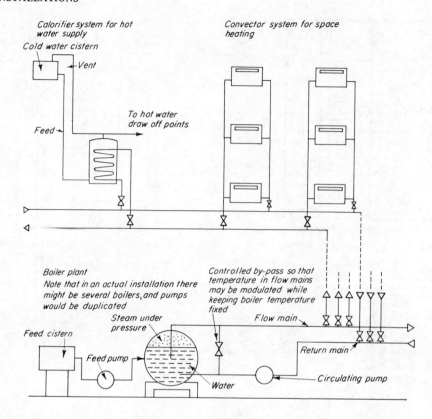

84 Schematic diagram of a high pressure hot water space heating system. Note that in the case illustrated the local hot water supply is heated by the main space heating flow. Where the points requiring hot water are scattered this can be economical in capital cost compared with a separate hot water supply circulation

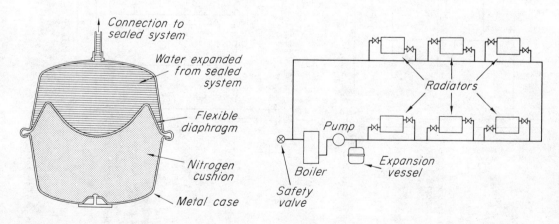

85 Pressure vessel for small high pressure hot water space heating system

86 Small high pressure hot water space heating installation

pipes. Diagram 84 shows a large scale high pressure system. The nitrogen pressurisation system now often used in preference to steam had enabled high pressure systems to be used in smaller installations and the development of small pressure vessels requiring no attention or power supply has extended the use of the system to the smallest installation. Diagram 85 shows the type of pressure vessel used. A metal container with a rubber diaphragm is pressurised with nitrogen and connected, on the other side to the nitrogen, to the pipework of the system. When expansion of water occurs as a result of the system warming up the nitrogen can be compressed via the flexible diaphragm. When the system cools the nitrogen will force the water back from the pressure vessel. Diagram 86 shows the form of installation resulting. There are many circumstances when the avoidance of feed and expansion cisterns and vent pipes, achieved by pressurisation, will be of considerable architectural advantage.

With the pumped circulation which is the rule in modern installations, there is considerable freedom in placing the boiler or furnace, and it is no longer necessary to have it at the bottom of the installation. In solid fuel installations considerations of fuel delivery and ash removal will restrict boiler positions to low levels in the building. Oil can, however, be pumped up to a small boiler feed tank, and with gas there is no difficulty at all in placing the boiler at high level if desired. If the boilers can be installed and the structural loading accepted, this arrangement may sometimes have considerable advantages in freeing ground-floor or basement area for other uses. The factors which must be taken into account in positioning boiler rooms include:

1 *Flue* Flues must normally be carried up to a level where the products of combustion will be harmlessly discharged. This often means that the flue must rise in the highest part of the building and will certainly limit its possible positions. Although it is possible, by means of induced draught fans, to force flue gases through lengths of horizontal flue where the boiler room is remote from the flue, there is clearly great advantage in having the boiler room at the foot of the flue.

2 *Delivery* Solid fuel fired boiler rooms require lorry access for fuel delivery and suitable provision must be made in all cases for replacement of the plant.

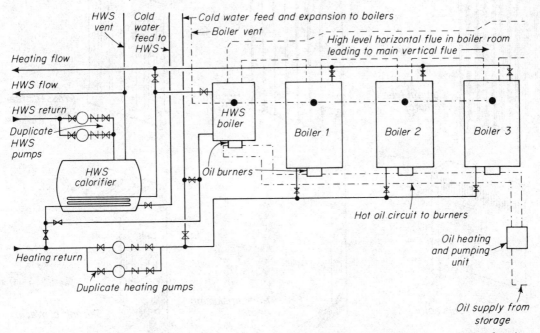

87 Schematic diagram illustrating the functioning of a typical low pressure hot water space heating and hot water supply boiler

THERMAL INSTALLATIONS

3 *Light and air* Air is necessary for combustion and to keep boiler room temperatures to a reasonable level. While this can be provided by ducts to subterranean boiler rooms far better environmental conditions can be provided by natural light and ventilation. There can be no doubt that good conditions in the boiler room contribute significantly to proper maintenance of the plant.

At its simplest a boiler plant consists of a boiler and pump. Larger installations appear to be a very complex mass of pipes and plant. In fact the basic principles involved are the same and the apparent complexity results mainly from the fact that in large installations it is usual, from considerations of maintenance and giving flexible heat output while using individual boilers to their full capacity, to divide the heating load between several boilers. Pumps are provided in duplicate, for maintenance and in case of failure; and where the building is divided into separate zones for control purposes a separate set of suplicate pumps may be required for each zone. Hot water supply will exhibit similar duplication of calorifiers and pumps. Diagram 87 shows diagrammatically the functioning of an oil-fired low pressure hot water space and water heating boiler plant.

The correct choice of heat-emitters is crucial for economy, amenity and comfort. Some types are shown in diagram 88. The main types available for use with hot water space heating are:

1 *Pipes* In early central heating installations large pipes were run around the building. Installations of this sort can still be seen, particularly in churches and assembly halls. The pipes were, however, awkward and obtrusive, and although pipe coils are still used in store-rooms and round roof-lights to avoid down-draughts, more efficient means of transferring the heat have been devised.

2 *Radiators* Present a much greater surface area for heat transfer, and the vertical arrangement of columns encourages convection currents. Most of the heat is given off as convection, but a useful if small proportion is emitted as radiation. To minimise wall staining, obstruction of useful wall space, and to provide the best thermal

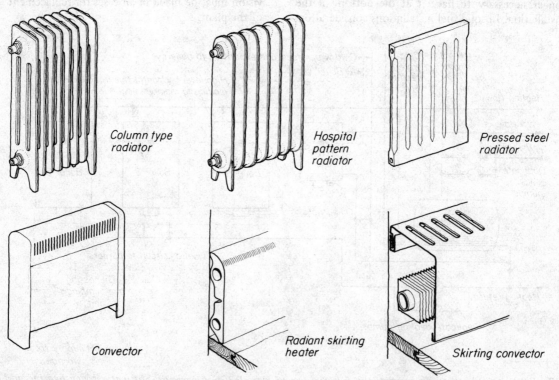

88 Some types of heat emitter suitable for use with low pressure hot water space heating

results, radiators are usually placed under windows. Pressed steel is now used for many radiators instead of cast iron, particularly for domestic applications. Steel radiators are light and often easy to clean. Control of individual radiators is usually by manual radiator valve, which does not permit of any fine adjustment. Valves which give a flow proportional to an indicated setting are, however, now available and thermostatic radiator valves are also obtainable although price may restrict their use. Systems employing radiators do not react swiftly to control because of the substantial thermal capacity of the radiators and the water contained in them.

3 *Convectors* These consist of gilled tubes at low level in a casing which encourages the movement of air over the gills by stack effect. The amounts of metal and water are considerably less than those contained in radiators and consequently an installation of this sort will react more swiftly to control. Individual convectors can be controlled by the same types of valves as radiators, or in some cases by manually operated dampers closing their inlet and output grilles. It is easy to build in convectors under cills, and in built-in furniture, and arrangements can be made to combine ventilation with convectors by drawing in air from outside rather than from the room itself. The amount of radiation from convectors is negligible. It is important for efficiency of convectors that the flow temperature of the heating installation should not drop significantly.

4 *Fan convectors (and unit heaters)* The heat transfer takes place in a gilled tube battery similar to that of a convector, but instead of relying on natural convection, air is blown over the heating battery by an electric fan. The output rate is usually high, and this type of heater is not usually suitable for very small rooms although manufacturers are developing smaller models. Individual control is, however, very easy, since the output is governed by the fan which can very easily be time-switched and thermostatically controlled. The delivery of air by fan assists in distributing the heat and there is considerable freedom in placing fan convectors. Advantage can often be taken of this to minimise pipe runs. Installations are often economical in first cost because of reduced pipe runs and small numbers of appliances required, and economical in running costs because of the closeness of control possible. In rooms where the heat requirement may change quickly, due perhaps to a sudden influx of people, fan convectors can give optimum comfort conditions, since some other systems which might provide better comfort in principle would not react sufficiently quickly to changing needs for heat. If fan convectors are placed on external walls it is possible by means of a duct through the wall and dampers to allow either the circulation of room air or the introduction of fresh air through the heater. This system can also be used for cooling when it is usually given the name fan coil.

5 *Skirting heaters* Convectors and special pipes are available shaped to take the place of a skirting board. There may be difficulties in making connections around doors, but in many cases the specially adapted pipes or convectors can give a neat unobtrusive solution.

6 *Panels* Continuous coils of pipe usually 12 mm in diameter and about 60 m long can be embedded in the construction of a building, in the bottom of the structural floor (ceiling heating) or more often nowadays in a screed on top of the structural floor (floor heating). Floor heating in particular maintains a very even temperature gradient, and gives off some 50 per cent of the heat by radiation. Flow temperatures are limited to 50°C, so that floor surface temperatures are kept low to avoid any discomfort. Floors are not normally warm to the touch. Boilers producing hot water supply and serving other radiators must have a flow temperature considerably higher than this. Diagrams 89 and 90 show a combined floor heating radiator and hot water supply installation in a house. Diagram 89 shows a mixing valve used to control the temperature of the water flowing in the underfloor panels. Diagrams 91, 92 and 93 explain the principles involved in more detail. The construction and insulation details of the screeds for hot water underfloor heating are similar to those for electric underfloor heating except that the thickness of the screed is standard at 50 or 62 mm since thermal capacity is not involved. Embedded panels react slowly to control. Systems of pipes forming part of a suspended ceiling provide a very much quicker reaction and a number of patented systems are on the market. Care must be taken

THERMAL INSTALLATIONS

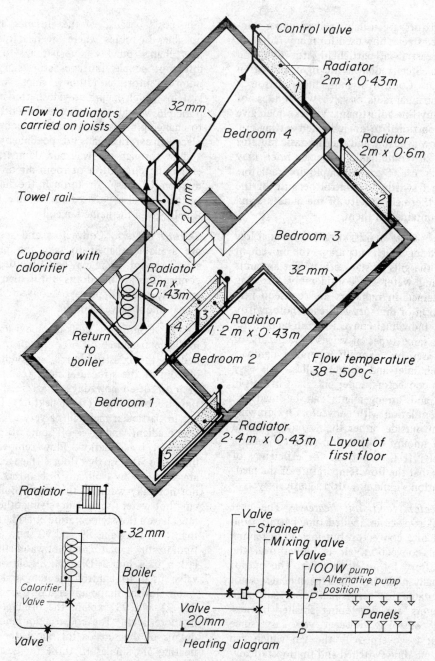

HOUSE HEATING: RADS ON FIRST FLOOR

89 *Axonometric view of pipe and radiator layout to first floor. Schematic diagram of heating installation in a two storey house having radiators on the first floor and underflow heating on the ground floor. Note the mixing value enabling the underfloor circuit to be controlled to a lower temperature than the radiator and calorifier circuits*

CONTROL OF HEATING INSTALLATIONS

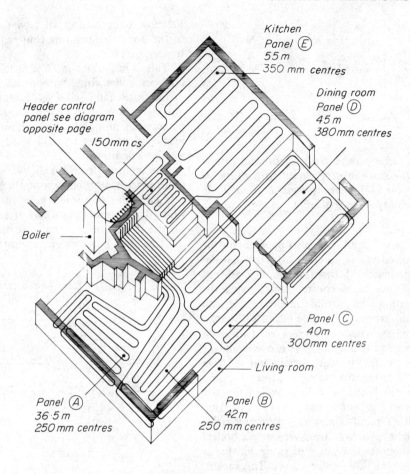

90 *Axonometric view of pipe work layout for underfloor heating (see diagram 89)*

that the rooms where this type of heating is used are sufficiently high to ensure comfort for the users of the building.

There is a wide variety of designs in most of the appliances described and many other devices (eg radiant panels, patent suspended heating ceilings, etc). New developments are constantly occurring. To obtain a comprehensive and up-to-date knowledge attention should be given to the contributions and advertisements in the technical journals.

CONTROL OF HEATING INSTALLATIONS

The great advantages of automatic control of heating installations are leading to rapidly increasing use of more and more sophisticated control systems. Small domestic heating installations which not long ago would have been controlled only by use of the flue damper are now fitted with thermostatic control of boiler firing, time-switch control of pump operation and thermostatic control of water flow tempera-

THERMAL INSTALLATIONS

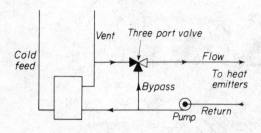

91 Shows a mixing valve giving a constant rate of flow round the space heating circuit with a variable flow temperature. This arrangement is usually the most appropriate for space heating

tures or radiator output. It is also possible, by the use of motorised valves, to close off particular circuits by time-switch at times when they are not required. In some cases control units, giving the user choices of several possible modes of use with variable timing, are being installed in small systems.

Control is exercised in respect of two main variables, time and temperature. It is now standard practice even in the smallest installation to have a time-switch giving overriding control of pump operation for the heating circuits (leaving the boiler in action providing hot water supply) or giving control both of pump and boiler firing (thereby giving overnight economy). Even very simple boilers now have thermostatic control of firing which can be used to govern flow temperatures. This means, however, that hot water supply and space heating will fluctuate together, which is clearly inappropriate. A better arrangement is that shown in diagrams 91, 92 and 93, which shows how a thermostatically-controlled three-port valve can, by allowing return water to recirculate, provide a circuit where

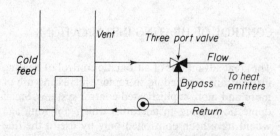

92 Shows a mixing valve used to give a variable rate of flow at a fixed temperature

the flow temperature can be varied independently of the boiler temperature (but not above the boiler temperature).

This is important in large installations to ensure efficient boiler firing without risk of condensation of flue gases and in all sizes of installation it enables hot water supply to be maintained at a fixed temperature by the direct boiler flow while the space-heating circuits can be modulated to suit prevailing weather conditions.

The method described above results in the whole of the space heating being controlled by the temperature prevailing at a single point in the building and while it is an economical and effective method the disadvantages are apparent — particularly in respect of larger buildings where very different conditions

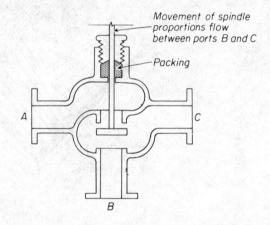

93 Diagrammatic section through a three part valve of the type appropriate to diagrams 67 and 68

can prevail in different parts. One method of attempting to overcome this is by use of external sensing devices, usually known as external compensators, which are sensitive to air temperature, usually wind, and in some cases to other external factors. This method assumes that the heat input required to the building can be directly related to the external conditions. Clearly sunshine on one facade can upset this assumption and buildings having systems of this sort will usually have to be divided into zones having different exposures. It is also assumed, however, even when the building is zoned, that the heat input required is entirely based on external conditions. This is often not the case as significant heat gains can arise from occupants electric lighting, machinery and other processes. In order to take these considerations into account in individual spaces local thermostatic control of the heat emitters in the room would be required. This is comparatively easy in the case of fan convectors where a simple thermostat can easily control the electric fan and thereby the heat output of the apparatus. Radiators and emitters where the flow of water has to be impeded present a more difficult problem but effective thermostatic radiator valves are available which can govern the water flow in the radiator in accordance with the needs of the room. The cost of thermostatic valves is considerably greater than that of ordinary radiator valves. On the other hand the closer control resulting will give fuel economy as well as better comfort conditions.

FLUES

The construction, insulation and fire-proofing of chimneys is dealt with in *MBC: Structure and Fabric*, but their functioning in terms of the discharge of effluent gases forms part of the scope of this volume. It is conventional to apply the term chimney to the structural elements while flue refers to the gas passages and to non-structural pipes (eg horizontal flues in boiler rooms or vertical flue pipes supported from some other structural member).

The primary purpose of flue is to carry the gases produced as a result of combustion together with entrained air and to discharge the necessary quantities of gas at a point where no nuisance or damage to health can result. In addition to this many solid fuel or oil fuel appliances require for their effective operation the 'pull' of air through the appliances which results from a proper flue draught.

The movement of gases up a flue is caused by the difference in density of hot flue contents compared with the colder outside air, which being heavier displaces the flue gases upwards. The forces generated are influenced by the height of the chimney and the temperature difference between the flue gases and the external air. Increases in either of these factors will increase the force available, known as the draught. The forces are quite small and may be increased or diminished or even overcome and put into reverse by the operation of the wind. Where the pattern of air movement (see chapter 3) round the building gives down draughts or produces higher pressures at the flue outlet than at the fire, or boiler then the down draught in the flue can result. This is normally overcome by carrying the flue up to a level which takes its outlet above the zone of varying air pressure. 1 m above the ridge level of a pitched roof or 2 m above a flat roof are figures sometimes quoted as likely to be satisfactory for small buildings, but topography, surrounding buildings and protrusions on the roofing, may affect conditions. Since the force moving the air is small, resistance to flow in the flue itself such as friction due to rough internal surfaces, (such as protruding cement joints between liners) sharp bends or changes of size or shape have a very significant effect in reducing draught and should be avoided. Maintaining the temperature of the flue gas is important and where possible flues should be internally rather than externally situated. This is particularly important in domestic situations where valuable heat savings can be made by internal flues, and in the case of oil and gas-fired appliances which operate intermittently and consequently allow the flue to cool. Apart from the effect in reducing draught, due to the reduced temperature of the flue gases, the cooling effect of flues may result in condensation which can attack the materials of the flue and run down, causing troubles at the foot, Where external flues are essential thermal insulation between the liner and the structure is very desirable. Leakage of air into the flue will cool the gases and dissipate the draught. Careful attention should be given to sound joints in the structure and also to properly sealed pipe junctions and soot doors. Diagram 94 shows a typical convection of a flue from a small boiler into a brick chimney.

For solid fuel and oil-fired appliances short flues (less than about 4 m) may not in any case develop sufficient draught to operate effectively. Tall flues, on the other hand (over about 10 m) may, particularly if the wind aids the draught, develop excessive draught. This effect can be dealt with by means of draught stabilisers which are counterbalanced flaps fitted at the base of the flue and weighted so as to open and admit air if the draught exceeds an appropriate value for the appliance. Draught stabilisers can also act as safety doors to cater for the risk of explosion with oil-fired installations which can suffer from sudden ignition of gas in the flue. Diagram 95 shows a typical stabiliser.

THERMAL INSTALLATIONS

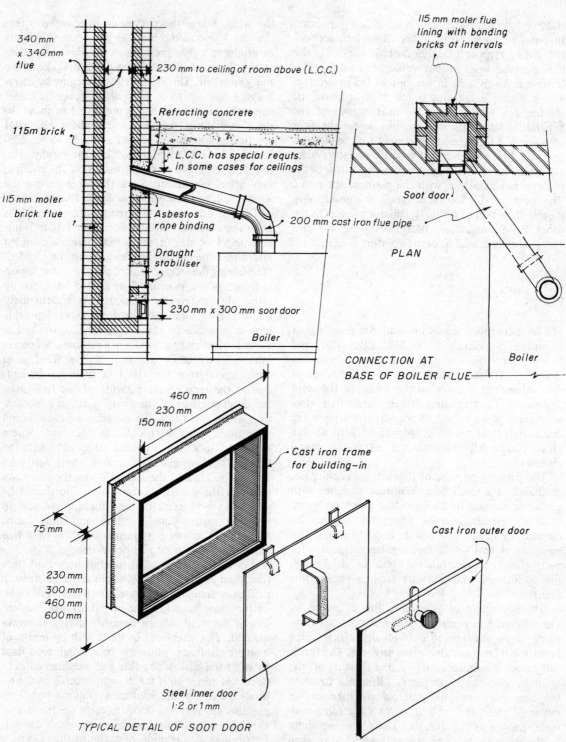

94 *Connection of flue from small boiler into brick chimney*

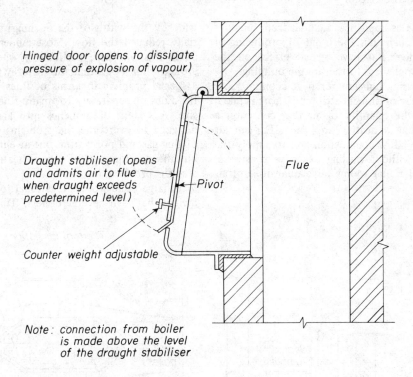

95 Draught stabiliser for oil fired boiler flue

Gas-fired appliances can normally operate without any flue draught and in many cases incorporate a hood which prevents fluctuations of flue draught from influencing the apparatus. Gas flues, therefore, act merely as a means of carrying away the products of combustion which in this case consist mainly of air and water vapour. Conditions for the siting and the discharge of gases are not so stringent and several interesting flue variations have been developed for them. Diagram 96 shows one such arrangement. Special unobtrusive terminals, including some in the form of ridge tiles for small installations, flues incorporated within partition blocks and of particular significance the balanced flue and SE duct. Balanced flue appliances have combustion chambers sealed from the interior of the building, but taking air from outside through a wall and also discharging the products of combustion at the same point. Inlet and outlet are so close that the air pressure on them is the same for any wind conditions. A typical arrangement is shown in diagram 97. The SE duct enables balance flue type appliances to be used away from external walls by means of a duct running through the building from bottom to top and vertical

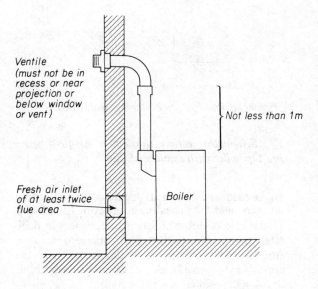

Simple flue arrangement possible with gas fired boiler (from diagram by Thomas Potterton Ltd.)

96 Simple flue installation with gas-fired boiler

THERMAL INSTALLATIONS

at both ends. Gas appliances take air from the duct and return is such that the input of producing of combustion does not have a bad effect on the operation of appliances further up the building.

With increasing concern for heat economy and the extensive use of draught stripping it care is taken not to limit the quantity of air that can enter a room containing a heating appliance. Enough air must enter to allow combustion and to supply the flue draught without causing excessive resistance. In the case of boiler rooms large quantities of air

wide by the width of the opening) which substantially reduced the flow. Most modern stoves overcome this problem, particularly when their doors are kept shut. For a time multi-storey flats presented a special problem in terms of flues, since carrying the flues up separately for many floors was expensive. A system of branches into a main flue was designed to overcome the problems of penetration of flue gas and noise from one dwelling to another, but the development of central heating has eliminated the problem.

In large boiler rooms it is usual to find the load divided between several boilers. Traditionally the

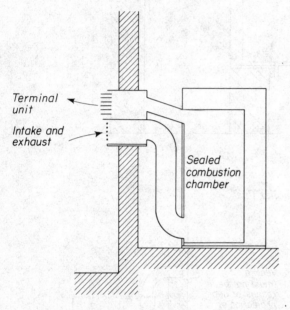

Balanced flue for gas boiler (or water heater)

97 *Schematic section through a gas-fired space heating boiler with a balanced flue*

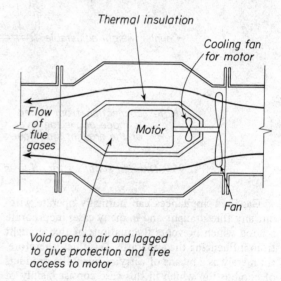

98 *Bifurcated fan used as induced draught for a flue with inadequate natural draught*

are needed and the areas required for ventilation are recommended in heating engineering handbooks. In the case of domestic heating appliances in habitable rooms, efforts have been made to duct air from outside to a point in the hearth near the appliance to provide air for combustion without drawing the cold air across the room.

In the traditional domestic open fire a large opening was provided to the flue and very large quantities of air passed from the room up the flue with consequent heat loss. Better practice was the use of a narrow throat (usually about 100 mm

flues from these boilers were collected into one single large diameter flue. The disadvantages of this in terms of flue efficiency are apparent and it is becoming increasingly usual to have separate flues carried up together grouped into a single structural chimney.

Small flues to gas-fired appliances often have no means for cleaning but flues to solid fuel and oil appliances should normally have access by means of soot floors so that inspection and cleaning is possible.

In cases where it is impossible to provide flues of adequate draught, such as inadequate height, or long horizontal runs of flue to reach chimneys remote from boiler rooms, induced draught fans may be employed to provide adequate 'pull' by mechanical means. In small and medium sized installations it is very usual to employ bifurcated fans. In this type of apparatus an electric motor, driving a fan fixed on the same spindle giving a simple and cheap arrangement. The motor is protected from the hot flue gases by dividing the duct as shown in diagram 98. In domestic installations particular care must be taken to avoid annoyance due to noise from the fan.

VENTILATION SYSTEMS

The need for ventilation and the range of conditions where natural ventilation will not suffice have been discussed in chapter 3.

Where, in industrial premises or kitchens the main consideration is the extraction of excess heat or fumes and close control of comfort is inappropriate, extract systems are used, with intakes preferably near to the sources of fumes, and fresh air is allowed to enter the space directly. This gives a cheap and effective method of preventing grossly uncomfrotable or unhealthy conditions from developing rather than the provision of comfort. Forcing air in rather than extracting would enable the air to be filtered and warmed but control of the distribution would be limited and liable to short-circuit when doors were opened. Where comfort is the air a combined system where both input and extract are controlled is usually employed. Diagram 99 shows the functioning of such a system.

When the system is required only for ventilation the extracted air will often be exhausted and the whole of the input will consist of fresh air, which in winter will have to be warmed before it is delivered to the rooms. In order to make warming-up in the morning more rapid provision for recirculating the air is very often made. Since it is necessary to warm the air there are advantages in using it as the means of heating the building and thereby eliminating any other form of heating installation. Air is not a very efficient heat transfer medium (see Table 48) and consequently it will be necessary, when heating is involved, to circulate more air than is required for ventilation. It is obviously undesirable to waste this air and it is also recirculated, sufficient fresh air for ventilation purposes being drawn in and mixed with the recirculating air.

The motive power for the distribution of air in ventilation systems is invariably provided by electrically-driven fans. There are three main types in use.

Propeller fans Suited for situations where there is no great resistance to air flow to be overcome.

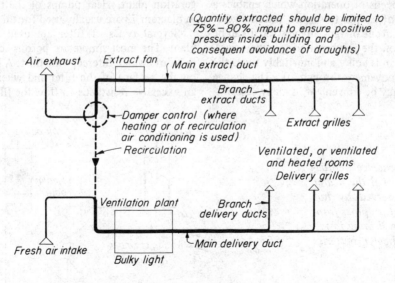

99 Schematic diagram showing the principles of operation of a typical mechanical ventilation plant

THERMAL INSTALLATIONS

The free intake and discharge condition of ventilation fans situated in wall openings giving direct in/out and out/in movement are eminently suited to this type of fan, which under these conditions can move large volumes of air economically and with very low installation costs. Short duct systems can also be served provided the resistance of the system is low.

Centrifugal fans Can develop pressures sufficient to drive air through air-treatment plant duct systems and are extensively used for this purpose. There are several types having impellers of different patterns giving various types of perform-

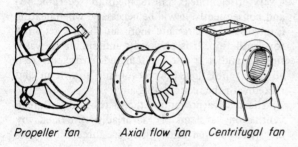

Propeller fan   Axial flow fan   Centrifugal fan

100 *Types of fan used for ventilation systems*

ance. Output from a fan can be varied by different motors and speeds of operation which enables a limited range of casings to serve a wide range of output. The output of fans already fixed can be easily varied on the job by altering pulley sizes. This type of fan is bulky and inevitably turns the direction air movement through 90°. The change of direction may be convenient in many cases.

Axial flow fans This type of fan is becoming popular. Efficiency is high, installation simple and the appearance neat, particularly in a line of ducting. Diagram 100 shows the three types of fan.

Except in the case of simple extract systems some treatment of the air for ventilation is usually required. This can involve merely filtering and warming the air. Diagram 101 shows a simple ventilation plant for this purpose. In cold dry weather the warming of the air in this type of plant will reduce the already low relative humidity still further and may cause timber shrinkage and cracking and complaints of dry, sore throats from the occupants of the building. Diagram 102 shows a more elaborate type of plant incorporating a spray chamber and two heater batteries which is capable, during the winter, of giving control of both temperature and relative humidity. The pre-heater enables incoming air to be raised to a predetermined temperature and after which it is humidified in the spray chamber and then heated to the desired final temperature, and relative humidity. This process can only be effective while the outside temperature is low, permitting the final heating to be effective in reducing the relative humidity. In summer these conditions do not apply but the plant can still be used to give fully conditioned air by cooling the water in the spray chamber. This requires the provision of refrigeration plant. Heat pumps of the types described in diagram 73 are usually used for this purpose.

Several types of filter are used for ventilation plant. The most simple are porous screens of paper or fabric which intercept the dust. A layer builds up on the surface of the filter and when this is giving an excessive resistance to flow the filter is replaced.

101 *Simple mechanical ventilation plant. Note that the heater battery can be served by hot water flow and return pipes from the normal space heating installation. Electric heater batteries may also be used.*

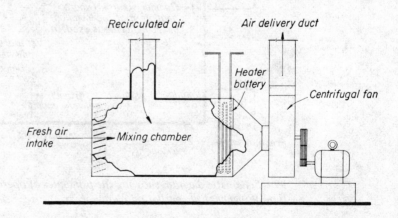

VENTILATION SYSTEMS

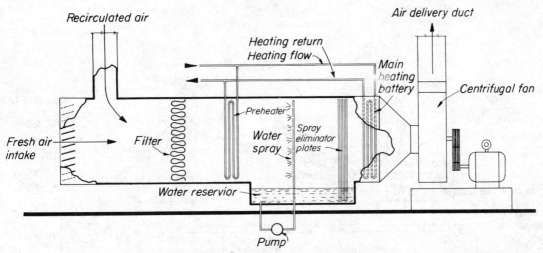

1 During Winter: The plant as shown can exercise complete control of air temperature and considerable control over relative humidity by correct use of pre-heater and main heater batteries

2 During Summer: If the water reservoir is chilled, cool air can be delivered at high relative humidity. If the air is chilled below the desired temperature and warmed again by the main heater battery, control of both temperature and relative humidity can be maintained

102 Air handling plant capable of maintaining control of temperature and relative humidity in summer and winter

In order to give more surface area of filter in a given space this type is often of a concertina shape. To overcome the inefficiency associated with the increasing resistance to air flow of the type of filter described a number of mechanisms have been developed. One example is an endless belt oiled wire mesh filter which is slowly and continuously rolled through an oil-bath thereby cleaning and reoiling the fabric.

A particularly efficient type of filter having a very low resistance to flow is the electrostatic type. The air is passed through a grid of wires charged at high voltage and the particles of dust are deposited as a result of electrostatic attraction.

The plant for air treatment is large and its placing warrants special consideration in planning. It is bulky, light, and requires to draw fresh air in and exhaust vitiated air. The top of the building may well be very convenient. Severe structural loads are not involved, and the occupation of space is less important than it might be at a lower level. Long ducts from a central plant may represent a severe planning problem which can be overcome in many cases by having a number of local plants rather than one central one. As insurance against total failure it may be desirable to have more than one plant. Sub-division of this sort may well assist in zoning the building for control purposes. When summer air-conditioning is required and a refrigeration plant must be provided, the siting consideration for this is very different. The plant is small but heavy, and produces considerable noise and vibration. There are advantages in accommodating it when possible at the bottom of the building. When refrigeration plants are used the heat must be dissipated. This is usually done by the external air, in small installations by an evaporative cooler where a fan blows air across heat-dissipated coils as water flows over them, or in larger installations by means of a cooling tower. In some cases ornamental fountains and pools have been used for the purpose.

Diagram 103 shows the relationships between the various plant areas required for an air conditioned building.

THERMAL INSTALLATIONS

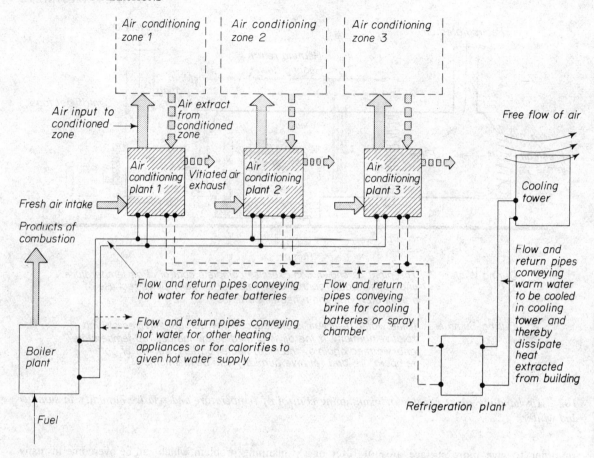

103 Schematic diagram showing the relationships between the various plant spaces required for air conditioning system

Air trunkings

Galvanised sheet steel is the material most commonly used for air trunkings. It is light, cheap, and easily fabricated. Builder's work may be used when this is convenient and when it can be made reasonably air-tight, and finished smoothly internally. CP 352 (1958) *Mechanical Ventilation and Air-Conditioning*, Table 8, tabulates a number of materials with their application and relative efficiencies, and gives details of the construction of sheet-steel trunkings.

The most efficient shape for trunkings is circular. Friction is reduced to a minimum, and the form of construction is inherently rigid. Rectangular trunkings, however, have many advantages; they are often more easily accommodated in buildings, and changes of cross-sectional area and junctions are more easily formed than in circular sections. Provided the shape of the rectangle does not become too elongated (maximum ratio of sides usually taken at 1 : 3) they give an acceptable aerodynamic performance.

For small systems such as lavatory extracts the use of PVC piping is becoming popular.

Input and extract devices for air distribution systems

When air is the heat distribution medium the problem of transfer to the environment is simplified in that the air is emitted directly into the rooms concerned. Care must, however, be given to the selection of a suitable grille or nozzle to give the right dis-

VENTILATION SYSTEMS

tribution of air without noise or other untoward effects. The simplest form of grille is the stamped plate. It is cheap but has limited free area (about 60 per cent), no directional control and risk of noise if the velocity of the air is high. The vane-type grille is a substantial improvement and there are many other special types giving particular distributions. It is even possible to input and extract through the same device at the same time. Nozzles can be used to direct jets of air in special directions with the minimum of trunking. Diagram 104 shows

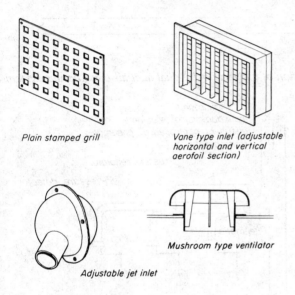

104 Simple input and extract devices for mechanical ventilation

these grilles together with an adjustable type often used in ships and aeroplanes. The form of extract grille is not usually critical and simple stamped or louvred types are generally used. Special patterns are made for use in floors, however, and the diagram shows a mushroom floor device suitable for use under theatre seats when a downward system of air distribution is adopted.

Special types of ventilation systems

In single trunking systems all the air delivered is at a single temperature and humidity and different conditions in various parts of the building can only be achieved by having separate plants serving the different zones. The large trunkings also present a problem especially where the branch serving a particular floor leaves the main vertical duct. If floor heights have been reduced to a minimum difficulties can arise with head-room. Systems which attempt to overcome these problems have been developed. They reduce trunking sizes by increased velocity. This requires circular trunkings to ensure airtightness and freedom from drumming, and some sound attenuation at the delivery points. There are two main methods of achieving local control. One is the *high velocity dual system* where every space is served by a duct carrying hot air and a duct carrying cold air. The trunkings deliver to boxes where the air is mixed and where the sound level is reduced by absorbent lining and baffles. A thermostatic device controls the balance of hot and cold delivery and a wide range of different conditions can be achieved in different rooms from the same plant and system. Diagram 105 shows the principles of this system and diagram 106 shows a typical room input unit. The *induction system* uses a single high-velocity trunking delivering air through a unit rather like a convector. The water temperature in the convector will be different to that of the air and local manual control of the water flow enables a range of different conditions to be achieved. The quantity of air delivered is limited to that required for ventilation. The additional air for effective heat transfer is entrained from the room air by the design of the unit. The trunking requirements for this system are more modest than any other and it is popular at present for office buildings. Its success is dependent on proper balance of the air input and water flow temperatures and zoning for different facades will normally be desirable. Diagram 106 shows a diagrammatic section through an induction unit.

Where rooms are too large to be served by induction units of the type shown in diagram 106 it is possible to use the same principle and achieve economy of main air trunkings by delivering conditioned outside air from a central intake plant to small subsidiary plants which take an appropriate volume of fresh air but also extract and recirculate air from the rooms they serve. Thus the main trunkings need only be sized to carry air for ventilation while the relatively larger trunkings required to achieve thermal control only run between the room served and the adjacently sited local plant. Local plants can be served with both chilled and

THERMAL INSTALLATIONS

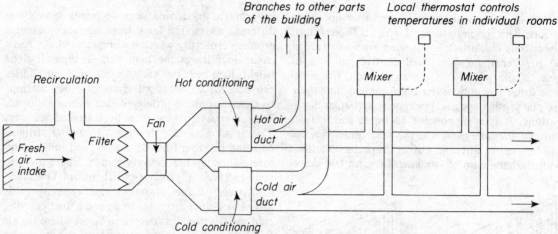

105 Schematic diagram showing principles of operation of a high velocity dual duct air conditioning system

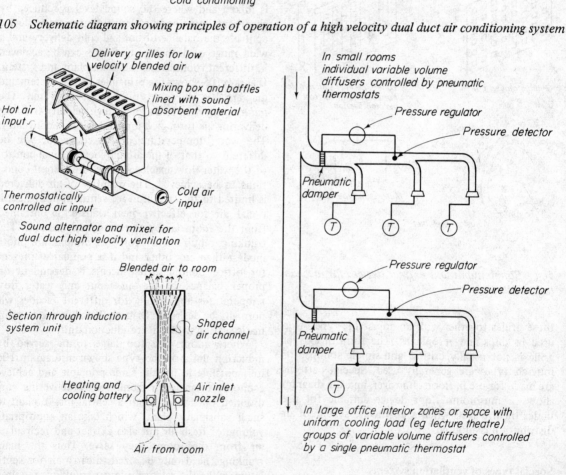

106 Typical room input unit for high velocity dual duct system and an induction unit

107 Variable volume air condition system using fan modulation

160

VENTILATION SYSTEMS

hot water and thus exercise local zone control. The system illustrated in diagram 112 is based upon this method.

Variable volume systems

There is increasing interest in methods of avoiding the need for both air and water systems and returning to the simplicity of the all air system. Apart from size of trunkings, one of the main difficulties with all air systems is the lack of control. In most cases the system has to be divided into zones, each with a separate plant, and within any one zone no individual room control is possible. Provided that adequate air for ventilation is maintained, regulating the supply of air to each room would enable temperatures to be controlled. In the normal air conditioning installation reducing the supply to one room would automatically increase the air supplied to other rooms, whether they needed it or not because the main fan could not easily be modulated. There are now at least two systems for overcoming this problem. The first depends on the use of fans with variable pitch blades, pneumatically controlled, which enable the air flow to the system to be automatically controlled from full flow to zero by pressure sensors in the duct work. With this method individual pneumatic thermostats can control air inout to rooms by damper, larger dampers will control pressure in main branches and pressure in the whole system will be regulated by fan adjustment. The second system employs a fan continuously driving air round a ring circuit of trunking from which room supplies can be diverted without requiring adjustment of fan operation. Diagrams 107 and 108 show sections of systems operating on these principles. In both systems, although control can be exercised between rooms, all the spaces served must be on the heating or the cooling sycles. Some heating and some cooling is not feasible.

Extract through lighting fittings

Where high levels of illumination are used considerable heat gains result. Levels of illumination of 500 lux present a significant problem, while at 1,000 lux the heat gain probably outweighs the losses even in cold weather. Since most of this heat is given off as convection from lighting fittings it is sensible to extract air in mechanically ventilated buildings through the lighting fittings rather than allow the heat to make its way into the room when it will represent a load that must be met by the refrigeration plant if comfort is to be maintained. It is possible not merely to extract the heat from the lighting fittings but also to recirculate it during winter so as to warm the incoming fresh air. In the summer the hot air from the lighting fittings would be exhausted to the outside air. Diagrams 109, 110 and 111 illustrate the principles involved. Diagram 130 shows a lighting fitting specially adapted for air extract.

These ideas have been substantially developed in recent years, particularly for large open office areas. Large volumes have relatively small rates of heat loss when compared with smaller ones and where high levels of lighting, often 1000 lux, are employed there is often no heating problem even in winter and the heat from the lights can maintain adequate temperatures. In the summer cooling will be needed. Diagram 112 shows part of a system of this sort. A local zone air handling plant takes fresh ventilation air from a main trunking, supplied by a central intake plant, mixes it with recirculated room air and delivers it to a large open office through a system of air trunkings in the ceiling and slot diffusers. The local zone plant draws air from the ceiling void which acts as a plenum chamber equalising pressures over the whole ceiling and enabling equally sized apertures in all the lighting

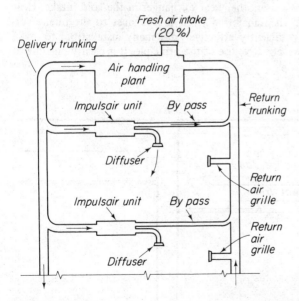

108 *Variable volume air conditioning system using ring trunking*

161

THERMAL INSTALLATIONS

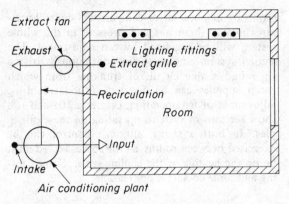

109 Air conditioning system not connected with lighting

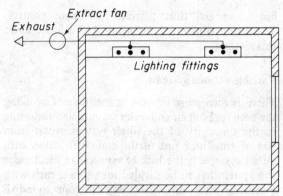

110 Air and connected heat being extracted via lighting fittings and exhausted

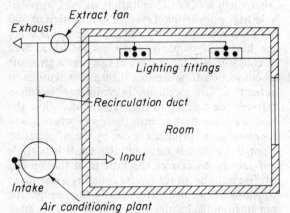

111 Lighting fittings forming part of extract system of air conditioning installation. Heat may be exhausted in summer or recirculated in winter

fittings to extract room air evenly throughout the room. The extracted air is then recirculated to give the volume flow required for adequate heating or cooling.

Very much attention is now given to energy conservation. Warm air exhausted from large air conditioned buildings as part of the ventilation process can waste considerable quantities of energy and a number of systems have been devised to save some of this. The simplest method is the 'run-around coil' shown in diagram 113 where water is circulated by pump from a heat exchanger in the warm exhaust to another heat exchanger in the cold intake. Heat is transferred from the exhaust to the intake sufficiently effectively for many installations to make effective use of the principle. It is clearly possible to

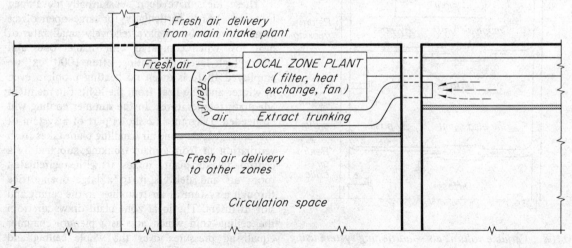

112 Local zone system for air conditioning

use a heat pump to improve the efficiency of heat recovery and diagram 114 shows this diagrammatically. A most ingenious method of achieving very high levels of heat recovery from exhaust to input trunkings is the heat recovery wheel. A disc, usually of a honeycomb of asbestos impregnated with lithium chloride, rotates passing through both ducts and transferring heat from the warm exhaust to the cold intake very efficiently. At present heat recovery wheels are expensive to buy.

In many buildings it is possible for one facade to require cooling while the remainder requires heating. In these circumstances it is uneconomic to input heat at the same time as extracting it from another part of the building and dissipating it via a cooling tower.

Diagram 116 shows a system, employing a heat pump, which enables the heat extracted from part of the building to be put back into other parts if they require it, or if there is no need for heat in the building to store it via the agency of large water cylinders, or finally if no further storage is needed to dissipate the excess heat via a cooling tower.

Trunking noise

Noise emanating from air systems must be kept to an acceptable minimum level. It can arise in two ways. First by vibrations from the plant being transmitted along the trunkings. This is avoided by providing in all systems a flexible canvas link joining the main trunkings to the fan. Second by noise generated

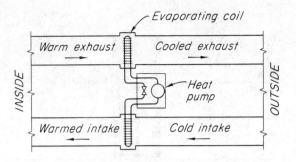

113 Run-around coil

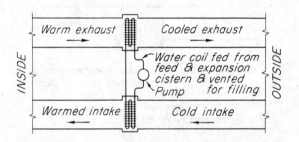

114 Heat pump transferring heat from exhaust to intake

in the trunkings or grilles by the passage of air. This noise can either emerge from the grille or may penetrate the wall of the duct even though there is no opening. To avoid the generation of unacceptable sound levels air velocity must be limited. It will be appreciated from the above that velocity will have

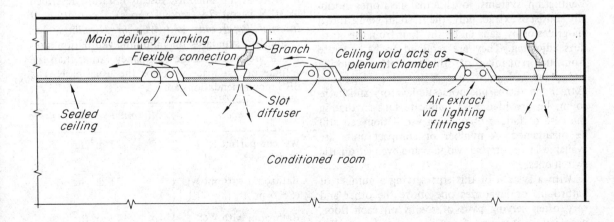

THERMAL INSTALLATIONS

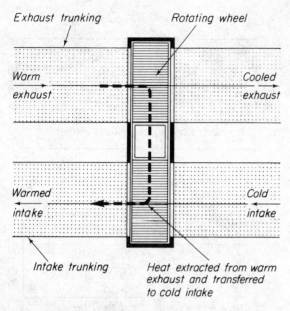

115 Heat recovery wheel

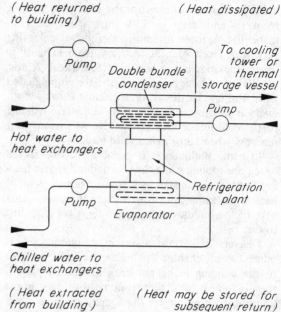

116

to be low near openings, can increase somewhat away from openings and increase still further if the trunking is enclosed behind sound insulating construction. Table 52 indicates a range of appropriate air velocities. In the case of rectangular sheet steel trunkings drumming may occur unless the flat sides are stiffened at appropriate intervals. High-velocity systems terminate in sound-attenuating boxes which also reduce the velocity at which the air is admitted to the room.

Ventilation of internal lavatories

Ventilation systems for internal lavatories normally consist of extract only, the fresh air being drawn via grilles or by gaps under the door from the corridors adjoining. They are required to be separate from the remainder of tthe ventilation trunkings and plant and must, except in cases where one extract fan serves one domestic lavatory and bathroom, be provided with duplicate fans so that in the case of failure of one fan ventilation can still be maintained. A number of compact units are available incorporating two fans and even automatic switch ones.

With a system of this sort serving a number of bathrooms or lavatories one above the other and very often serving pairs of rooms on each floor, penetration of sound from one room to another could occur in a quite unacceptable way. This is avoided by serving the rooms by means of branches from the main trunkings called shunts, which should be more than 1 m long and incorporate two bends. Diagram 117 shows a detail of this type of installation. The arrangement is also effective in preventing smoke penetration to other spaces served in the case of a fire in one room. Since the trunkings involved are usually small plastic pipes can form a very neat and satisfactory way of forming them.

Work at the Building Research Station described in BRS Digest 78 (second series), *Ventilation of Internal Bathrooms and WCs in Dwellings,* has established standards for minimum ventilation rates which are based on the occupancy rather than the volume of the space which is the usual basis. The BRS recommendations are:

Space	Minimum ventilation rate
WC compartment (3-4.5 m^3)	21 m^3/hr
Bathroom without WC (5.5-7 m^3)	21 m^3/hr
Bathroom with WC	42 m^3/hr

VENTILATION SYSTEMS

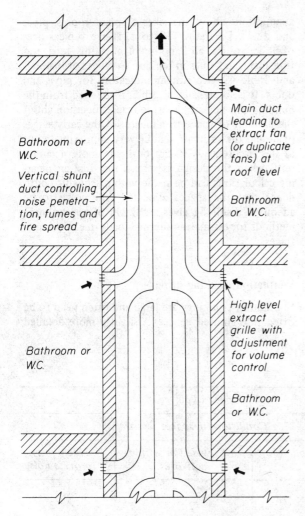

117 Shunt system for the ventilation of internal bathrooms and lavatories

It is recommended that the ductwork is designed to cater for 20 per cent more ventilation to allow for balancing. A velocity of 3 m/s is usual for this type of installation.

The main trunking is often kept to a fixed size for the whole height of building served and in all cases extract grilles will be fitted with dampers to allow the system to be balanced.

Control of fans for individual components can be by means of the light switch, preferably with a time delay switch to ensure 20 minutes running after use. Fans serving more than one room must be kept in permanent operation.

Table 50 Typical air change rates

Ventilation space	Air changes/hour
Boiler rooms	15–30
Garages: parking	6 (minimum)*
repair	10 (minimum)*
Laundries	10–15
Kitchens	20–60*
Sculleries and wash-ups	10–15*
Banking halls	6
Cinemas and theatres	6–10
Dance halls	10–12
Restaurants	10–15
Offices	4–6
Bathrooms	6*

Note: These rates are for ventilation rather than heat transfer, which may in some cases require higher rates. Extract rates will be less than input except in cases marked * where extract is the standard and negative pressure in space is desired.

Table 51 Methods of making approximate estimates of thermal and ventilation loads

Heating

Assume an average U-value for the whole of the skin (including windows) and an average ventilation rate for the whole volume.

Estimated heat requirement =

$$\left[\begin{array}{l} (U\ value \times\ area) + (volume \times air\ change\ rate \times \\ \quad\quad volumetric\ specific\ heat\ of\ air) \end{array} \right]$$
$$\times\ temperature\ difference$$

For a single-storey building (U-value of walls and roof 1.1 W/m^3/deg C) with 50 per cent glazing the total requirement may vary from about 35 W/m^3 for 50 m × 50 m on plan to 60 W/m^3 for 10 m × 10 m on plan.

For a multi-storey building 20 m × 80 m on plan with 50 per cent glazing and similar walls and roof the total requirements may vary from 30 W/m^3 for two floors to 27 W/m^3 for 12 floors.

Cooling

Perimeter zones (up to 7 m from windows)
    25 per cent glazing 120 W/m^2
    60 per cent glazing 180 W/m^2

Interior zones (more than 7 m from windows)
    75 W/m^2

Note that perimeter zones are greatly affected by orientation and fenestration and all areas by lighting levels and occupancy.

THERMAL INSTALLATIONS

Concern is sometimes felt about the heat losses resulting but natural ventilation would also result in heat loss and there is no means of eliminating it. Since the ventilated rooms are often totally internal, however, no fabric losses can take place. There are several sources of heat gain. Occupants, lighting, hot water supply and heating flow and return pipe passing by contribute heat, sometimes flues may be adjacent and drawing off bath water raises the room temperature. It may be found that internal lavatories become uncomfortably hot. The problem should be borne in mind during design. If necessary ventilation rates may have to be augmented to keep the temperature rise to within acceptable limits.

Approximate sizing of heating and ventilating pipes, trunkings and plant

In the early stages of building design, before a proper engineering analysis of the heating and ventilation installation is possible, it is necessary to make assumptions about the sizes of plant rooms, pipes and ducts. Diagrams 118 to 121 give typical sizes of boiler rooms, air handling plant rooms, refrigeration plant and flues in relation to ranges of loadings and Table 49 give similar indicators for pipes and ducts. It is, therefore, possible, if the load from the building can be estimated, to reach a decision about the amount of space to allocate. In the early stages of design the estimate will be very approximate but as design proceeds it can be made more accurate until when the design is settled the final engineering calculations can be made with reasonable confidence that the spaces allocated will prove to be adequate. Table 51 gives a number of approximate methods for estimating thermal loads in early design stages.

Ventilation trunking sizing

Ventilation trunkings are large and often need to be fitted into limited spaces. A slightly more detailed

Table 52 Speed of air in ducts, typical values

Type of duct	Speed of air m/s		
	Conditions in spaces served		
Conventional systems	Low background noise	Crowded or with activities	Machine or process noise
Diffusers or grilles (assume 60% free area)	1.7	2.5	3
Branches opening to diffusers	2	3	4
Main distributing ducts	5	7	10
	Relationship of duct to space served		
High velocity systems	In ceiling	In ceiling of adjoining corridor	In concrete casing	Generally
Distributing ducts	8	12	30	20

* High velocity systems deliver to sound-attenuating and mixing boxes, details of which must be obtained from manufacturers. Diffusers and branches to diffusers data is not appropriate.

High velocity ducts are normally circular (better air flow, reduced noise, less leakage). In approximate sizing arithmetic can be saved by consulting duct sizing charts.

Extract from spaces served by high velocity systems is at conventional velocities.

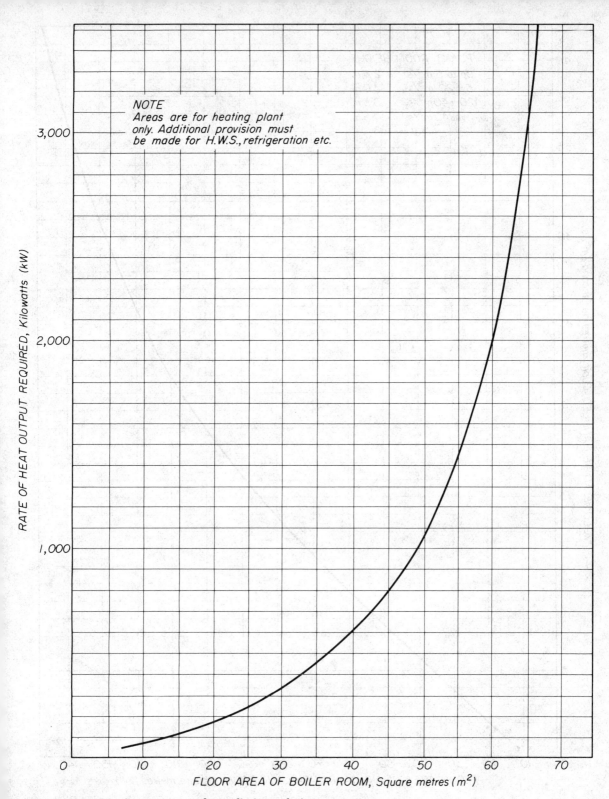

118 *Typical boiler room areas for preliminary design use*

Reproduced by courtesy of E.W. Shaw, MIHVE, FRSH, MIPlantE, Senior Lecturer in Heating and Ventilating, National College for Heating, Ventilating, Refrigeration and Fan Engineering

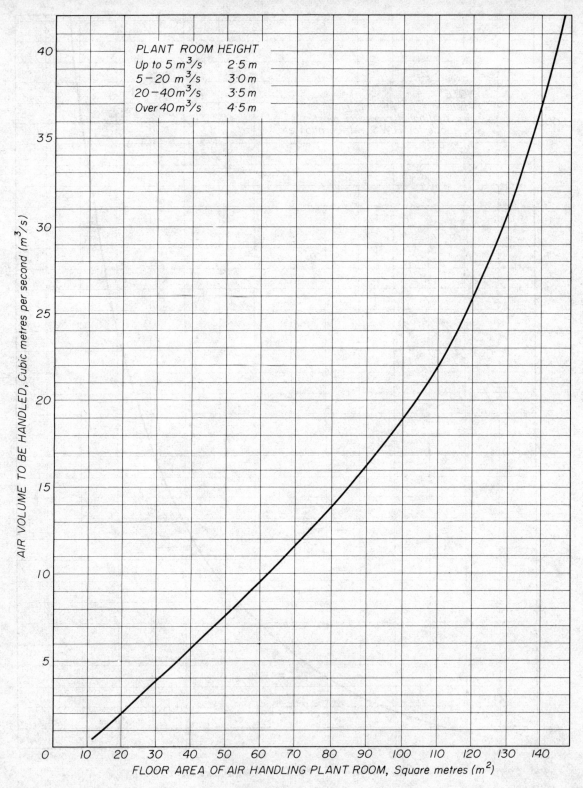

119 *Typical air handling plant sizes for preliminary design use*

Reproduced by courtesy of B.E. Lawrence, AMInstF, AMIHVE, AMInstR, Lecturer in Air Conditioning, National College for Heating, Ventilating, Refrigeration and Fan Engineering

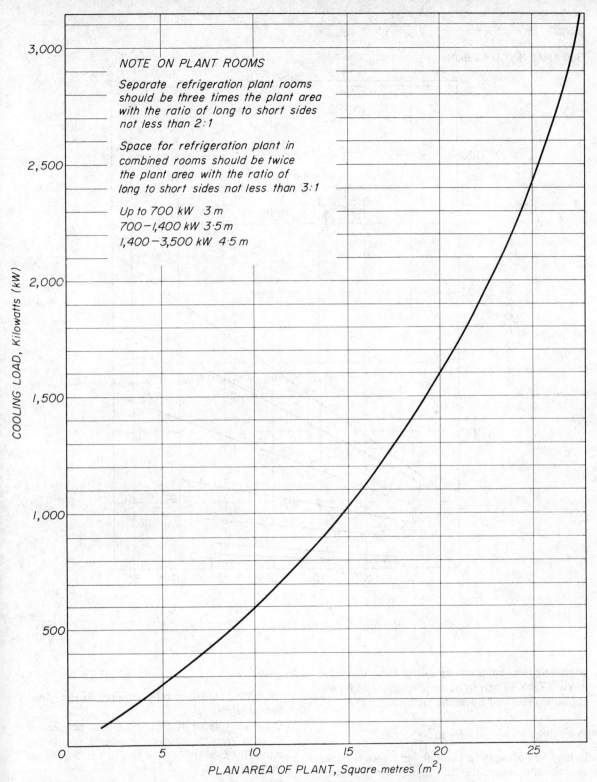

120 *Typical areas occupied by refrigeration plant for preliminary design use*

Reproduced by courtesy of B.E. Lawrence, AMInstF, AMIHVE, AMInstR, Lecturer in Air Conditioning, National College for Heating, Ventilating, Refrigeration and Fan Engineering

THERMAL INSTALLATIONS

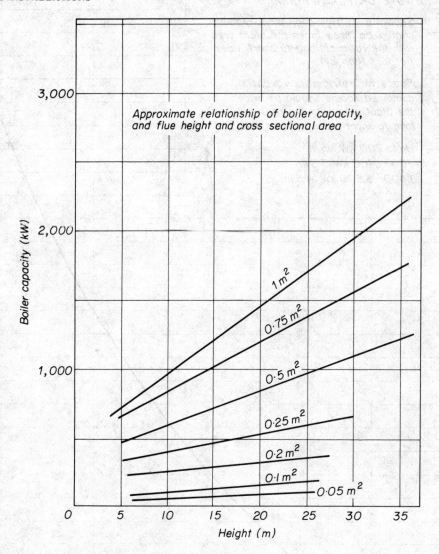

121 Flue sizes for solid fuel and oil fired boilers

procedure for approximate sizing than the value given in Table 49 may be of use in design. The following notes described a possible method.

Procedure for approximate sizing

It may be necessary in early design stages to make approximate estimates of sizes. The following method is suggested for this purpose. It is usually helpful to sketch a three-dimensional diagram of the proposed layout to assist in the sizing process.

Method

1 Establish ventilation rates required in each area. See table 50.
2 Calculate volume of air to be moved per *second* in each area to achieve the desired ventilation rate. This is likely to take one of the two forms shown:

Volume (m³/s) =

$$\frac{\text{Number of people} \times \text{rate per hour per person}}{3600} \quad \text{(see Table 6)}$$

VENTILATION SYSTEMS

or

Volume (m³/s) =

$$\frac{\text{Volume of space} \times \text{number of air changes per hour}}{3600} \quad \text{(see Table 50)}$$

or

$$\frac{\text{Heat to be dissipated (W)}}{\text{Volume} \times \text{specific heat of air (1.3kJ/m)}}$$

(from people, lighting, processes, etc)

× final temperature differences desired between inside and outside (°C).

3 Size grilles by deciding the proportion of the volume of air (2 above) which will pass through each grille or diffuser and divide by the speed. This will give the free area. (Where manufacturers' literature is available volume flow through grilles may often be directly established from figures quoted.)

4 Size branches to diffusers and main trunkings by adding the volumes of air passing through grilles and trunkings served by the section under consideration and dividing sum by the air speed given in Table 52.

Note In duct layouts, sharp bends should be avoided and in the case of rectangular ducts the ratio of short to long sides should not exceed 1 : 3.

In recent years there has been a rapid increase in the use of completely prefabricated plants known as packaged plants. This is particularly so in the field of air handling where packaged plants capable of being installed on flat roofs without protecting buildings are frequently employed. Access for maintenance and repair is from the roof itself thus saving circulation space, and the plants themselves are extremely compactly designed. For a given duty a packaged plant will be much smaller than the corresponding plant room. Several manufacturers produce ranges of sizes and their literature should be consulted. Diagram 122 shows a typical roof mounted packaged air conditioning plant.

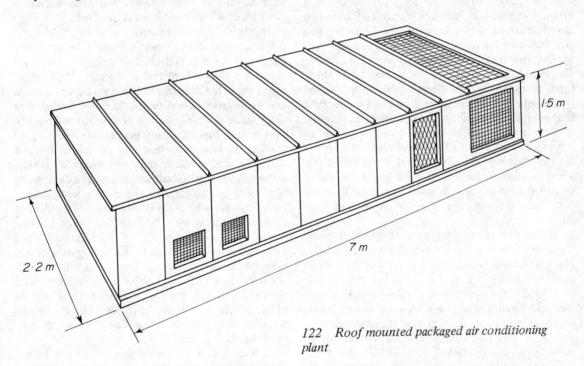

122 *Roof mounted packaged air conditioning plant*

8 Electric lighting

Since the end of the nineteenth century virtually all buildings have been provided with electric lighting installations for use at night. The source of light was usually the tungsten filament lamp, known as the gls (general lighting service) lamp. This type of lamp gives light as a result of raising the temperature of a thin filament by means of an electric current to such a degree that light is emitted. SInce the lamp is small and very bright it gives bright reflectors and sharp shadows. The colour of the light is not particularly compatible with daylight and considerable heat is given off by the lamp. The development of the fluorescent tubes has provided another source of light for use in buildings, with very different characteristics. They consist of a long tube containing mercury vapour with filament at each end. In use an electric current passes between the filaments and gives rise to ultra violet radiation which causes coating (phosphorus) on the inside of the tube to glow. This type of lamp gives very much more light (about three times) for a given electrical consumption and is consequently both cheaper to run and reduces the heat input to the building. The chemical composition of the phosphor can be varied to obtain a range of colours of light varying from north light and colour matching through daylight, natural, white and warm-white to specific colours. Control gear is necessary to run the tube and it is more expensive in initial cost than the tungsten lamp of equivalent light output.

The advent of the fluorescent lamp has provided a source of light which is compatible with daylight, cheap to run and does not impose such a considerable heat gain on buildings. As a result of this it is possible to have electric lighting installations which supplement daylighting and can in extreme cases provide the only source of light in a windowless environment. The implications of this for the design of buildings are considerable.

Although its necessity has been traditionally unquestioned natural lighting is associated with some serious disadvantages. The depth to which natural light can penetrate is limited and consequently the depth of buildings lit in this way is governed. This results in long external walls and in limited use of restricted sites as well as high heat loss. Glazing itself makes a significant contribution to heat loss, presents a cold surface on the interior of the building and may be subject to condensation. In the summer the heat of the sun may give rise to overheating. At night the dark area of windows normally requires a covering. Where specially close control of lighting is needed the variations in natural lighting will often mean that electric lamps must be used. Sky glare may result from large windows and where the effect of direct sunlight could be disadvantageous special control devices may be required. The north light roofs of factories are an example of this. Patches of sunlight can affect proper vision and make accidents more likely where machinery is involved.

It might be thought that, as a result of these disadvantages, electric lighting would be used very frequently to the exclusion of daylight. There are many buildings where this is the case. Department stores are often designed in this way. Some factories are deliberately designed to exclude daylight, in many more the contribution of daylight to illumination is very limited. Many people, however, regard daylight as very important and desirable. It seems that this desire stems not only from discernible differences between artificial lighting and daylighting but also to a considerable degree to a wish to have visual contact with the exterior and to be aware of changes in the weather and also of other external activities. In terms of quality of illumination it must be borne in mind that while electric lighting comes from overhead the sideways light from windows will often give better light on vertical surfaces and a better three-dimensional quality to objects than will be obtained from the overhead lighting. Rooflights would suffer from the same disadvantages as overhead electric lighting.

It is clear that in the design of modern buildings it cannot automatically be assumed that windows are needed. A decision about the best form of illumination in the light of the considerations described will be needed and if daylighting is selected considerations beyond simple illuminations must be borne in mind.

In relation to electric lighting it will be necessary to decide what part the lighting will play:

1 *By night*
Electric lighting will be the sole illumination.

2 *By day*
Sole source of illumination (windowless building). Main source of illumination with windows for view or to give better modelling (small windows can make a considerable contribution to better three-dimensional appearance).
To supplement daylight where rooms are too deep for adequate natural lighting (known as PSALI, permanent supplementary artificial lighting installation).

It should be noted that, in addition to providing overall illumination in a way similar to daylighting electric light can be used to provide dramatic contrast effects; it is also frequently used as an emergency lighting or for security purposes at night. Rarely used spaces may be provided with minimal electric lighting sufficient for movement and safety; any work being carried out by means of portable lamps.

TERMINOLOGY

The terminology and quantities used for lighting are not in common use and prior to dealing with criteria for design and estimation some definitions are desirable:

Phenomenon	Term and symbol	Definition
Luminous intensity	Candela (cd)	Basic SI unit. Defined terms of a 'black body' of fixed area, temperature and pressure.
Luminous flux (flow of light)	lumen (lm)	Flux emitted by unit solid angle by source of intensity one candela.
Illumination (light falling on surface)	lux lx	1 lumen/m^2

Phenomenon	Term and symbol	Definition
Luminance (light emitted by unit area)	apostilb asb	1 lumen/m^2 (not an SI unit but convenient in use since illumination (1 lux) x reflection factor = luminance (apostilb)
	candela/m^2 (cd/m^2)	Equal to π apostilb
Reflection factor (sometimes called reflectance)	ρ	Ratio of reflected luminous flux to incident luminous flux

CRITERIA FOR DESIGN OF ELECTRIC LIGHTING

When it has been decided that an electric lighting installation is required and what its objectives are it is necessary to establish a number of detailed design criteria. These can be divided into two main groups covering quantity and quality of light.

Quantity This is normally defined in terms of the amount of light falling on a horizontal plane at working level known as the working plane. Quantities are given in terms of 1 ux. The Illuminating Engineering Society's Code, is an authoritative source giving illumination levels suitable for a wide range of applications. Some typical values are:

	Illumination level lux	Limiting glare index
Offices	500	19
Drawing offices		
General	500	16
Boards	750	16
Auditorium and foyers	100	–
Shops	500	19
Living rooms		–
General	50	
Reading	150	
Sewing	300	

ELECTRIC LIGHTING

Quality Several factors can exercise a critical influence on the success of lighting installations apart from the proper level of illumination on the working plane. Glare is an important consideration. Excessively bright areas in the field of view can give rise first to discomfort and in acute cases to disability in seeing. Glare is a complex phenomenon dependent to some degree on the observer but also governed by the brightness, size, number, distance away and direction of the light sources. Methods have been developed to evaluate glare and it is possible to calculate a glare index for proposed installations. Suitable values for the glare index for a range of applications are given in the IES Code. In order to design lighting installations in relation to glare it is necessary to know the distribution of light output from lighting fittings. The British Zonal Method (BZ) exists for this purpose. Light fittings are divided into 10 categories based on the direction of light output. Diagram 123 (c) shows the principles of this classification.

An observer is not likely to be conscious of bright sources of light 55° or more above the horizontal (see diagram 123 (a)) and some ranges of lighting fittings have been designed giving complete cut-off of light within this angle. Diagram 123 (b) shows the geometry of this cut-off. With this type of fitting, glare is not likely, but fittings may have to be close together to give adequate distribution of light.

It is very possible to have a room with adequate light on the working plane, low glare levels but which gives an unsatisfactory and gloomy appearance. This will usually be due to unsatisfactory luminance of the surroundings in relation to the task. Hence the term luminance ratio has come to represent a range of ratios of the luminances of task, immediate surroundings and background and suitable ratios which result in interiors which appear cheerful and brightly lit have been established. The Bodmann ratio of 10 : 4 : 3 for task, immediate surroundings and walls and ceiling has been widely accepted as giving satisfactory conditions.

In recent years it has been increasingly appreciated that meeting all the criteria given alone would not necessarily give lighting conditions which permitted good appreciation of three-dimensional form (modelling). Although they are not yet widely used for design some concepts have been developed to take this into account. Scalar illumination is the light falling on a point from all directions (not merely on to a plane). Vector illumination is used to describe the angles at which the difference of illumination on a point is at its greatest. It has magnitude and direction. The system is not used, at present to any great extent in design and it is important to note that in a room well lit in terms of the Bodmann ratio described above the effect of reflected light from walls and floors will be to give less harsh

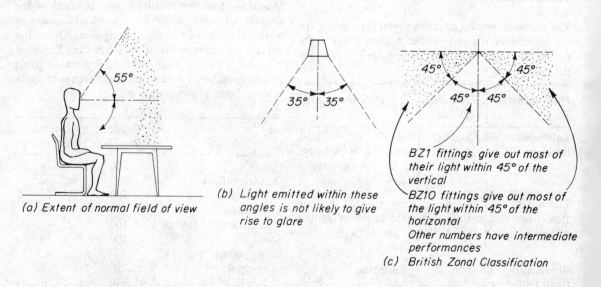

123   BZ classification and rough geometrical limit of glare

modellings than would be the case with a flow of light solely from overhead.

Lighting fittings

It is convenient to regard lighting fittings as falling into three categories: decorative fittings which are intended to be seen rather than to give optimum lighting distribution and performance; general utility fittings which are intended to give economical and effective illumination; and special fittings, usually with special optical arrangements such as reflectors or lenses to give highly directional light. It is the general utility type of fitting which is of present concern (diagrams 125 and 126).

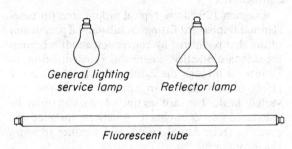

124 Typical electric lamps

All lighting fittings must be strong enough in construction to withstand handling and erection to sustain the lamps and any associated control gear (particularly the case with fluorescent lamps which require ballasts which are heavy and perhaps starters) and to be cleaned without damage. Access and electrical and thermal insulation is necessary for electrical cables and earthing must be catered for. The temperatures likely to be reached in operation must be considered. Fluorescent tubes particularly are susceptible to temperature and are inefficient in operation at very high and low temperatures. Ventilation of the fitting is consequently very usual. Completely closed fittings present another problem also in that because of thermal expansion and contraction of the air inside they cannot be completely sealed and dust and often flies may gain access, with consequent unsightliness and need for frequent cleaning.

From the optical point of view the fitting should obscure the lamp from direct view to reduce glare and should present, if it is to be directly viewed, a larger surface area of lower brightness. In addition

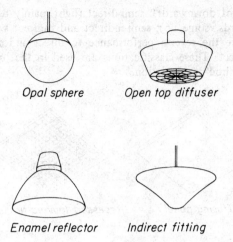

125 Typical fittings for tungsten lamps

the shape and arrangement of louvres and openings, if any, will be devised to give a particular distribution of light in upwards, sideways and downwards directions. The BZ system of classification has been described. There is in addition a traditional descriptive classification of lighting fittings into direct

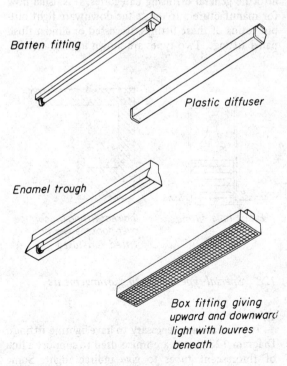

126 Typical fittings for fluorescent lamps

ELECTRIC LIGHTING

(light downwards); semi-direct (light mainly downwards, some up); semi-indirect and indirect which have the reverse performance to direct and semi-direct. These classifications are used in the lumen method of calculation.

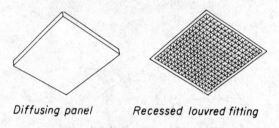

Diffusing panel   Recessed louvred fitting

127 Typical panel fittings (usually with fluorescent tubes)

Diagrams 125 and 126 show typical fittings for tungsten and fluorescent lamps. The enamel reflectors and trough fall into the direct category, the diffuser and louvred fittings into the semi-direct category and the plastic diffuser and opal sphere into the general diffusing categories. It is usual now for manufacturers to quote the downward light output ratios of these fittings. Recessed or almost flush panel fittings. Two types are shown in diagram 127.

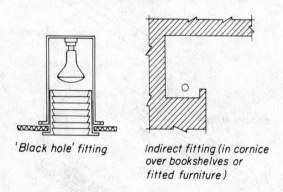

'Black hole' fitting   Indirect fitting (in cornice over bookshelves or fitted furniture)

128 Special types of lighting arrangements

It is not always necessary to have lighting fittings. Diagram 128 shows a cornice used to support a line of fluorescent tubes to give indirect light. Some fittings have been devised to give light while remaining completely concealed. The 'black hole' fitting shown in diagram 128 is an example of this type.

Other factors of significance in electric lighting design

Heat gain

For comparatively low levels of electric lighting in buildings with good natural ventilation present little problem. As lighting levels increase, however, heat input from the lighting fittings must be taken into account. 500 lux using fluorescent lighting is often taken as the level at which the problem becomes critical. Heat emitted from lighting fittings is partly in the form of radiation and partly in the form of convection.

Diagram 129 shows typical proportions for tungsten and fluorescent fittings. A substantial proportion of the ehat is emitted by convection and it is becoming standard practice, where this problem exists, to extract air through the lighting fitting. This has the effect of removing a large proportion of the convected heat. The hot extracted air can then be exhausted, or as in many American installations, partially recirculated during cold weather to warm the incoming air.

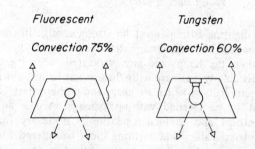

129 Emission of heat from lighting fittings

Diagram 130 shows a lighting fitting adapted for extract.

The radiant heat cannot be dealt with in the same way and must be balanced by an input of cold air from an air-conditioning plant. This adds considerably to the expense of the building. It will be noted that tungsten lamps not only have a higher heat input for their light output but also emit a

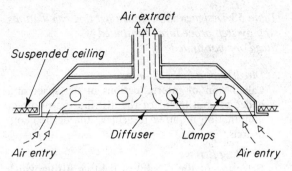

130 Lighting fitting combined with air extract

higher proportion as radiation. It follows that their use in the circumstances under consideration is inappropriate.

Where air is to be extracted through lighting fittings the design of the lighting installation must be integrated with that of the ventilator system. In view of the ducts running from lighting fittings suspended ceilings are almost inevitable with this arrangement.

Cost

Tungsten lamps and fittings are cheaper than comparable fluorescent fittings but in order to obtain a particular level of lighting it is possible that more will be required. For a given level of illumination more electricity is required by tungsten installations and replacement of lamps is more frequent than for fluorescent lamps. Running costs are therefore higher for tungsten installations. This factor is, however, influenced by the periods for which the installation is in use. It is apparent, therefore, that for installations which receive little use tungsten will be more economical.

Nursery schools which have adequate daylighting are closed before electric lighting is necessary and are not much used in the evenings because of the inappropriate size of furniture are an example of this. Lighting in cupboards, ducts and other occasionally used spaces is another.

On the other hand, where the lighting installations will be used for long periods, fluorescent installations are likely to give the most economical solution. In the case of air-conditioned buildings the cost of extracting heat from the lighting can be an important cost factor and fluorescent lighting is indicated.

ESTIMATION OF LIGHTING LEVELS

Lumen method

This is a simple way of estimating the number and size of lighting fittings needed to give adequate light on the working plane. The effect of reflectors from walls and ceilings is taken into account but the appearance on luminance ratios are not. *Interior Lighting Design,* published by the British Lighting Council, contains a detailed survey and the data required for the use of this method.

Basic formula:

$$N = \frac{A \times E}{F \times UF \times MF}$$

or $E = \dfrac{N \times F \times UF \times MF}{A}$ or $F = \dfrac{A \times E}{N \times UF \times MF}$

where:
- N = number of lighting fittings
- E = illumination level required, 1ux
- F = 'average through life' lamp flux, lumens
- UF = Utilisation factor. An experimentally determined factor taking into account shape of room, reflection factors of surfaces and the performance of lighting fittings. Tables are given by manufacturers of fittings. To use them the room ratio must be calculated:

$$\frac{\text{length} \times \text{width}}{\text{height of lightings above working plane} \times (\text{length} + \text{width})}$$

- MF = maintenance factor. An allowance for reduced light emission due to dirt. Depends on frequency of cleaning. Usually taken as 0.8 in clean interiors.

Note: In some specialized cases additional factors must be taken into account (eg absorption of light by dust in the air in industrial buildings). To ensure even illumination the spacing of fittings should not exceed 1½ x mounting height above the working plane (1.25 in the case of louvred or recessed fittings).

Luminance design

Luminance design is a means of taking into account the appearance of all wall, ceiling and floor surfaces and selecting lighting fittings not only to give light on the working plane but also on to the walls and ceiling. For designers of buildings and interiors the technique seems particularly appropriate and a step-

ELECTRIC LIGHTING

by-step method of carrying out this type of analysis is provided together with a worked example. In order to use this method it is necessary to ahve the manufacturer's data on a range of possible shapes and sizes of electric lighting fittings.

See Tables 53, 54, 56 & 57 and worked examples.

Both the above methods are related to even, overall levels of lighting; where lighting distribution is to vary over a room or where it is desired to achieve areas of more intense illumination the point-by-point method is used. The calculation is involved and requires knowledge of the light distribution from the fitting; it is beyond the present scope. Detailed descriptions of the method are contained in most texts on electric lighting.

Estimations of the output of permanent supplementary artificial lighting installations (PSALI) are usually made by the lumen method. The level of lighting required is, however, no longer the same as that quoted in the IES code. The lighting level provided must be related to the natural lighting conditions which prevail. The problem is discussed in BRS Digest 76 (second series), *Integrated Daylight and Artificial Light in Buildings*.

Where the formula

$$\text{Level of supplementary lighting} = 10 \times \text{The average daylight factor over the area supplemented} \times \frac{\text{Sky luminance (cd/m}^2\text{)}}{3}$$

is given as a guide to the levels of supplementary lighting required. In most working interiors a level of 500 lux gives reasonable satisfaction.

Table 53 Designed appearance lighting calculations, step-by-step procedure

A routine for determining the number and type of lighting fittings required to provide a desired luminance pattern or balance of illumination in a space with given internal finishes.

The procedure is in four main stages:

1 *Development of luminance specification*

Specification of desired internal finishes and definition of the task to be carried out in the space. From these basic decisions a complete luminance specification is developed.

Table 53 Designed appearance lighting calculations, step by step procedure–continued
Stage one continued

2 *Direct and indirect illumination*
Calculation of interreflections in the space and hence establishment of the distribution of direct illumination required to achieve the desired conditions.

3 *Lighting fittings*
Selection of the type(s) of lighting fitting which will provide a suitable light distribution.

4 *Number and layout of fittings*
Determination of the number of fittings required.

STAGE ONE DEVELOPMENT OF LUMINANCE SPECIFICATION

1 Select surface finishes, colours and reflectances for the working plane and for the other room surfaces, and enter in colums 1, 2 and 3 respectively of Table A.

2 Establish the reflectance of the task (from manufacturers' literature or published data, or by comparison with shade cards of known reflectance, etc). Enter in column 3 of Table A.

3 Establish the illumination level required on the task (from IES code) and enter in column 4 of Table A.

4 Calculate the task luminance (luminance = illumination × reflectance) and enter in column 5, Table A.

5 Complete the specification by determining *either* luminance *or* illumination of the room surfaces. As a guide in working environments, use either Bodmann's luminance ratios or Jay's illumination ratios, which give the desirable levels on other room surfaces in terms of the task luminance or illumination. The recommended ratios are:-

	Task	Immediate surround	Walls	Ceiling
Bodmann *(luminance)*	1	0.4	0.3	0.3
Jay *(illumination)*	1	1	0.5 to 0.8	0.35 to 0.8

Table 53 Designed appearance lighting calculations, step by step procedure–continued

6 Having specified reflectance and luminance, calculate illumination; *or* having specified reflectance and illumination, calculate luminance. Complete Table A for all room surfaces. (Luminance = reflectance × illumination).

Table A

Surface	1 Finish	2 Colour	3 Reflection factor	4 Illumination	5 Luminance
Task					
Immediate surround					
Floor					
Working plane					
Walls					
Ceiling					

Notes

1 *Floor cavity reflectance* should be used instead of reflectance when working plane is above floor level. It is roughly the weighted average reflectance of floor and work surfaces, minus 0.1.

2 Where walls are of different reflectances, use the weighted average. Reflectance of uncurtained windows may be taken as 0.1.

DIRECT AND INDIRECT ILLUMINATION

STAGE TWO DIRECT AND INDIRECT ILLUMINATION

7 Calculate the *room index:*

$$\frac{\text{width} \times \text{length}}{(\text{width} + \text{length}) \times \text{height}}$$ or use nomogram.

Note: height is measured from ceiling to working plane or floor, as appropriate.

8 Using the value of room index just calculated, select the appropriate multiplying factors from table 54 and enter them in table B. Also enter luminances from table A in the first column of table B. When room index does not coincide with tabular values use linear interpolation. To check that the multiplying factors have been entered correctly, the sum of all the factors in one column of table B should be unity.

9 For each pair of surfaces, work out the indirect illumination on one due to light reflected from the other (indirect illumination–source luminance × multiplying factor). Enter the results in table B, and sum the indirect illumination on each surface.

Note that the values of indirect illumination in one row of the table are arrived at by multiplying the factors in that row, in turn, by the luminance at the beginning of the row.

Table B

Light coming from	Luminance (from col. 5, Table A)	Light falling on					
		Floor or working plane		Walls		Ceiling	
		Mult fac	Ind illum	Mult fac	Ind illum	Mult fac	Ind illum
Floor or working plane							
Walls							
Ceiling							
Total indirect illumination on each surface							

Table 53 Designed appearance lighting calculations, step by step procedure—continued

10 Obtain the direct illumination required on each surface by subtracting indirect illumination)Table B) from total illumination (table A). Complete table C below.
11 If the direct illumination required on any surface has a negative value, the luminance specification developed in steps 1 to 6 cannot be met because that surface is too brightly illuminated. In such a case, return to stage 1 and adjust the specification, either by altering reflectances of the room surfaces (step 1) or by choosing different luminance or illumination ratios (step 5).

N.B. Changing the task, luminance or illumination alone will not help.

STAGE THREE LIGHTING FITTING SPECIFICATION

12 Work out the ratio R =
$$\frac{\text{Direct illumination on walls}}{\text{Direct illumination on floor or WP}}$$
13 If lighting fittings are to be suspended below the ceiling, calculate a new room index for this stage of the procedure, measuring height from the place of the fittings to the floor or working plane. Otherwise use the room index found in step 7.
14 Select the appropriate BZ classification for the lighting fittings from table 55, using the values of R and of room index from steps 12 and 13. Linear interpolation should be used for intermediate values of room index.
15 From table 56 obtain the direct *ratio* d for this BZ classification and room index, and note its value once again, interpolate for intermediate values of room index.
16 Calculate the ratio C =
$$\frac{\text{Direct illumination on ceiling}}{\text{Direct illumination on floor or wp}}$$
17 Calculate the required flux fraction ratio, F = C x x (see note).
18 Choose a lighting fitting of the required BZ classification and flux fraction ratio from manufacturers' literature. Note the downward light output ratio (DLOR) given by the manufacturer.

Note: Flux fraction ratio is the ratio between the amount of light emitted from the fitting above the horizontal to the amount below the horizontal. Some manufacturers give this value directly; in other cases the flux fraction ratio is obtained by dividing the percentage of light emitted upwards by the percentage emitted downwards.

STAGE FOUR NUMBER AND LAYOUT OF FITTINGS

19 Determine the total lamp flux required from the formula
$$\text{Flux} = \frac{\text{Area} \times \text{direct illumination on working plane}}{\text{Direct ratio} \times \text{DLOR} \times \text{maintenance factor}}$$

A good average value for the maintenance factor is 0.8; for information on the more precise determination of maintenance factors see IES Technical Report No. 9, *Depreciation and Maintenance of Interior Lighting*.

Table C

1	2	3	4
Surface	Total illumination (from table A)	Indirect illumination (from table B)	Direct illumination required (Col 2–Col 3)
Floor or working plane			
Walls			
Ceiling			

DEVELOPMENT OF LUMINANCE SPECIFICATION

Table 53 cont/d...

20 Establish the number of lamps required by dividing the total lamp flux obtained in step 19 by the light output of the type of lamp appropriate to the fitting chosen. (Use the *lighting design lumen* figures given by the lamp manufacturer.) Hence obtain the number of fittings required.

21 Sketch out an acceptable layout of fittings. The number may have to be increased to give a pattern suited to the space.

22 Check the spacing/mounting height ratios for the proposed layout and ensure that they do not exceed the values given for the appropriate BZ classification in Table 55. If they do, the layout should be modified until satisfactory spacing/mounting height ratios are obtained. Failure to do so will result in excessive variation in the illumination on the working plane.

23 Carry out a glare check on the proposed installation, using either the Mears' circular calculator or the tables in IES Technical Report No. 10. Ensure that the glare index for the proposed installation does not exceed the Limiting Glare Index given in the IES Code for the task which is being performed in the space.

Table 53 based on method developed by Peter Jay and Dennis Coomber.
Worked example by Dennis Coomber of Peter Jay and Partners.
Reproduced by courtesy of Peter Jay and Dennis Coomber.

Luminance design

Worked example

Required: a general lighting scheme for a computer room. No special requirement for local concentrations of light. Room dimensions: 12.0 m x 6.0 m x 3.0 m high. The working plane is taken as 0.85 m above floor level.

The step-by-step procedure operates as follows:

STAGE ONE: DEVELOPMENT OF LUMINANCE SPECIFICATION

Step 1 Preferred surface finishes, colours and reflectances are entered in Table 53A, which then appears as below:

Step 2 The task may be taken as reading and writing, and for task reflectance we take 0.8 (white paper).

Step 3 From the IES code, task illumination is to be *600 lux*.

Step 4 Therefore task luminance = 600 x 0.8 = *480 apb*.

In the example, the working plane is of more interest than the floor. Before going on to step 5 we therefore work out the floor cavity reflectance and enter it in Table 53A. Assuming that worktops etc account for one-third of the total floor area, the floor cavity reflectance will be
$\frac{1}{3}$ x 0.45)+($\frac{2}{3}$x 0.30)−0.10 = 0.15+0.20−0.10 = 0.25

Step 1

Surface	Finish	Colour	Reflection factor	Illumination	Luminance
Task					
Immediate surround	Work surfaces	Light teak	0.45		
Floor	Semtex vinyl asbestos tiles	Roman grey	0.30		
Working plane					
Walls	Plaster	Pale blue	0.50		
Ceiling	Acoustic tiles	White	0.70		

Table 54 Luminance design calculations

Room index	Multiplying factors			
	On to ceiling or floor, from floor or ceiling	On to ceiling or floor from walls	On to walls from ceiling and floor	On to walls from walls
k_r	$_{cf}M_{cf}$	$_{cf}M_w$	$_wM_{cf}$	$_wM_w$
.500	.1998	.8002	.2000	.5999
.525	.2129	.7871	.2066	.5868
.550	.2257	.7743	.2129	.5742
.575	.2384	.7616	.2190	.5621
.600	.2508	.7492	.2248	.5505
.625	.2630	.7370	.2303	.5394
.650	.2749	.7251	.2357	.5287
.675	.2866	.7134	.2408	.5184
.700	.2980	.7020	.2457	.5086
.725	.3092	.6908	.2504	.4991
.750	.3201	.6799	.2550	.4900
.775	.3307	.6693	.2594	.4813
.800	.3411	.6589	.2636	.4728
.825	.3512	.6488	.2676	.4647
.850	.3610	.6390	.2716	.4569
.875	.3707	.6293	.2753	.4493
.900	.3800	.6200	.2790	.4420
.925	.3892	.6108	.2825	.4350
.950	.3981	.6019	.2859	.4282
.975	.4068	.5932	.2892	.4216
1.0	.4153	.5748	.2924	.4152
1.05	.4316	.5684	.2984	.4031
1.10	.4471	.5529	.3041	.3918
1.15	.4618	.5382	.3095	.3811
1.20	.4758	.5242	.3145	.3710
1.25	.4892	.5108	.3192	.3615
1.30	.5020	.4980	.3237	.3525
1.35	.5141	.4859	.3280	.3440
1.40	.5257	.4743	.3320	.3360
1.45	.5368	.4632	.3358	.3283
1.50	.5474	.4526	.3395	.3211
1.55	.5575	.4425	.3429	.3142
1.60	.5672	.4328	.3462	.3076
1.65	.5765	.4235	.3494	.3013
1.70	.5854	.4146	.3524	.2952

Table 54 Luminance design calculations—continued

	Multiplying factors			
Room index	On to ceiling or floor, from floor or ceiling	On to ceiling or floor from walls	On to walls from ceiling and floor	On to walls from walls
k_r	$_{cf}M_{fc}$	$_{cf}M_w$	$_wM_{cf}$	$_wM_w$
1.75	.5940	.4060	.3553	.2895
1.80	.6022	.3978	.3580	.2840
1.85	.6101	.3899	.3607	.2787
1.90	.6177	.3823	.3632	.2736
1.95	.6250	.3750	.3656	.2688
2.00	.6320	.3680	.3680	.2641
2.10	.6454	.3546	.3724	.2552
2.20	.6577	.3423	.3765	.2470
2.30	.6693	.3307	.3803	.2394
2.40	.6801	.3199	.3839	.2323
2.50	.6902	.3098	.3872	.2256
2.60	.6998	.3002	.3903	.2194
2.70	.7807	.2913	.3933	.2135
2.80	.7171	.2829	.3960	.2079
2.90	.7251	.2749	.3987	.2027
3.00	.7326	.2674	.4011	.1978
3.10	.7397	.2603	.4035	.1931
3.20	.7464	.2536	.4057	.1886
3.30	.7528	.2472	.4078	.1844
3.40	.7589	.2411	.4098	.1804
3.50	.7647	.2353	.4117	.1765
3.60	.7702	.2298	.4136	.1729
3.70	.7755	.2245	.4153	.1694
3.80	.7805	.2195	.4170	.1661
3.90	.7854	.2146	.4186	.1629
4.00	.7900	.2100	.4201	.1598
4.25	.8006	.1994	.4237	.1527
4.50	.8103	.1897	.4269	.1462
4.75	.8190	.1810	.4298	.1404
5.00	.8270	.1730	.4325	.1350
5.25	.8343	.1657	.4350	.1300
5.50	.8410	.1590	.4373	.1255
5.75	.8472	.1528	.4394	.1212
6.00	.8529	.1471	.4413	.1173

Reproduced by permission of the Illuminating Engineering Society, from IES *Monograph*, No. 10.

ELECTRIC LIGHTING

Table 55 Values of the ratio 'R'

Classification	Associated S/H_m ratio	Room Index (k_f)									
		0.6	0.8	1.0	1.25	1.5	2.0	2.5	3.0	4.0	5.0
BZ1	1 : 1	0.242	0.223	0.212	0.202	0.196	0.189	0.185	0.182	0.179	0.177
BZ2	1 : 1	0.323	0.295	0.278	0.264	0.255	0.244	0.237	0.233	0.228	0.224
BZ3	1.25 : 1	*	0.362	0.344	0.329	0.319	0.307	0.299	0.294	0.288	0.284
BZ4	1.25 : 1	*	0.457	0.432	0.413	0.400	0.384	0.375	0.368	0.360	0.355
BZ5	1.5 : 1	*	0.530	0.512	0.496	0.484	0.471	0.462	0.457	0.451	0.448
BZ6	1.5 : 1	*	0.780	0.760	0.746	0.738	0.733	0.734	0.739	0.751	0.764
BZ7	1.5 : 1	*	1.006	0.977	0.958	0.947	0.942	0.943	0.951	0.967	0.987
BZ8	1.5 : 1	*	1.213	1.170	1.140	1.122	1.110	1.108	1.115	1.135	1.155
BZ9	1.5 : 1	*	1.568	1.435	1.379	1.352	1.308	1.293	1.289	1.296	1.312
BZ10	1.5 : 1	*	2.345	2.048	1.854	1.743	1.627	1.573	1.547	1.531	1.535

*Room index 0.6 impossible with standard fitting spacing for classifications BZ 3 to 10.

Table 56 Direct ratios for BZ classifications

Classification	Associated S/H_m ratio	Room Index (k_f)									
		0.6	0.8	1.0	1.25	1.5	2.0	2.5	3.0	4.0	5.0
BZ1	1 : 1	0.553	0.642	0.703	0.755	0.792	0.841	0.871	0.892	0.918	0.934
BZ2	1 : 1	0.481	0.576	0.643	0.703	0.746	0.804	0.840	0.866	0.898	0.918
BZ3	1.25 : 1	*	0.525	0.593	0.655	0.702	0.765	0.807	0.836	0.874	0.898
BZ4	1.25 : 1	*	0.467	0.536	0.602	0.652	0.722	0.769	0.803	0.847	0.876
BZ5	1.5 : 1	*	0.430	0.494	0.558	0.608	0.680	0.730	0.766	0.816	0.848
BZ6	1.5 : 1	*	0.339	0.397	0.456	0.504	0.577	0.630	0.670	0.727	0.766
BZ7	1.5 : 1	*	0.284	0.338	0.395	0.442	0.515	0.570	0.612	0.674	0.717
BZ8	1.5 : 1	*	0.248	0.299	0.354	0.401	0.474	0.530	0.574	0.638	0.684
BZ9	1.5 : 1	*	0.203	0.258	0.312	0.357	0.433	0.492	0.538	0.607	0.656
BZ10	1.5 : 1	*	0.146	0.196	0.252	0.301	0.381	0.443	0.492	0.566	0.620

*Room index 0.6 impossible with standard fitting spacing for classifications BZ 3 to 10.

LUMINANCE DESIGN CALCULATIONS

Step 5

Surface	Finish	Colour	Reflection factor	Illumination	Luminance
Task	Paper	White	0.80	600	480
Immediate surround	Work surfaces	Light teak	0.45		
Floor	Semtex tiles	Roman grey	0.30		
Working plane			0.25		
Walls	Plaster	Pale blue	0.50		144
Ceiling	Acoustic tiles	White	0.70		144

Step 6

Surface	Finish	Colour	Reflection factor	Illumination	Luminance
Task	Paper	White	0.80	600	480
Immediate surround	Work surfaces	Light teak	0.45	600	270
Floor	Semtex tiles	Roman gray	0.30		
Working plane			0.25	600	150
Walls	Plaster	Pale blue	0.50	288	144
Ceiling	Acoustic tiles	White	0.70	206	144

Step 5 Using Bodmann's luminance ratios, the luminance of both walls and ceiling should be 0.3 times the task luminance. We therefore enter 0.3 × 480 = 144 asb in column 5 of Table 53A, in each case.

Obviously the illumination on the work surfaces and on the working plane will be the same as that on the task, since we have assumed that all three are at the same height above the floor. The respective luminances are therefore:

    Immediate surround: 0.45 × 600 = *270 apb*
    Working plane: 0.25 × 600 = *150 apb*

Enter these figures in column 5 of Table 53A.

In this case the luminance ratio for the immediate surround is 270/480 = 0.56; this is higher than Bodmann's recommendation, but would still be quite acceptable. (The ratio could be reduced by reducing the reflectance of the work surfaces, and if this were done the floor cavity reflectance would also be reduced.)

Step 6 Calculate the illumination on walls and ceiling, and complete Table 53A. We have:

Walls: Illumination = Luminance/Reflectance
                      = 144/0.5 = *288 lux*
Ceiling: Illumination = 144/0.7 = *206 lux*

ELECTRIC LIGHTING

The calculations in stage two will use only the luminances of the ceiling, walls and working plane, and we do not need to complete the line in table 53A, dealing with the floor.

The completed table 53A is shown in Step 6.

STAGE TWO: DIRECT AND INDIRECT ILLUMINATION

Step 7 Since the working plane is 0.85 m above floor level we take height as (3.00−0.85) = 2.15 m. Therefore

$$\text{Room index} = \frac{12 \times 6}{(12+6) \times 2.15} = 1.86$$

Step 8 Consider first the multiplying factor for indirect illumination on the ceiling due to reflection from the floor. From table 54, this factor is

0.6101 for room index 1.85
and 0.6177 for room index 1.90

By linear interpolation, the factor for room index 1.86 is

$$0.6101 + \frac{1.86 - 1.85}{1.90 - 1.85} \times (0.6177 - 0.6101)$$
$$= 0.6101 + \frac{0.01}{0.05} \times 0.0076 = 0.6101 + 0.0015 = 0.6116$$

Using a similar interpolation process for the other factors, and taking luminances from table 53A, we have the following entries in table 53B.

Step 8

Light coming from	Luminance	Light falling on to					
		Working plane		Walls		Ceiling	
		Mult factor	Indirect illum	Mult factor	Indirect illum	Mult factor	Indirect illum
Working plane	150			.3612		.6116	
Walls	144	.3884		.2777		.3884	
Ceiling	144	.6116		.3612			

Step 9

Light coming from	Luminance	Light falling on to					
		Working plane		Walls		Ceiling	
		Mult factor	Indirect illum	Mult factor	Indirect illum	Mult factor	Indirect illum
Working plane	150			.3612	54	.6116	92
Walls	144	.3884	56	.2777	40	.3884	56
Ceiling	144	.6116	88	.3612	52		
Total indirect illumination			144		146		148

LIGHTING FITTING SPECIFICATION

As a check on the multiplying factors, note that the column totals are 1.0000, 1.0001 and 1.0000 respectively.

Step 9 The indirect illumination on the working plane due to reflection from the walls is

144 (wall luminance) × 0.3884 (MF) = *56 lux*

and so on. Completing table 53B we have:

Step 10 Now completing table 53C.

Step 11 None of the direct illumination values is negative, so the luminance specification derived in step 5 is feasible. Since some direct illumination is required on the ceiling, we shall use suspended rather than recessed lighting fittings.

STAGE THREE: LIGHTING FITTING SPECIFICATION

Step 12 Ratio R = 142/456 = 0.311 (using direct illumination values from table 53C).

Step 13 Suppose the lightiny fittings are suspended 0.30 m below the ceiling. Then the new room index for stage 3 will be

$$\frac{12 \times 6}{(12 + 6) \times (2.15 - 0.30)} = 2.16$$

Step 14 From table 55 we have the following values of R:

Room index	2.0	2.5
BZ3	0.307	0.299
BZ4	0.384	0.375

So by linear interpolation, for room index 2.16, the R-values are

BZ3: $0.307 - \frac{2.16 - 2.0}{2.5 - 2.0} \times (0.307 - 0.299)$

$= 0.307 - \frac{0.16}{0.5} \times 0.008 = \mathit{0.304}$

BZ4: $0.384 - \frac{2.16 - 2.0}{2.5 - 2.0} \times (0.384 - 0.375)$

$= 0.384 - \frac{0.16}{0.5} \times 0.009 = \mathit{0.381}$

Clearly a BZ3 fitting is closest to the requirement.

Step 15 By linear interpolation in table 56, the direct ratio of a BZ3 fitting at a room index 2.16 is

$0.765 + \frac{2.16 - 2.0}{2.5 - 2.0} \times (0.807 - 0.765) = \mathit{0.778}$

Step 16 Using direct illumination values from table 53C,

Ratio C = 58/456 = *0.127*

Step 17 Flux fraction ratio F

= C × d
= 0.127 × 0.778 = *0.099*

Step 18 Select a surface-mounting BZ3 fitting with flux fraction ratio about 0.1 from manufacturer's catalogue. The computer room will be air-conditioned and fluorescent fittings will be preferred so as to keep down the heat gain. The DLOR of such a fitting is found to be about *0.40*.

Step 10

Surface	Total illumination	Indirect illumination	Direct illumination
Working plane	600	144	456
Walls	288	146	142
Ceiling	206	148	58

ELECTRIC LIGHTING

STAGE FOUR: NUMBER AND LAYOUT OF FITTINGS

Step 19 The total lamp flux required is

$$\frac{(12 \times 6) \times 456}{0.778 \times 0.4 \times 0.8} = 132{,}000 \text{ lumens}$$

Step 20 Lighting design lumen figures for high efficiency fluorescent tubes are:

$$\begin{aligned}&1.8 \text{ m, } 85 \text{ watt: } 5550 \text{ lm}\\&1.5 \text{ m, } 65 \text{ watt: } 4400 \text{ lm}\\&1.2 \text{ m, } 40 \text{ watt: } 2600 \text{ lm}\end{aligned}$$

Dividing these figures into the total lamp flux we find that we require 24 1.8 m 85-watt tubes, *or* 30 1.5 m 65-watt tubes, or 51 1.2 m 40-watt tubes to meet the specification. Assuming that twin-tube fittings are used, we have a choice between 12 1.8 m, 15 1.5 m or 26 1.2 m fittings.

Steps 21 and 22 Consider first the use of 12 1.8 m 85 watt twin fittings; the natural arrangement will be three rows with four fittings in each. Mounting height is 1.85 m above the working plane (step 13); the spacing between rows will be 2.00 m. This gives a spacing/mounting height ratio of 1.08 compared with a maximum of 1.25 for a BZ3 fitting (table 55 or 56) and is therefore satisfactory.

Each fitting will be about 1.90 m long, and the clear space between fittings in the same row will therefore be 1.10 m. The value of the spacing along rows is taken from the centre of one fitting to the centre of the next, and is accordingly 3.00 m. This results in a spacing/mounting height ratio of 3.00/1.62, which is excessive, and we therefore reject the idea of using 1.8 m fittings.

Consider instead the use of 15 1.5 m 65-watt fittings, arranged in three rows of five. The spacing between rows remains the same as before and is satisfactory. Each fitting is 1.60 m long, and there is about 0.80 m clear space between them. The spacing/mounting height ratio along the rows is therefore $(1.60+0.80)/1.85 = 1.30$. This is close to the limiting value, and by reducing the clear space between fittings to 0.70 m we can get a satisfactory spacing/mounting height ratio of 1.24.

It would also be desirable to increase the spacing between rows slightly by moving the outer rows towards the walls. With 2.20 m between fittings a ratio of 1.19 would be obtained.

Step 23 Carrying out a glare check by use of the Mears' circular calculator gives a final glare index of about 14.

A more detailed calculation using tables in IES Technical Report No. 10 is as follows:

Initial glare index	19.9
Correction for downward flux	+3.2
Correction for luminous area	−6.1
Correction for mounting height	−0.8
Correction for 0.3 floor reflectance	−2.8
Final glare index	13.4

The IES code sets a limiting glare index of 19 for computer rooms, so that the proposed installation is completely satisfactory.

Table 57
Lamp Flux. 'Average through life' lumen output (Typical values, 240V 25°C)

5'0" 80 W Fluorescent Tubes

Northlight	3250
Daylight	4500
Natural	3500
White	5000
Warm-white	5000

Incandescent Lamps

40 W	330
60 W	600
100 W	1200
150 W	2000
200 W	2750
300 W	4400
500 W	8000
750 W	12800
1000 W	18000
1500 W	28500.

UTILITY SERVICES

9 Water supply

PRIVATE WATER SUPPLIES

While public utility water mains can serve the majority of buildings, it is sometimes necessary for private water supply arrangements to be made. Expert advice from specialist firms can be obtained, and the local authority public health department will require to be satisfied as to the suitability of the supply.

It is important to note that where a private source water supply has to be provided it is usually also necessary to make private arrangements to dispose of sewage, and care must be taken to avoid pollution of the water source.

For human consumption it is best that the water come straight from the ground, rather than from a stream or pond which is exposed to probable pollution. A dug well of big enough diameter to admit a man and his spade, or a borehole of small dimensions made by mechanical means, and just big enough to admit the necessary pump or suction pipe are possible methods. In either of these the upper part should be lined to exclude surface water, as this is liable to be polluted. It is best to leave the lower part unlined if it will stand safely without collapsing, but lining is usually needed through clay zones, and some form of porous or perforated liner may be needed in water-bearing strata which will not stand up without support. Most modern wells are lined with pre-cast concrete cylinders, and bore-holes with stell tubing.

When a well is being dug the water level must be held down by pumping to enable the well digger to work. The excavated material is shovelled into a bucket and hauled out with a rope. Great care must be taken to ensure the safety of the digger both from falling objects and from collapsing sides of the excavation. Boring operations do not involve de-watering. For most modern requirements boreholes are the most practical and convenient, but dug wells are useful in situations where the water-bearing strata does not yield water freely, and where the larger peripheral area of the well is therefore an advantage.

Dug wells for domestic or farm use are commonly 1 m to 1.5 m in diameter, and boreholes 150 to 200 mm. For small requirements where the strata is coarse-grained gravel or soft, fissured rock, driven tube-wells are often successful. These are usually of steel tube, 30 to 50 mm diameter, screwed together and having a point and perforations at the lower end.

The usual consumption of water approximates to:

0.09–0.18 m^3 per day per person
0.36–0.45 m^3 per day per ordinary house
0.14 m^3 per day per dairy cow
0.05 m^3 per day per dry cow.

For market garden and pasture irrigation in season, 12 mm of 'rain' per week is generally considered sufficient provision.

Diagram 131 shows how rain soaks into porous ground and finds its way through layers of rock or sand to feed springs, wells, boreholes and rivers. Underground 'rivers' are very unusual, and the water ordinarily lies in myriads of tiny crevices in rock layers and between particles of sand. This forms a vast underground store of water, rather like a tank filled with both water and pebbles. Usually there is a movement of water through these cavices to an outlet. A simple example of this is a gravel-capped hilltop overlying clay: springs flow at the junction with the clay, and water may be obtained from wells dug in the gravel down to the face of the clay. If the catchment area is small, and the gravel clean and readily porous, the water will run out quickly and the springs and wells will fail in prolonged dry weather. If the area is extensive and the gravel interspersed with fine-grained material, the rate of flow will be diminished and springs and wells will continue to yield small supplies even in dry times. The same things happen in the more complex conditions shown in the later diagrams. An 'artesian' borehole is one which pierces a layer

WATER SUPPLY

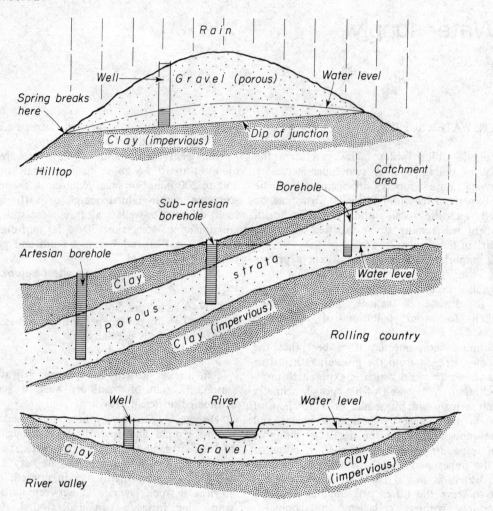

131 Wells and boreholes for private water supply

of clay and enters a lower porous zone from which water rises as a 'gusher' above ground level. A similar borehole where water rises part way but does not reach surface is called 'sub-artesian'.

Pumps

There are very many types of pumps; broadly they fall into three main categories:

1 The displacement pump in which one or more plungers move backwards and forwards or up and down, each drawing water past a valve into a cylinder and then forcing it out past another valve.

2 The rotary pump, which has meshing 'gear' wheels, or a rotor and a stator, which are pressed together in such a way that water is pocketed, passed through the pump body, and forced out as the meshing elements meet.

3 The centrifugal pump, in which a rotating impeller flings water to its periphery, where the kinetic energy is converted to pressure energy by a volute or set of diffuser vanes.

All these types are obtainable in sizes to meet almost every requirement, but each has its advantages and its limitations, and selection must take into consideration all aspects of the application—

the height the water is to be elevated, the cleanliness of the water, the form of power available to work the pump, the amount of noise that can be tolerated, and the length of life required of the mechanism.

Most hand-operated pumps are of the displacement type, having a plunger which is lifted and lowered alternately, or pushed backwards and forwards in a cylinder. Another variety, called the 'semi-rotary', has a metal plate which fits diametrically across a round chamber and is oscillated with a backwards and forwards semi-rotary movement; the plate is fitted with valves, so is the chamber, which is divided so that the pump has the equivalent of two cylinders. This is a popular type for raising water from a well to an overhead tank.

Displacement pumps are also operated by windmills or water wheels, gearing being provided to produce fairly slow speeds, usually between 15 and 50 reciprocating cycles per minute.

Windmills are erected on towers which are usually between 6 and 12 m in height. To span times of no wind it is necessary to provide ample storage. The pump units can be small, as they will be operating continuously when the wind is blowing.

The plungers in displacement pumps are fitted with 'cups' made of leather or fabric, or operate through packed glands. Usually these pumps give many years of reliable service, the cups or packing being renewed at intervals at low cost for materials.

Rotary pumps of the gear type are convenient and inexpensive but rely on close metal-to-metal clearances, and if the water they handle is gritty they wear quickly and become noisy and inefficient. They are particularly useful for pumping oil and chemicals, where cleanliness is assured. They usually rotate at speeds of 1000 rpm or more.

Some metal-and-rubber rotaries are useful and reliable for small water-supply duties and have good self-priming characteristics. These pumps usually run at 1450 rpm—a useful electric motor speed, and are quiet in operation.

As both 'displacement' and 'rotary' pumps are positive in action they give approximately constant volume at any fixed speed, irrespective of the pressure.

Centrifugal pumps usually run at 1450 or 2900 rpm and are often direct-coupled to electric motors. Ordinary centrifugal pumps have no ability to self-prime; the pipe system must be completely filled with water and all air must be released before the pump is started. But there are a number of devices which can be incorporated in the design to provide self-priming; most of them include circulation of part of the water from the delivery side, back to the suction side through a venturi tube. This draws water up the suction pipe and dispells the air in the system. Centrifugal pumps are not 'positive' in action and the volume delivered varies widely at different pressures, even at constant speed.

Most of these centrifugal pumps can drawn up water 4.5 to 6 m vetrically; by special design as much as 9 m can be attained, but this is unusual (about 10 m is the theoretical maximum 'suction'). The total head these pumps can produce depends on the peripheral speed of the impeller. Most small single impeller pumps give total heads of between 10 and 20 m. Larger single-stage pumps give up to 60 m. To attain greater heads several pump units are assembled in series on one spindle and are driven by the same motor.

Another type of 'centrifugal' is the 'water-ring' pump. This has a different form of impeller which runs with close clearances between the facings of the casting sides; it pulls water in through one set of holes in the sides of the casting and forces it out through another. It is essentially a clean water type of pump and has a relatively low power efficiency; but it is useful because it will draw water up about 8.5 m, has good self-priming characteristics, and gives total heads of 45 to 100 m per impeller stage when running at 2900 rpm. The volume does not vary with pressure changes as much as with a conventional centrifugal pump.

Most of these types of pumps can be made in shallow-well and deep-well forms. In every case the actual pumping unit must be either submerged or else located within 'sunction limit' of the water level, that is to say, for practical purposes not higher than about 4.5 m above water level. Two convenient forms of pump now used in deep wells are boreholes are:

(a) The reciprocating 'displacement' pump with its cylinder located at or below water level, suspended from ground level, and having its operating rod extended above ground and worked by a crank mechanism driven by an engine, electric motor or windmill.

(b) The 'submersible' electric pump: this has a special electric motor designed for working under water. To its rotating shaft are attached one or more centrifugal pump impellers. The water passes through each of these in series and extra pressure is added by each to build up sufficient for delivering water to the desired height. The pressure given by each impeller is a function of its peripheral velocity, hence, as most of these pumps are designed to go down relatively small boreholes, several impellers are generally needed. Pumps are available for raising over 1.3 l/s from 100 mm boreholes, and 25 l/s from 200 mm ones, and pressures up to 300 m are attained with standard pumps.

Three other special types of pump deserve mention; each is limited in its application.

1 *The hydraulic ram* This device derives power from falling water. Water is taken from above a dam, piped to the ram at a lower level, and discharged through a 'pulse-valve' to waste below the dam. Water flowing through the ram washes the pulse-valve to a closed position, thus arresting the flow suddenly and causing a hydraulic shock; this shock jerks a small amount of water past a delivery valve into the delivery main. After the shock has passed the pulse-valve drops open and water flows again. This pulsing action is allowed to go on continuously day and night, and a hydraulic ram correctly installed can give trouble-free service for many years. Variations in design allow special types of ram to be operated by dirty water, and to pump clean water from a well or other pure source.

2 *The jet-pump* This is a device for raising water from deep wells and boreholes. An ejector (or 'jet') is hung below water level and power water is delivered down a 'drive-pipe' to it; this water enters the ejector near the bottom and is directed upwards through a nozzle. The water-jet from the nozzle rushes through a venturi-throat and creates a vacuum which draws in fresh water through a foot-valve below the device. This fresh water and the power water flow together up a riser-pipe to ground level. Generally, a centrifugal pump is fitted at ground level with its inlet connected to this riser-pipe: from its outlet some of the water goes to the reservoir and the rest is returned to the jet as power water. Correct proportioning of pressures, heads, volumes and pipe-sizes is essential to successful operation. This device is sometimes convenient, but is necessarily of low mechanical efficiency as power is absorbed in pressurising about twice as much water as is delivered to the reservoir.

3 *The air-lift* This uses compressed air. The air is piped to a foot-piece below water level; thence the bubbles rise and carry water up with them. It is essential for the foot-piece to be adequately submerged, and about 50 per cent submergence is common; that is to say, if water level is 15 m below the discharge point, the foot-piece must be 15 m below water (viz 30 m below the discharge point). Airlifts cease to function below about 35 per cent submergence. The higher the percentage submergence the less volume of air is needed to raise each gallon of water, and the higher the required air pressure. Correct proportioning of pipe sizes, depth of foot-piece, air volume and pressure is essential. Airlifts have a low power efficiency, but they are particularly useful in handling water containing sand, mud, or raw sewage, since it does not have to pass through any moving mechanism and wear and tear is limited to abrasion on the stationary pipe system.

Purity

It is frequently necessary to obtain information on the purity of a water supply to assess its suitability for drinking purposes, for steam-raising or other industrial processes where excessive hardness may be a disadvantage, or to forecast its action on metals such as galvanised tanks or pipes. Samples are usually collected in containers supplied by the analyst, with instructions for taking the sample.

The analyst's report will give a number of analytical figures together with an opinion on its suitability for the purpose in mind. The interpretation of the results will be influenced by the source of the water and other considerations but it will give the following figures:

Total solids This will comprise both dissolved and suspended solids, organic and inorganic. The former are mainly derived from animal and vegetable debris, and the latter from the earth.

Water from chalk or from salt-bearing strata may contain over 1000 parts per million, whereas water from upland sources may contain less than 50 parts per million. (Note carefully the units employed, as it is customary nowadays to express results in parts per million, but some analysts still use parts per 100 000.)

Ammonia More than 0.08 ppm of albuminoid ammonia in water from a shallow well, coupled with a similar amount of free and saline ammonia, usually indicates pollution by sewage, but the same amount of free and saline ammonia in water from a deep well would be due to another cause.

'Oxygen absorbed' This is a measure of the amount of organic matter in a water, mainly derived from the decay of animal and vegetable matter. Generally speaking, the figure for a very pure water would be less than 0.5 ppm, up to 1.5 ppm would be reasonable pure, and anything over 2.0 ppm would be classed as impure, but these figures may be doubled if the water is surfaced water from an upland source.

Nitrates More than 5 ppm of nitrates indicates than the source may be liable to pollution, although at the time of sampling it may be satisfactory.

Nitrites The presence of nitrites normally indicates recent sewage pollution.

Chlorine The combined chlorine is an indication of the salinity of the water. A very salty water is unpalatable. Chlorides may also be derived from urine.

Hardness is of two kinds: that which is removed by boiling is termed 'temporary', and that which is not, 'permanent'. The former is deposited as fur in pipes and boilers, unless the water is softened or steps taken to prevent its deposition. Both kinds of hardness are due mainly to salts of calcium and magnesium, and the hardness of water is usually measured in degrees Clark. Waters usually vary from 0–50 degrees of hardness, anything over 20 is hard, less than 10 would be regarded as soft. On this basis a soft water would not be likely to cause serious deposition of fur in 20 years.

pH This measures the acidity or alkalinity of the water. Soft acid waters are derived from hard insoluble rocks or from peaty upland, they have a pH less than 7.0 and may corrode pipes and tanks unless passed through a cylinder packed with limestone to neutralise the acidity. Waters with a pH of more than 7.0 are alkaline and are less likely to attack metals.

Water may also contain bacteria, some derived from soil. These latter grow at ordinary temperatures and normally do not acuse disease, although their presence in very large numbers is not desirable, and those which normally flourish in the intestines of animals, which grow at blood heat, may include disease-producing species. The latter are termed 'coliform', and their presence indicates pollution of animal origin.

If the water is intended for drinking purposes, for use in a dairy or in the manufacture of preparation of foodstuffs a bacteriological examination is necessary, in addition to the chemical analysis. Samples are taken in sterile bottles supplied by the analyst.

Treatment

Water treatment normally consists of three processes:

1 Filtration
2 Sterilisation
3 Softening

1 *Filtration* is normally done through sand, either by simple percolation (slow filter) or under pressure (rapid filter). It may be necessary to add a small amount of some chamical, such as alumina, to assist filtration. The filters become clogged in time, and have to be washed back with clean water. The process removes suspended matter from the water.

For water which is to be used for baths, flushing wcs, etc. no further processing will be required in many cases. Diagram 132 shows a simple filtration and storage unit from which water can be pumped up to a storage cistern in the building itself.

2 *Sterilisation* Before water can be consumed by humans, it must be sterilised. In large installations and in public supplies chlorine is added to the water by means of special plant. In small installations this is not feasible. It consists of a very fine filter capable of being cleaned and impregnated with silver which has a bacteriological effect upon the water. Diagram 133 shows such a filter. It is usually used at the kitchen sink to give additionally filtered and sterilised water at this point

WATER SUPPLY

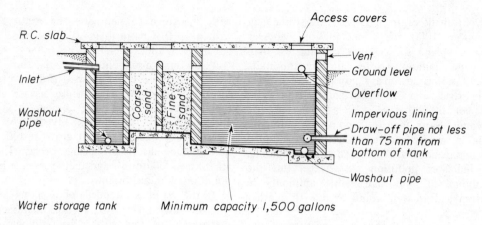

132 Simple filter and storage cistern for small private water supply installation

while the remainder of the installation delivers the water from the simple filter described above.

3 *Softening* Normal hard water is not a risk to health but has a number of disadvantages. In hot water pipes and boilers scale may be deposited and

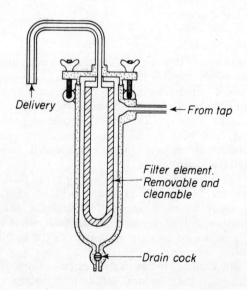

133 Sterilising filter for producing potable water in a private water supply installation

in domestic use scum is deposited and soap does not lather well. Some industrial processes may be affected. There are several ways in which water may be softened, base exchanged, lime-soda and inhibitors.

Domestic water softeners belong to the first type. The water is passed through a medium called a 'zeolite', which converts the calcium salts in the water to sodium salts. From time to time the zeolite medium requires regenerating by passing strong brine through it. A modern development of this process employs synthetic resins, etc, in place of the natural zeolites, and water of comparable purity to distilled water can be obtained for certain industrial processes where water of high purity is necessary. This is expensive and unnecessary for domestic water.

For industrial plant, water may be softened by the addition of lime to remove temporary hardness, or with lime and soda to remove both temporary and permanent. The process may be a batch process whereby tanks of water are treated with the required amount of lime or lime soda, the sludge allowed to settle and the water run off to use, or the addition may be made continuously by a special plant. In either case the disposal of large volumes of sludge is necessary.

To prevent the deposition of fur in boilers by hard waters, certain complex phosphates (such as 'Calgon') may be added which prevent the formation of scale. These compounds are growing in numbers, and can be regarded as inhibitors rather than true softeners, although the effect may be the same.

MAINS WATER SUPPLY

The vast majority of buildings take their water from public water supplies. This supply will be suitable for drinking although in some areas it may be hard and require softening for some uses and in others the use of particular pipe materials may be prohibited because of rapid corrosion due to the nature of the water. The adequacy of a mains supply will depend on the size of the water mains, the pressure of water in them (expressed in terms of 'head', ie the height to which the water in the main would rise in a vertical pipe) and the demand on the main. The water undertaking should be consulted about a supply both to the building and for the building operation at a very early stage.

The desirable minimum size for water mains is governed by firefighting requirements. A 75 mm diameter pipe fed from both ends and a 100 mm diameter pipe fed from one end are considered satisfactory. In towns a grid of pipes served by two trunk mains enables supplies to be maintained even when individual sections have to be shut off. Diagram 134 shows a typical arrangement.* An adequate head of water in the main is necessary to provide an adequate flow and to raise water to the top of buildings. A minimum head of 30 m is desirable for firefighting purposes

*Supplies to isolated buildings will normally be by single pipe.

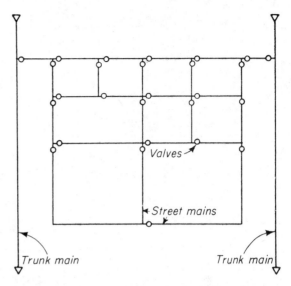

134 Typical urban water main grid. Note that two trunk mains supply the grid and that any pipe can be taken out of service without cutting off the supply to the others

and a maximum head of 70 m is thought appropriate to limit wastage and noise in pipework. The head is provided by siting service reservoirs at suitable heights above the buildings being

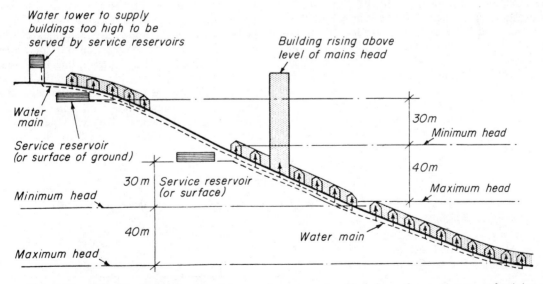

135 Principles of distribution of water from service reservoirs. Note how the maximum and minimum heads of water are achieved by siting the reservoir at a level higher than the buildings that it serves. In practice the slopes of the ground may make close compliance difficult. While water is being drawn off pressures will fluctuate due to friction in the pipework. Water to the tall building will have to be pumped

WATER SUPPLY

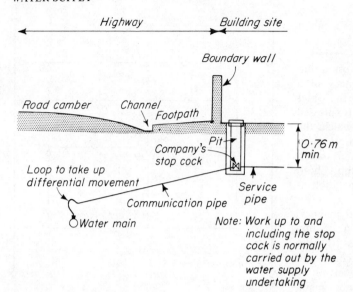

136 Typical details of the connection (by communication pipe) from the town water main and the service pipe on the site. This part of the installation is often carried out in the early stages of building to provide a supply for construction

served. Diagram 135 shows the principles involved. It also shows a water tower built to give a supply to buildings which cannot be served from a normal service reservoir, and a building rising above the mains head which will require its own system of pumps to raise water to the top. It is clearly not always possible to arrange service reservoirs in ideal relationship to the buildings they supply and while it will be rare to find heads of over 70 m many areas will have less than 30 m. The full head of water from the reservoir is only available when no flow is taking place in the mains. At most times the actual head at a given point will be reduced by the friction losses due to flow. This will vary with demand from day to day and from time to time. Water undertakings carry out a continuous programme of pressure measurement and will usually be able to give information about the pressure variations to be anticipated at any point.

Although a position beneath the footpath has often been advocated, water mains, in common with other main services, are usually to be found under the carriageway. Connection of the main to the site is almost invariably made by the water supply undertaking which has the right to dig up the road for this purpose. An ingenious box-like mechanism bolted to the main enables the connection to be made without interrupting flow in the main itself. Diagram 136 shows the general arrangement for a typical domestic connection where the communication pipe and the service pipe would be 12 mm in diameter. The minimum cover of 0.76 m below ground for the service

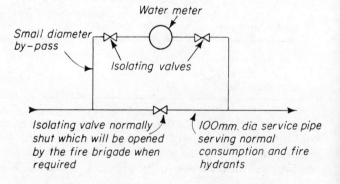

137 Method of installing water-meter of size appropriate to normal consumption where the service pipe has to be of 100 mm diameter for fire-fighting purposes

pipe to avoid risk of freezing will be noted. This part of the work is normally carried out at the commencement of the building contract so that a supply of water for building can be obtained by the contractor. In large buildings or where sanitary appliances are supplied directly from the service pipe a larger diameter may be required. In some cases where street hydrants are not close enough to the building fire brigade hydrants will be required on the building site necessitating a 100 mm diameter service pipe.

The charge for water in most buildings is based on a proportion of the rateable value or on an annual charge per sanitary appliance irrespective of the amount of water used. In some cases, however, particularly in industrial buildings, the water may be metered and the charge based on actual consumption. The installation of a meter, often in a pit near the boundary, presents no special problem except in the case of the large diameter pipe put in to serve fire hydrants, which is very much in excess of the diameter that would be appropriate to the normal flow. If, due to the distance away of the building, it is not practicable to separate the pipework for normal consumption a suitable meter size may sometimes be achieved by the use of a by-pass. Diagram 137 shows a possible arrangement.

COLD WATER STORAGE AND DISTRIBUTION

The pipework from the company's stop-cock onwards will usually be installed after the carcase of the building is complete. Provision for the service pipe to enter 0.76 m below ground and to rise inside away from the external wall is usually made by building in drainpipes at the time of laying the foundations and ground floor slab. Some features of pipe distribution within the building are common to all installations since they deal with avoidance of damage from frost. Water supply pipes may not be built in to external walls and should not pass through or near the eaves where cold draughts are difficult to avoid and if they are not within the heated volume of the building pipes should be lagged.* Apart from these considerations the size and pattern of distribution of pipework will depend very much on the provision for water storage. Not all countries require water storage and it is possible in some parts of the world to have all parts of the water

*A stopcock and a drain cock to enable the installation to be shut off and emptied must be provided immediately the service pipe enters the building, and drain cocks to empty any section of pipe not drained by the main drain cock or by draw-off points.

Table 58 Merits and demerits of water storage in buildings

Advantages of water storage	Disadvantages
1 Provides a reserve against failure of the mains supply.	1 Space and support must be provided for the storage cistern. (In high buildings a proportion of storage at ground level is becoming usual to avoid loads on the building and to save space at high level.
2 Sudden demands are met from the storage cistern which then fills slowly, thereby making the demand on the main more even. This results in: (a) economy of water mains and in size of service pipe; (b) reduced possibility of mains pressure dropping to nothing, which could lead to 'back-syphonage' of water from sanitary appliances into the main.	2 Storage cisterns may become dirty (particularly if not provided with a cover). Drinking water is, however, supplied direct from the main.
3 Reduced pressure on the installation which minimises noise and wastage and enables economical appliances (eg hot water cylinders of reasonable gauge) to be used.	3 Because of the reduced pressure distributing pipes have to be larger.
4 Heating and hot water supply apparatus can be vented to the storage cistern, thereby minimising safety valve requirements.	

supply installation directly under mains pressure. In Britain hot water supply installations must be supplied from storage (except instantaneous heaters) and in many parts of the country all sanitary appliances must be supplied from storage except for taps supplying water for cooking and drinking, which come directly from the main.

The main advantages and disadvantages of water storage may be tabulated:

In many cases the water undertaking will require storage to be provided and there will be no question of taking any decision about this. If storage authority some thought should be given to whether it is a desirable provision in any case. The main factors which would indicate the provision of storage are:

1 *Economy* If the stroage cistern can be accommodated without expense (say in the roof space) and the service pipe from the main is long in relation to the distributing pipes a system employing a small diameter service pipe and storage may be the more economical.

2 *Noise* The reduced pressure from storage cisterns as compared with mains pressure will usually give quieter operation, reduce water hammer and particularly ball-valve noise. (Substantial water storage is not essential to achieve these objects which could be accomplished by a small 'break tank' but the two are normally associated.)

3 *Reserve for mains failure* Some areas are subject to gross pressure fluctuations and even the cessation of flow at times during normal operation. Most places enjoy very consistent supply. The consequences of mains failure must, however, be taken into account. In some industrial situations provision of water storage could form a well worth while insurance against disruption of manufacturing processes.

Tables 60 and 61 give standards for water storage and sizes of typical cisterns. For domestic premises 0.23 m^3 actual capacity of storage is required for hot and cold and 0.12 m^3 actual capacity for hot water supply only. These quantities could be achieved respectively by galvanised sheet steel cisterns of 0.32 m^3 nominal capacity (0.92 m x 0.61 m x 0.59 m high) and 0.18 m^3 nominal capacity (0.61 m x 0.61 m x 0.48 m high).

Diagrams 138, 139 & 140 show a typical domestic water supply installation. 138 shows the service pipe (subject to mains pressure) serving the storage cistern and kitchen sink with stopcock and drain-cocks. Diagram 139 shows the cold water distributing pipes (pipes subject to pressure from cistern only) and Diagram 140 the hot water distributing pipes. Diagram 140 shows in addition the flow and return pipes to the hot water cylinder that would be required if hot water supply is to be achieved from a domestic boiler. Where water storage is employed only for hot water, it is possible to supply all the cold taps from the service pipe. This may result in some saving of pipework, particularly if 12 mm pipe is used throughout, which is very usual. If a 12 mm pipe is used from the communication pipe some restriction of flow may be experienced if several fittings are used simultaneously. Noisy operation of ball valves and taps is very likely with this arrangement. Diagram 141 shows a typical installation. The diameter of the service pipe at entry, 25 mm, should be noted. This is to allow for simultaneous use of all taps without restriction of flow.

The pipe sizes shown in these diagrams, although typical of domestic installations, relate to the building shown and are not necessarily applicable in other cases. The size depends on flow required, length of pipe and head from cistern to draw-off (see Table 66 for pipe-sizing methods). The general principles demonstrated in these diagrams apply to larger and more complex installations which, with appropriate sizing, will be similar in their nature.

The connection of a bidet presents a special problem since the points of input are below the flood level of the appliance and consequently there is a serious risk of back syphorage. This is usually overcome, as shown in diagram 142 by providing a separate cold water distributing pipe for the bidet and taking the hot water distributing pipe from a vented pipe at least 2 m above the bidet. If the water supply fails there should be no possibility of drawing water from the bidet to any other sanitary appliances.

Diagram 143 shows the details of installation and insulation of a typical cold water storage cistern. The cistern should be included within the insulated perimeter of the building. If it is installed above an insulated ceiling, the ceiling insulation should be omitted within the area of

COLD WATER SUPPLY AND DISTRIBUTION

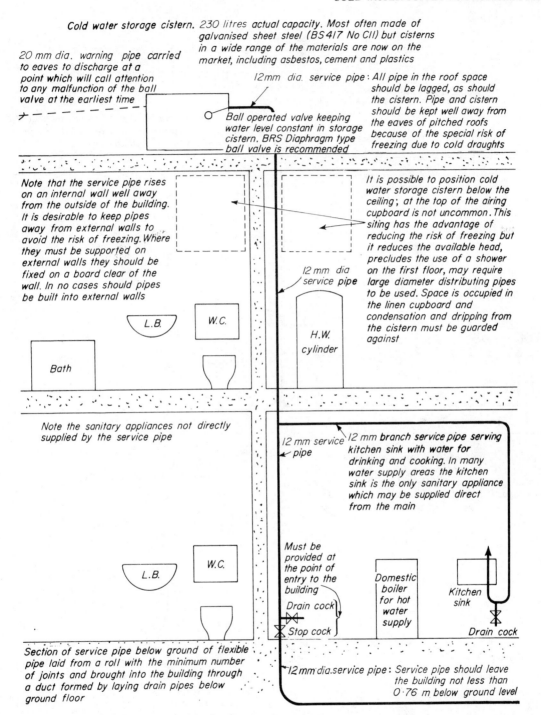

138 Typical service pipe installation to a dwelling where storage for hot and cold water is provided. Note particularly: Pipe rising on an internal wall to avoid frost. Stop cock and emptying plug. Kitchen sink served direct from main. (All other taps from storage)

WATER SUPPLY

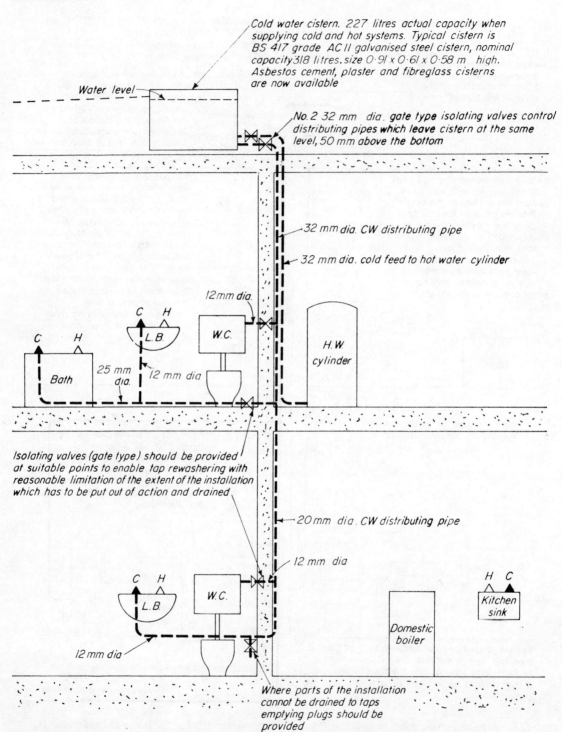

139 Typical cold water supply distributing pipe system. (To be read in conjunction with diagram 138

COLD WATER SUPPLY AND DISTRIBUTION

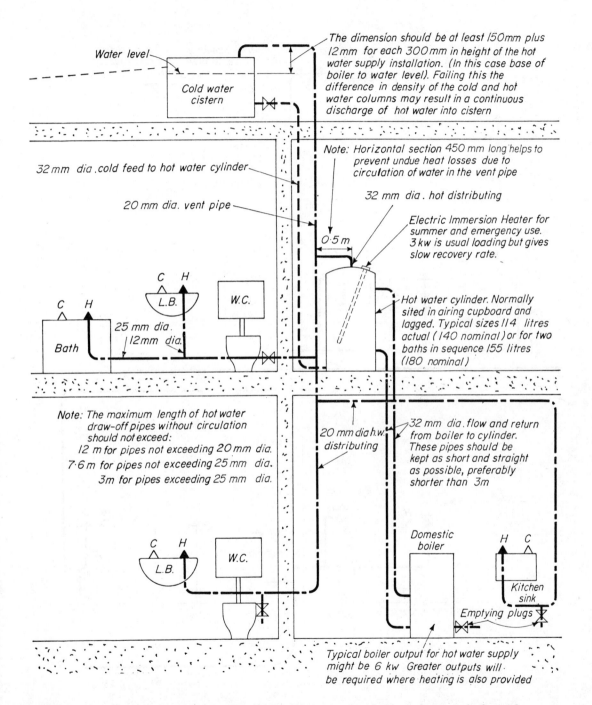

140 *Hot water supply distributing pipes. (Water heating is provided by a domestic boiler with flow and return pipes leading to a patent indirect cylinder)*

WATER SUPPLY

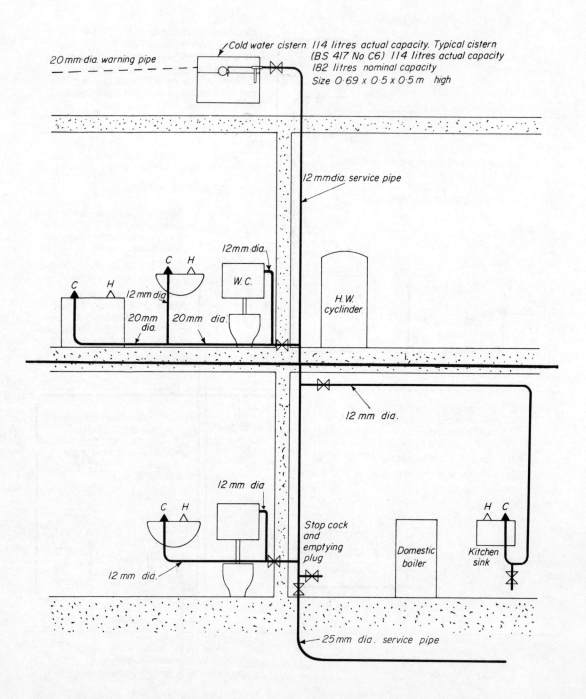

141 *Typical service pipe installation to a dwelling where storage is required for hot water supply only*

COLD WATER SUPPLY AND DISTRIBUTION

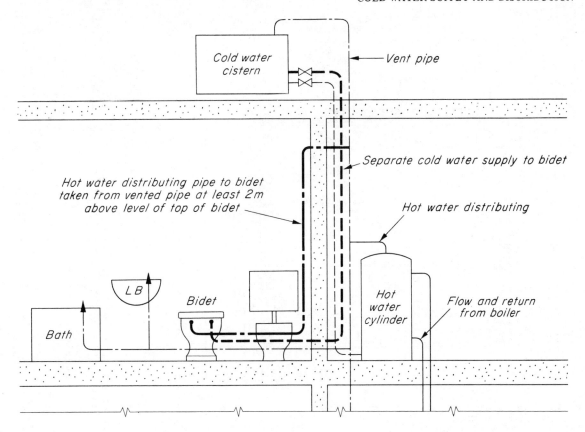

142 Connection of bidet to hot and cold water supply

the cistern insulation so that heat from the building may penetrate easily. There is some risk of condensation forming on the sides and bottoms of cold water cisterns. This problem is particularly likely to arise in a linen cupboard if hot water cylinder and cold water cistern are both installed there. Trouble can be avoided if an asbestos-cement sheet is fixed below the cistern. The drips will be intercepted and will then evaporate.

The traditional material for cold water cisterns has been galvanised sheet steel. There is, however, rapidly increasing use of other materials, particularly plastics, for use as domestic cisterns.

Very large storage cisterns are often built up from panels of iron or steel about 1 m square. In some cases the cisterns are formed of concrete lined with bitumen.

Cold water storage cisterns often present difficulties of placing. They may well be too large to go upstairs and through doors or hatches. The problem is particularly acute when new cisterns have to be fitted in place of old ones which were placed before construction was complete. In these cases several cisterns may be linked together to provide the storage required. The supply and the draw-off should be at opposite ends of the group to ensure a flow of water through all the cisterns.

It is usual in large installations to provide at least two cisterns with separately valved ball valves and outgoes so that one may be put out of action for cleaning or repair without affecting the supply to the building. Diagram 144 shows a typical arrangement.

WATER SUPPLY

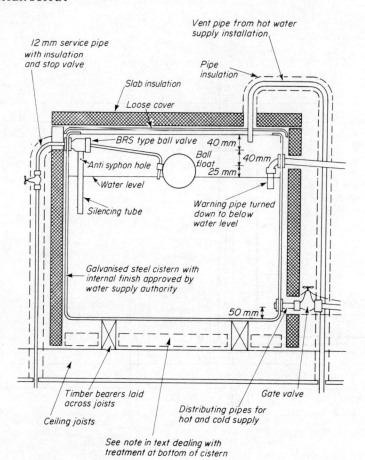

143 Typical cold water supply cistern

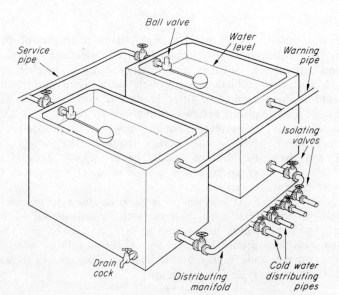

144 Typical water storage installation with two linked cisterns

COLD WATER SUPPLY AND DISTRIBUTION

Supplies to buildings rising above the level of the mains head

New problems arise in high buildings. It will probably be necessary to pump water up to the top of the building since the head in the main is not likely to be adequate and if the building is very tall it may be necessary to divide the distribution into zones to keep the water pressure within reasonable limits. It is not satisfactory to have pumps in continuous operation, nor is it practicable to have them switch on and off for every draw-off of water. Diagram 145 shows one means of overcoming this problem by means of a pneumatic vessel. A cushion of air under pressure is maintained in the top of a pressure cylinder; when a tap is opened the air is able to expand by forcing water out of the cylinder and through the pipework. This process can continue until the water level drops to a predetermined point, when the pumps will be switched on to raise the level again. Drinking water is drawn off from the pressure vessel, although it will be noted that within the reach of the mains head the drinking water is supplied direct from the main. Precautions are taken to ensure purity. The air pumped in is filtered to prevent dust and insects gaining access and the capacity of the vessel is kept reasonably small to prevent stagnation. For dwellings the volume of water between high and low water levels would be no more than 4.5 litres per dwelling. For flats of up to about 15 floors a simplified system is possible. Diagram 146 shows the arrangement. It differs from the previous system principally in supplying minor drinking water draw-offs by means of an enlarged section of service pipe above the level of the highest flat. This enlarged section, coupled with the special air vent above, enables water to flow to drinking water taps without the pump being operated except when the whole enlarged section becomes empty. As soon as this occurs a float switch brings the pumps into operation until the pipe is full. The ball valve prevents overflowing into the storage cistern unless the level there drops when the pumps will also be brought into operation by a float switch. It will be noted that in this case also the drinking water taps are supplied direct from the main where this is possible. In buildings which rise above 15 floors zoning to reduce the pressure in the distributing pipes at the bottom of the building is necessary. The usual zone height is about 30 m. Diagram 147 shows one method of dealing with the problem. The main water storage is now at ground level. This avoids a load of perhaps hundreds of thousands of kg at the very top of a high building. The systems for drinking water supply and cold water distribution are now completely separate. Pumps raise the water in stages of 60 m in both cases, but with cold water cisterns distributing to stages of 30 m. Drinking water is distributed on the same system using storage cisterns but in this case the cisterns are sealed against insects and dust, the warning pipe and air vent are fitted with filters for the

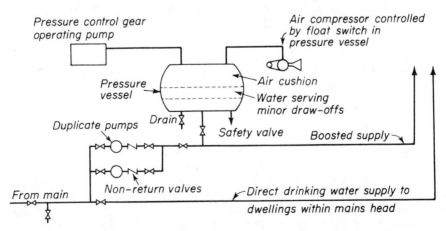

145 *Pneumatic Booster for raising water in buildings extending above the level of the mains head*

WATER SUPPLY

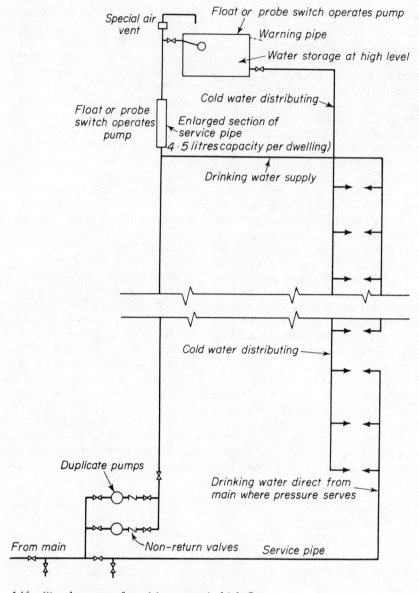

146 *Simple system for raising water in high flats*

same purpose and stagnation is avoided by limiting the cistern capacity to 9 litres per dwelling. Diagram 148 shows typical cistern arrangements. Pressure-reducing valves can have the same effect in reducing pressure as intermediate cisterns and are increasingly used. MO 6 HL Design Bulletin 3, Service cores in high flats, Part 6, Cold Water Services, gives a detailed account of the problem and several examples of solutions.

The temperature in internal pipe ducts is likely to be at least several degrees above ambient temperature because of heating and hot water supply pipes, flues, etc. Cold water passing through long lengths of pipe may be noticeably warmed. This can become unacceptable for drinking water, particularly in office buildings where long

HOT WATER SUPPLY

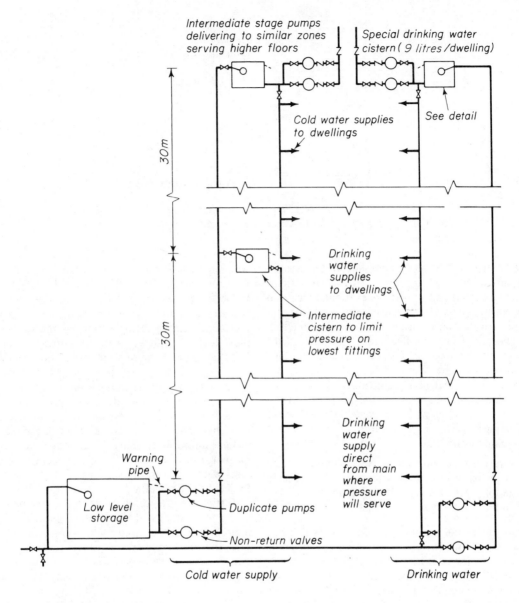

147 System for raising and distributing water in high buildings. Note: the ground level main storage, separate drinking water circuiting and the possibility of raising the water up another stage of the building

length of pipe in warm ducts may be serving drinking fountains which give rise to only a small flow. In such cases an insulated flow and return circuit serving the drinking water points with water pumped through a drinking water cooler is desirable.

HOT WATER SUPPLY

Hot water can be produced by a wide variety of appliances involving the whole range of available fuels. The methods can most usefully be classified

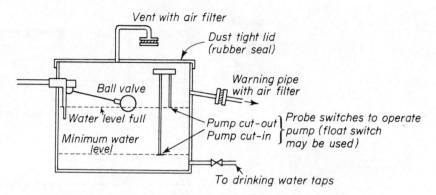

148 *Storage cistern for drinking water supply. Note: filtration of air entering. Sealing of top. Limited volume to ensure fresh water (4.5 l per dwelling)*

into central and local systems and instantaneous and thermal storage appliance. In central systems water is heated usually in the boiler room by a cheap fuel which is also being used for space heating and distributed to taps throughout the building; in local systems the water is heated (often by electricity or gas) adjacent to particular appliances and to groups of appliances. One of the main criteria of choice between these systems is comparative economy of the installation and running costs. In order to keep water consumption within reasonable limits 'dead legs' of pipework serving hot water taps are governed to maximum lengths.

12 m for pipes not exceeding 20 mm diameter
7.6 m for pipes not exceeding 25 mm diameter
3 m for pipes exceeding 25 mm diameter
1 m for pipes serving spray taps

This limits the amount of cold water which has to be run to waste before hot water is delivered. The reason for the length limit is water saving, but it also serves to keep within bounds the time delay and consequent inconvenience which would result from the use of very long pipes. Where, in central systems individual appliances are too far from the central plant to allow the use of single pipes to serve the taps, a circuit of pipe-work (called a secondary circulation) with either gravity or pumped flow can be used to maintain a constant flow of hot water which can be quickly drained off through a short length of delivery pipe serving taps.

It will be readily appreciated that a central system which has to have a secondary circulation will require a substantial pipe installation and that the heat losses from the pipework, even where reduced by lagging, will be continuous, including the summer months. Local systems, while they save pipework, usually being served by the cold water pipe system, and eliminate heat losses from distributing pipes, require a number of heaters, which often use a more expensive fuel than a central system. It is apparent, therefore, that large and continuous hot water demands, particularly hot if they are close to the central plant, are best dealt with by central installation, while scattered hot water points, particularly if the use is very intermittent, can often be more economically served by local water heaters. In many cases it will be difficult to decide which method is more economical.

Instantaneous water heaters are normally restricted to use as local heaters because of the limited flow of hot water which can be produced. The thermal capacity of water is high and to raise 8 litres of water per minute 55 deg C (the flow for a lavatory basin tap and the usual temperature increase required) requires a heat input of 17 kW. This is a rate equivalent to the full space heating requirement of a sizeable house. The output which can be achieved from an appliance of reasonable size is, therefore, very limited. A large gas instantaneous heater may deliver a flow of between 10 to 20 litres per minute, while the flow from a 3 kW electric spray tap may be only 1 or 2 litres per minute. Diagram 149 shows an instantaneous gas water heater. Gas instantaneous heaters are used for the whole hot water supply of small dwellings, but apart from this, instantaneous water heaters are used

HOT WATER SUPPLY

locally for the specific appliances that they serve.

Thermal storage heaters use a vessel to accumulate hot water. It is possible with this form of heater to use a modest heat input, while sudden heavy draw-off can be catered for by the stored hot water. It is necessary to balance the capacity of the heater and the heat input (recovery rate)

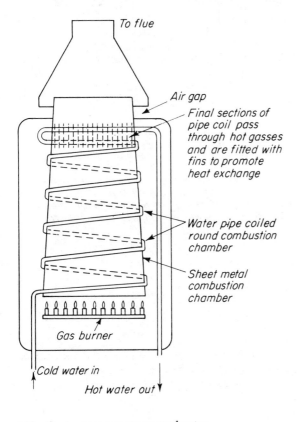

149 Instantaneous gas water heater

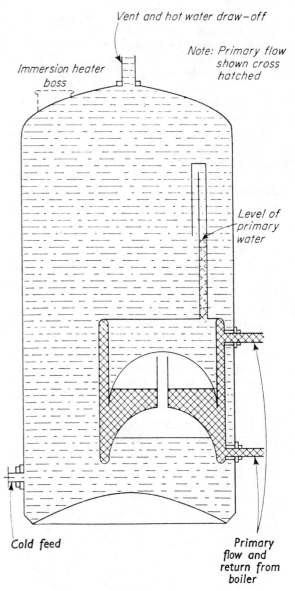

so that the likely demands can be met. Diagram 140 shows a typical central thermal storage system for a house. The water is heated in a boiler on the ground floor and the hot water accumulates by thermo-syphon action in the cylinder on the first floor. Fresh cold water is introduced into the cylinder from a cistern at high level and makes up any water drawn off from hot taps. At one time a decision had to be taken whether the system should be 'direct' (water passing through the boiler can also be drawn off from the taps) or 'indirect' (water circulating through the boiler

150 'Primatic' type of patent indirect hot water storage cylinder. The primary flow is shown cross hatched. The air bubbles which separate it from the water in the cylinder can be observed. Expansion and contraction of the water and even the effects of pump operation can be accommodated without allowing the primary flow to mix. Filling and venting, however, take place through the cold feed and vent

211

WATER SUPPLY

and heating the cylinder as a separate system from the water drawn off at taps). The direct system is subject to furring or corrosion of pipes, while the indirect system, although universal in large installations, is expensive for small houses. The problem has been overcome by the development of patent indirect cylinders which effectively separate the boiler flow (primary circulation) from the stored hot water while allowing the boiler to be filled and vented through the cylinder without the need for an additional cistern and pipework. Diagram 150 shows a section through a patent indirect cylinder of this type, the 'Primatic', demonstrating how the primary flow is kept from contact with the stored water by means of entrapped air pockets.

The system shown is provided with an immersion heater for use when the boiler is out of action or during the summer, when the boiler, which often also supplies space heating, may be turned off. The usual loading for immersion heaters used in this way is 3 kW. This gives a very slow recovery rate. The 3 kW heater will take some three hours to raise 150 litres of water 55 deg C. This slow rate of heating a cylinder may cause considerable annoyance and often contrasts unfavourably with the very much quicker heating-up achieved by the boiler, which will invariably have a higher rate of heat output. This is particularly the case when heating is also provided by the boiler. Except in the coldest weather the boiler will have unused capacity which can contribute to water heating and will thus speed the recovery rate of the cylinder.

In large buildings, particularly blocks of flats, several indirect type hot water cylinders or pressure type water heaters may be required. The separate cold water supply to an individual hot water cylinder coupled with hot water draw-off from the top of the cylinder means that, even in the case of failure of the water supply, the cylinder cannot be emptied. A similar performance is required from installations with more than one cylinder. Diagram 151 shows how this may be achieved by taking the connections to the cylinder from the distributing pipe at a level higher than that of the cylinder served.

Diagram 87 shows the arrangements in the boiler room for water heating and water storage in large installations with pumped secondary circulation. The term 'calorifier' is usually used in this case to describe the hot water storage vessels.

Diagram 84 shows how hot water supply may be achieved using the heat from the boiler room

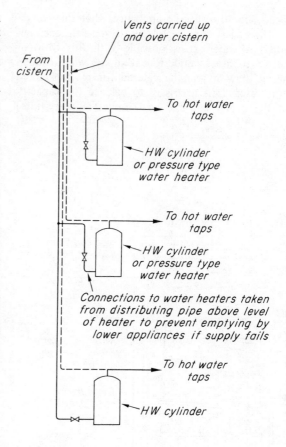

151 Cold water supply to several hot water cylinders ensuring that the cylinders cannot be emptied in the case of water supply failure

but without any separate hot water secondary circulation. An indirect cylinder is heated by water from the space-heating pipework. The cold water storage and distribution arrangements are similar to the domestic ones in Diagram 140. In the summer, when the heating is not operating, an immersion heater in the cylinder is used. This is a particularly economical arrangement when small groups of sanitary appliances are widely scattered in a building. It is quite feasible to have several hot water supply systems of this sort in different parts of the building served from one boiler room.

HOT WATER SUPPLY

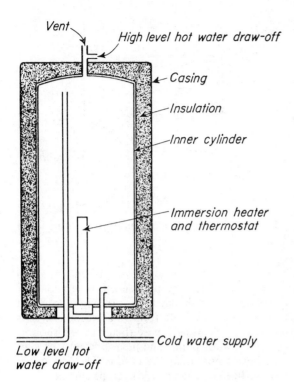

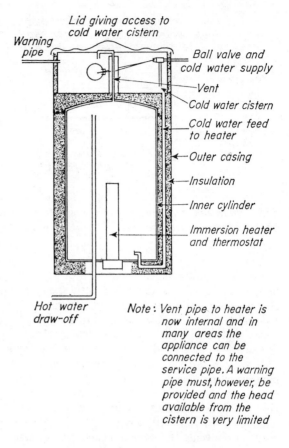

Note: Vent pipe to heater is now internal and in many areas the appliance can be connected to the service pipe. A warning pipe must, however, be provided and the head available from the cistern is very limited

Note: The delivery of hot water is controlled by taps on the distributing pipes served by draw-off points. Several taps may be served

152 Pressure type electric water heater. Requires vent pipe carried back to cold water cistern (as in diagram 140). Can supply many taps at levels both below and above the heater

153 Cistern type of electric thermal storage heater. Convenient for use in flats since it may be served from the main and requires only a warning pipe, not a vent. May be used to supply many taps. The pressure at the taps will usually be low and all pipework must be kept below the level of the cistern

154 Free outlet type electric water heater. Neither vent nor warning pipe is required. Venting takes place through open delivery arm. The admission of cold water makes hot water overflow down delivery tube. Limited to one point, or two if swivel delivery arm is used

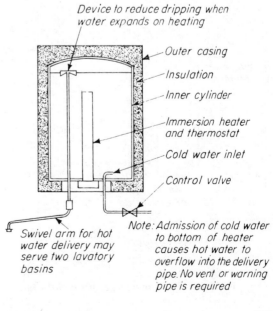

Note: Admission of cold water to bottom of heater causes hot water to overflow into the delivery pipe. No vent or warning pipe is required

213

WATER SUPPLY

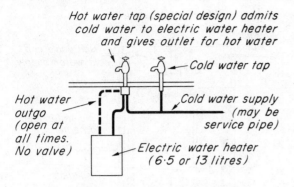

155 Undersink water heater with free outlet (can be connected to service pipe. No vent or warning pipe required)

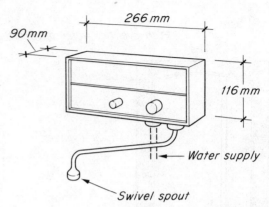

157 Heatrae 'Carousel' instantaneous electric water heater

Where no central boiler plant exists or where the sanitary appliances are not conveniently sited gas or electric water heaters may be used. Diagrams 152, 153, 154 and 155 show various types of electric water heaters. The pressure type heater (152) is essentially similar to the hot water cylinder and requires the same connections, including a vent pipe which must be carried back to the cold water cistern. In blocks of flats this can give rise to expensive pipework and the cistern type water heater (diagram 153) has been developed to obviate the need for the vent. This type of heater can normally be supplied by the service pipe rather than a special supply from the cistern. No vent is required but a warning pipe must be provided. In operation this type of heater is similar to the pressure type except that the head available from the integral cistern is limited. It will usually be difficult to use showers and all draw-off points and pipework must be kept at a lower level than the cistern. The free outlet type of heater (diagram 154) can be supplied from the service pipe provided its capacity is less than 13.5 ℓ (larger capacities may be permitted but only if effective anti-syphon devices are fitten). This type of heater is particularly convenient since no vent or warning pipe is required but only one outlet can

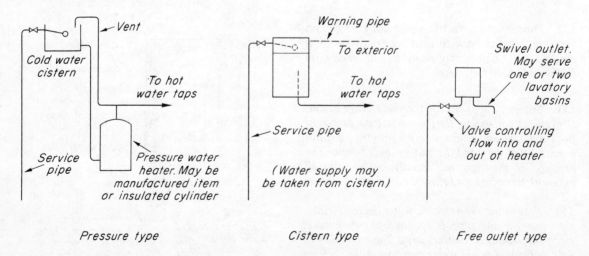

156 Water supply arrangements for various types of water heater

HOT WATER SUPPLY

be served. This outlet is often made to swivel so that two lavatory basins may be served. By the use of special tap fittings this type of heater can be adapted to be fitted below the level of the appliance served. Diagram 155 shows the arrangement. Diagram 156 shows schematically the piping requirements for pressure, cistern and free outlet type water heaters. An instantaneous heater is shown in diagram 157. The electrical loading for this type of heater is usually 3 kW and the output of hot water only 1 or 2 litres per minute. The outlet normally delivers a spray and is used only for washing hands or for simple showers.

There are gas appliances to match each of the electrical heaters. Some of the smaller ones may be permitted to discharge their products of combustion into rooms which, if acceptable, saves the expense of a flue. (See chapters 7 and 16 for gas flue considerations.) Table 59 sums up the main types of electric and gas water heaters together with notes on their piping and flue requirements.

Table 59 Types of gas and electrical water heaters

	GAS		ELECTRICITY	
	Type of appliance	*Special requirements*	*Type of appliance*	*Special requirements*
Instantaneous	Geyser Single point (may be able to provide boiling water) Multipoint	(Flue not normally required if heater is in a room) Flue needed 75–125 mm	Electric spray taps (3 kW) (limited flow)	
Thermal storage	Storage heaters* Small 1–5 galls (serve single basin and sink) Multipoint 12–100 galls Cistern multipoint	Vent pipe and flue required Flue and warning pipe required	Free outlet (1 or 2 sanitary appl.) sometimes from main Pressure* (may be built up by use of immersion heater and copper cylinder) Cistern (must lie above highest draw-off point)	Sometimes connected directly to service pipe. Vent pipe required Warning pipe required
Supplementary	Circulator (fixed) adjacent to and provides heat for storage cylinder) May be used to supplement boiler	Usually installed in cupboard where flue is required	Immersion heater (provides heat for storage cylinder to supplement boiler as for summer use)	

* Must be normally supplied from cistern.

10 Sanitary appliances

The use of water in buildings for sanitary and other purposes is made possible and convenient by the provision of sanitary appliances which are of appropriate form and have water supply and outgo arrangements sized and placed to assist the function which the appliance serves. All sanitary appliances should have surfaces which are smooth and easily cleaned; the surface and preferably also the body of the material should be impervious; the shape should be appropriate to its use and free from crevices or features which would make cleaning difficult. It is important that water supplies should run no risk of contamination from their proximity to the foul water in the appliance, particularly where direct supply from the mains is involved. In most cases it will be found that appliances are so designed that taps discharge above the highest flood level of the water so that they cannot become submerged and make back syphonage into the supply piping possible.

MATERIALS

The main materials used for sanitary appliances are:

Ceramics

Give very durable glazed surfaces in a wide range of colours. There are three main classes, the significant properties of which, from the point of view of selection of sanitary appliances, may be summarised:

Earthenware Relatively cheap.

Fireclay Strong and resistant to knocks: widely specified for public buildings and institutions.

Vitreous china Unlike the other materials which have a glaze on a permeable body, vitreous china is completely impermeable.

Most designs of sanitary appliances are made in one material only and if a particular design is desired the material of manufacture may be inevitably established.

Cast iron

Cast iron may have a white or coloured vitreous finish fired on. Mainly used for very large appliances which would be impracticable to manufacture and too heavy to move when made in ceramics.

Pressed steel

Used in recent years in cases where cast iron might previously have been employed. Finish can be of vitreous enamel. Noisy when used for sinks: sound-deadening materials are used on underside. One-piece drainers and sinks have great hygienic advantages when compared with separate sinks and drainers.

Stainless steel

The natural surface of stainless steel, although perhaps less satisfactory visually than vitreous enamels or ceramics in good condition, is able to withstand impact and abrasion. One-piece sinks and drainers have similar hygienic advantages to pressed steel.

Plastics

Perspex White and coloured *perspex* is increasingly being made into sanitary appliances. Its surface is liable to scratching.

A wide range of plastics are used in WC seats, cisterns, etc.

Terazzo

Special designs not involving large numbers of units in the materials described above, involve considerable expense because of the nature of the manufacturing processes and the patterns and moulds required. Special appliances can, however, be made reasonably simply from terazzo and very large items such as plunge baths can be formed in situ in this way.

SANITARY APPLIANCES

APPLIANCES

WCs

A water-closet consists of a pan containing water and receiving excrement and a device for providing a flush of water. Normally the pan and flushing device are separate but in some modern fittings they are coupled together. The pans are almost invariably of ceramic material.

There is a very long history to this type of appliance. However, the first device having features clearly similar to present practice was developed by Sir John Harrington in the reign of Elizabeth I. Flushing water closets did not, however, come into general use until piped water supplies became available. Many developments in design took place in the nineteenth century. One culminated in the valve closet

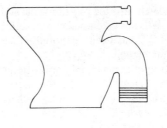

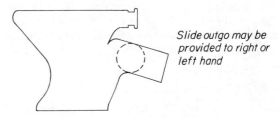

Slide outgo may be provided to right or left hand

159 WC pan with S-trap and P-trap with alternative right- or left-hand side outgoes

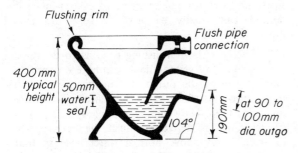

158 Section through wash-down WC pan. P-trap

which was highly effective and quiet in operation but has been completely superseded because of the problems of water supply (see *Flushing cisterns*, page 218), maintaining the mechanism and because a wooden casing was required which was difficult to keep clean. Some closets of this type are still in operation. The non-mechanical developments led to the type of pan most widely used at present: the *wash-down* (Diagram 158 shows a section through a WC pan of this type with a P-trap. Diagram 159 shows other outgo configurations which may aid planning and neatness of pipe installation). Clearing the contents of this type of pan depends on a powerful flush of water. A method which ensures clearing of the pan and does not require such a fierce flush is the syphonic closet shown in diagram 160. In this the operation of the flush causes a flow in the outgo which, due to the shape of the outgo, induces a

syphonic action which removes the water and other contents from the pan. The pan is refilled and the seal re-established by the last of the flush.

Although universally accepted the form and particularly the height of WC pans does not accord with medical opinion, which regards as low a level as possible as physiologically desirable. A compromise between established practice and the ideal was the health closet which sloped downwards from front to back, theoretically giving a lower sitting position and support to the thighs. The user, however, tended to slip backwards, thereby fouling the back of the pan. Seats with stops to prevent sliding were developed but this type of closet is rarely employed.

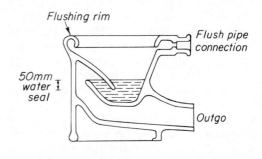

160 Section through syphonic WC pan

SANITARY APPLIANCES

Recently detailed research into the physiological and hygienic problems of WCs has taken place in America and a new pattern of WC has been proposed which may be produced commercially.

The floor space around and behind conventional WC pans is difficult to clean. To make this easier, particularly in institutions and hospitals, where hygiene is important, *corbelled washdown closets* are available. Two types are made, one requiring a heavy wall into which the corbell extension of the pan is built and the other which may be used with light partitions consisting of a metal frame fixed in the floor screed and the thickness of the wall and projecting from the wall to support the WC pan.

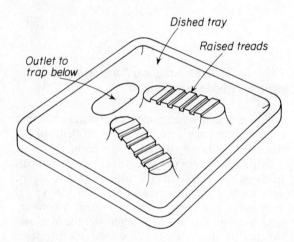

161 Squatting or eastern closet

The *Squatting* or *Eastern* closet (diagram 161) is rarely, if ever, installed in this country, although its use in Europe is widespread. It is flushed in the same way as a conventional WC. It is not suitable for old or infirm people but it gives a good physiological posture and when installed in a tiled enclosure enables very easy cleaning by means of a hose. It is very surprising that some of these closets are not provided in public lavatories, where they could give very much more hygienic conditions than are usual at present.

A number of particular features in addition to the points generally desirable for sanitary appliances should be borne in mind when selecting WC pans.

1 *Flush* The flush should scour the whole inner surface of the pan and should be sufficiently powerful to discharge the contents of the pan. In most cases the water is distributed round the pan by a flushing run but in some designs the run is dispensed with. The performance in this respect of both wash/down and syphonic pans should be observed before selection.

2 *Water area* The surface area of the water in the pan should be as large as possible so that all excrement falls directly into the water rather than on to the dry sides of the WC, where it would tend to stick, even when the WC is flushed.

3 *Back* The back of the inner surface of the pan should be vertical or nearly so to avoid fouling.

WC seats The traditional wooden seats have been replaced by plastic ring seats which are very much more hygienic. Seats with flat undersides are desirable and open-fronted rather than complete ring seats are best for all applications. Cover-flaps, if desired, can usually be provided from the same hinges as the seat.

WC flushing cisterns Flushing cisterns to deliver a fixed quantity of water quickly can have very simple plug control of the outgo, but in this country the type known as water waste preventers (WWP) are invariably used. They employ a simple syphonic type of mechanism to deliver the flush, the delivery tube of which rises above normal water level so that dripping, a continuous overflow into the pan with consequent waste of water, cannot take place. The earliest satisfactory form of WWP employed an inverted cast iron cone in a cast iron cistern which was raised and then dropped into the water, thereby causing it to rise inside the cone and to overflow into the flush pipe, giving rise to syphonic action which discharged the contents of the WWP. The action of the cone was usually assisted by allowing it to fall into a well in the bottom of the cistern, which ensured that water would rise inside the cone and gave rise to the characteristic form of this type of cistern. The fall of the heavy cone gave rise to noise and this and the neater form which can be achieved have resulted in the use of piston-operated flush mechanism shown in diagram 162. The flushing cistern may be positioned at the traditional high level (some 2 m to bottom of the cistern) or at the currently more popular low level (about 1 m to the top of the cistern). In each case a flush pipe connects the cistern to the pan. For high-level suites a galvanised telescopic pipe in which the top section can

SANITARY APPLIANCES

be pushed into the bottom section to suit the actual height is normally used. For low-level suites an L-shaped pipe, often finished with vitreous enamel, is cut to suit. Some closets are available which are bolted directly to the pan and described as close-coupled. A variety of materials can be selected for flushing cisterns. Plastics is widely used at present, as are ceramics. Vitreous enamelled steel is also available. Cisterns are available in 14 litres, 11.5 litres and 9 litres capacity. Water supply undertakings govern the maximum size which may be used on their supply, which is often 9 litres. In order to conserve water a type of cistern ahs recently been developed which will deliver either a 4.5 litre flush or a 9 litre one. The 4.5 litre flush is intended to be used when no solid matter is in the pan.

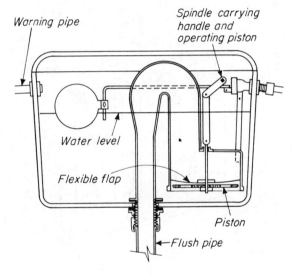

162 *Piston operated water waste preventer*

The water supply to flushing cisterns is invariably by a 12 mm ball valve which should have an isolating valve to control its supply in case of failure of maintenance. An 18 mm diameter warning pipe is also required. Normally the supply- and warning-pipes are on opposite sides of the cistern but special patterns with bottom entry for neatness of pipework are available. The appropriate position for discharge of the warning pipe often presents a problem in multi-storey buildings and internal lavatories. Common warning pipes are not popular because of the difficulty of tracing the source of the flow. Clear sections at each flow can sometimes minimise this difficulty. High-level cisterns with the warning pipe discharging over its own pan have been used. Recently a device has been marketed which discharges the flow from the warning pipe into the bath by means of a specially designed overflow fitting.

In some cases, usually women's lavatories in industrial buildings, consecutive flushes must be catered for more quickly than a normal cistern would be expected to refill. This can be catered for by the use of a flushing trough. This consists of a galvanised sheet steel trough running right across the WC compartments with one ball valve and warning pipe. Each WC has a flush pipe and a flushing device delivering 9 litres of flush. The quantity of water contained in the trough is very much greater than in a single cistern and consecutive flushes are possible.

Flushing valves which deliver a fixed quantity of water are popular in some countries. In Britain most water supply authorities condemn their use because of the risk of unnoticed leakage into the pan, the possibility of back syphonage and the possibility of maladjustment giving excessive quantity for the flush. To supply a single flushing valve a 32 mm diameter pipe is likely to be required in contrast to the 12 mm pipe required for a cistern.

Slop hoppers and housemaids' closets

Slop hoppers are appliances similar to high-level WCs where the pan is adapted to support a bucket or pan on a metal grid. Bib taps can be provided for filling or flushing the bucket and buckets can be emptied into the hopper and the contents flushed away. The 100 mm diameter outgo avoids problems of blockage. A sink may be associated with the slop hopper and sinks are sometimes used for disposal of excrement in hospitals where bedpan washers are not available.

The water supply for the flushing cistern is 12 mm diameter, for the taps 18 mm or 25 mm diameter hot and cold and the outgo is 100 mm diameter.

Urinals

Urinals present the most critical nuisance problem found in sanitary appliances. They are installed in male communal and public lavatories. Their cost compares unfavourably with a high-level WC suite but cleaning is very much simplified, some space can be saved and more users can be dealt with than

219

SANITARY APPLIANCES

would be the case with WCs. Traditionally a urinal consists of a slab with a channel at its foot or a series of 'stalls' which include projecting dividing pieces. Ceramic materials are normally used. Flushing is provided at the rate of 4.5 litres per stall at about 20-minute intervals from an automatic flushing system and distributed through a system of sparge pipes or spreaders. Diagrams 163 and 164 show slab and stall urinals. SInce frequent use may be anticipated in situations where urinals are installed and flushing is at intervals, the flow from urinals is very concentrated relative to other sanitary appliances.

fluids often employed, special care must be taken in selecting trap materials and urinal traps can be obtained in fireclay and in vitreous-enamelled metal. In expensive installations devices are often provided to minimise splashing of feet and legs. They take the form of raised ceramic screens next to the channel or glass panels at low level. The usual width of urinal stalls is 0.61 m but it is questionable whether full use can ever be made of an installation based on this spacing.

One-piece urinals of modern materials, such as stainless steel, obviate joints in units of up to 3 or 4

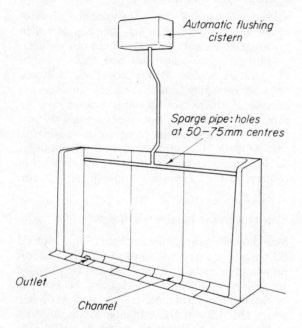

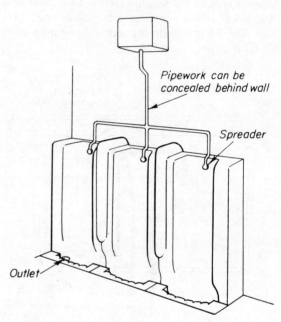

163 Slab type urinal with automatic flushing cistern, sparge pipe and channel

164 Stall type urinal

In addition the faces of slab stall are wetted by urine which is not immediately washed away. These circumstances tend to give rise to deposits in the outgo and to smell in the room containing the urinals. In addition the joints may leak and the channel lower than floor level may conflict with structural requirements. The whole urinal may have to be raised on a step to accommodate the channel. An outlet with a grating to catch cigarette ends, etc, opening to a 62 mm or 75 mm diameter trap is normally provided to every 6 or 7 stalls. Because of the concentrated flow and the nature of the cleaning

stalls but are not widely used. Bowl urinals are, however, in places where reasonably responsible use can be anticipated, tending to be widely used. Diagram 165 shows a range of urinals of this type. They are substantially cheaper than stalls, have a smaller surface area for wetting and drying, and have no cement joints. The inverted trap arrangement shown in the diagram retains some of the flushing water and thereby dilutes the first part of the fresh flows. The level of fixing requires careful thought and small boys may be inconvenienced by this type of appliance.

SANITARY APPLIANCES

a pop-up waste should be used rather than a plug and chain. The flushing rim and spray could become submerged if the waste became blocked and in view of the nature of their use bidets represent a particular risk of contamination for water supply. Special water supply provisions are usually called for to avoid any possibility of contamination of water. Normally the problem can be satisfied by taking special connecting pipes for both hot and cold from the distributing pipes at least 2 m above the bidet although in some cases separate distributing pipes or even a local cold water cistern may be called for. Bidets are treated as waste appliances for drainage purposes. Diagram 142 shows the connection of a bidet into a domestic water supply installation.

Baths and showers

Cast iron finished with vitreous enamel is the most usual material for baths. Glazed fireclay baths are available and have some advantages but they are expensive, heavy and cold to the touch unless they have been warmed by hot water. Their use is therefore confined to institutions where use is fairly continuous. In recent years enamelled pressed steel, fibreglass reinforced plastics and perspex have been employed and the use of perspex is becoming well established. For special designs in the past baths were sometimes cut from marble. Terrazzo is a more economical material likely to be used at present for in situ special baths. Both marble and terrazzo suffer from the same limitations as fireclay.

The traditional roll-edge, free-standing bath still

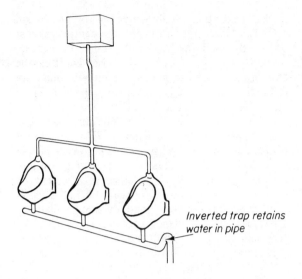

165 Range of bowl type urinals with inverted trap discharge

Bidets

Bidets (diagram 166) are rare in this country although their use is thought to be increasing. They are of ceramic materials and are provided in association with WCs for perineal washing. 12 mm diameter hot and cold valves control the temperature and rate of water supply which can in most cases be directed either round the flushing rim (thereby warming it if desired) or to an ascending spray. Outgo is via a 32 mm diameter waste and trap. It is desirable that

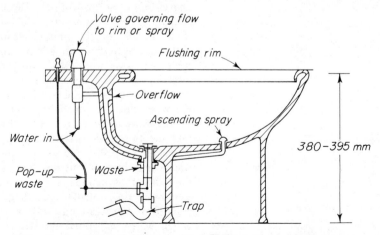

166 Section through bidet showing flushing rim, ascending spray and pop-up waste

SANITARY APPLIANCES

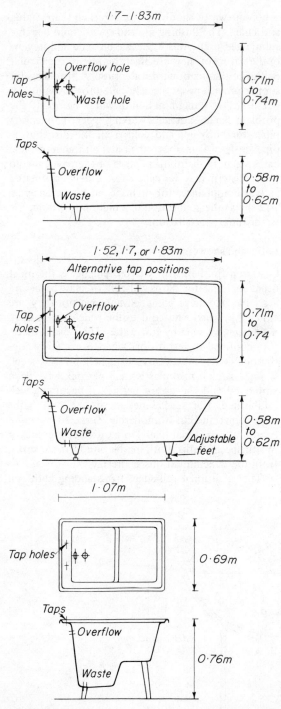

167 Typical baths
Top: Traditional roll top free standing
Centre: Magna type   Bottom: 'Sitz' bath

used in institutions, is shown at the top of diagram 167. The Magna type shown at the centre is almost universally used for present-day domestic installations. Its flat top with straight raised edges can be built against walls, although watertight joints are difficult to achieve, particularly when the bath stands on a wooden floor, and a cover panel can be simply installed to conceal the underside and feet. The 'sitz' bath (lower part diagram 167) is economical of space and water and has advantages for old people since the user can maintain a normal sitting position. All these baths can be obtained in a variety of colours and with a variety of special details, such as hand-grips, soap recesses and decorative features. There is a clear tendency to use smaller baths for economy in space and hot water.

Water supply to baths is usually by means of hot and cold pillar taps at the end of the bath or on the long side near the wall. In domestic circumstances 18 mm diameter taps are used but in institutions 25 mm diameter taps are normal to increase the speed of filling. In some cases valves are used instead of taps with the water delivered to the bath by a spout. Provided both hot and cold water are supplied from a storage system and back syphonage cannot occur, it is possible to have an inlet spout near the bottom of the bath. This is described as a steamless inlet since hot water introduced at low level will give off less steam than would be the case if the water fell from a tap. Special inlet fittings can be obtained combining hot and cold valves, spout, diverter valves and hand-shower fittings.

Wastes are 38 mm diameter in domestic situations and 50 mm in institutions. Deep seal traps will sometimes require pockets to be left or cut in the floor because of the limited clearance under the bath. An overflow is normally provided. At one time these overflows were taken through outer walls to discharge into the open. The cold draughts, entry of insects and inconvenience if water flowed out of the pipe which attended this method, are avoided by the current system of connecting the overflow into the trap as shown in diagram 168. Plugs and chains are normally used to control outflow. Pop-up wastes may be used but they increase the length of the waste thus making the accommodation of the trap more difficult. The standing waste is a tube rising from the waste outgo to overflow level down which water will flow. By raising or turning the tube the contents of the bath can be discharged into the waste. This device is more easily cleaned than

SANITARY APPLIANCES

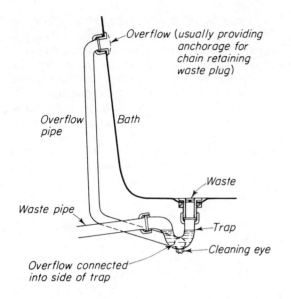

168 Bath overflow. Delivery into trap below water level prevents cold draughts and ingress of insects

Two types of delivery heads are used for showers, the traditional 'rose' type delivers water through a disc pierced with holes and is frequently fixed overhead. The adjustable umbrella type spray shown in diagram 169 has come into use relatively recently. This type of spray consumes very much less water than the rose type and is conventionally installed at chest level. These points avoid the previous disadvantages of showers which were high rates of hot water consumption and discharge of water inevitably

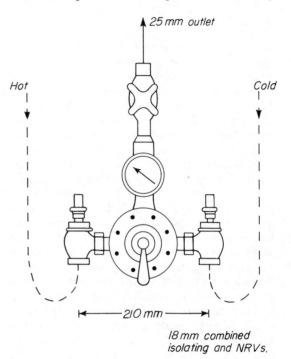

170 Thermostatic mixing valve serving a range of showers

the concealed overflow but is not so neat in appearance and does not appear to be widely used.

Many baths have hand or fixed shower fittings. Operation of the shower may cause some splashing and plastic curtains falling into the bath are often used to control this. Where showers are used independently of baths they are usually formed by a shallow ceramic tray 1 m square in a tiled compartment. A 35 mm diameter waste discharges water from the tray.

169 Adjustable 'umbrella' type shower spray

overhead, which is very unpopular with women users. The hot and cold water supplies to shower sprays must be similar in pressure and taken from storage cisterns. The simplest form of control is by manual adjustment of valves on hot and cold supplies. Fluctuation of pressure in one of the supplies due to the opening of other taps can alter the temperature of the shower and may even cause scalding. Mixing valves provide a superior method of control. Non-thermostatic valves normally have anti-scalding provisions which shut off the hot water supply if the cold fails. Thermostatic valves automatically adjust themselves to give a fixed temperature of flow.

223

SANITARY APPLIANCES

Diagram 170 shows a thermostatic mixing valve with associated isolating valves. In addition to their use in individual shower compartments ranges of shower sprays can be used for walk through installations. Single mixing valves can deliver water to a range of shower sprays. A minimum 1.5 m head is necessary to operate a shower spray and this will be increased if a mixing valve is employed. This requirement can influence the placing of the cold water cistern in domestic buildings.

Wash basins

A wide variety of sizes and designs are available. Diagram 171 shows the features of a typical basin which may range in size from barely large enough to wash the hands to 500 mm x 640 mm or more. Basins are normally made of ceramic materials although others such as plastics are available and used where weight is important. Water supply is controlled by 12 mm diameter pillar taps for hot and cold. Waste outgo are of 32 mm diameter and incorporate an integral overflow as shown in the diagram. Control of outgo is usually by plug and chain but pop-up wastes are available. Self-draining recesses to hold soap are formed in the shelf. This type of basin has disadvantages. The overflow and plug and chain waste are not easy to keep clean, the position of the taps makes it difficult to wash the hands in running water and in addition the flows may be too cold or too hot. To overcome these problems a type of basin has been introduced having only one tap which delivers a spray of water at the right temperature for washing, has no overflow or plug and chain. This basin is very well suited to hand washing but cannot be filled with water for the other uses which form an important part of domestic requirements. Its use is therefore usually confined to communal and public lavatories. Basins can be supported in a number of ways. Some types can be obtained with lugs for building into walls so that they cantilever from the wall. Brackets built in or plugged to the wall give the same type of support. A ceramic pedestal support is often used and it conceals the pipework although making access more difficult. A metal frame fixed to the wall and having legs and rails for towels may also be used.

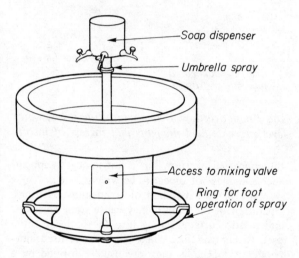

172 Wash fountain for industrial hand washing. (Current water supply byelaws require a separate spray and tap for each place at the fountain)

Rectangular wash basins can be joined into ranges by means of cover strips. A space between each basin is neater and easier to clean and allows more comfortable use of the basins. In industrial situations economical and easily maintained hand-washing arrangements can be made by means of long troughs with spray taps or circular wash fountains which employ an umbrella spray. Diagram 172 shows a ceramic wash fountain. Stainless steel models are available. In common with most other sanitary appliances the taps serving lavatory basins must discharge above the flood level of the appliance to obviate any risk of back syphonage. Diagram 173 shows the nature of the clearance required.

171 Section through typical lavatory basin

SANITARY APPLIANCES

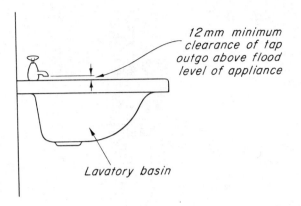

173 Water supply to obviate back syphonage risk

Sinks

The 'belfast'-type sink emerged as the most successful of a variety of ceramic sinks. A range of sizes can be obtained. 610 mm x 460 mm x 205 mm is popular for domestic use. Diagram 135 shows a belfast sink. This type of sink is usually equipped with a draining board most often made of wood. This is not easy to keep clean, particularly at its point of junction with the sink and pressed steel sinks with integral drainers which give no lodgement for dirt have largely superseded the belfast sink for domestic and similar use. Stainless steel and vitreous enamelled steel are popular and vitreous enamelled cast iron is available. Patterns with single and double drainers and single and double sinks can be obtained. One disadvantage of the sheet steel sinks is the noise which crockery makes. They are supplied with sound-deadening material supplied to the underside. Teak sinks reduce crockery breakage and their use has been popular in large kitchens but the hygienic advantages of stainless steel are leading to its widespread use in these circumstances too.

The water supply to 'belfast' sinks is by bib taps usually 12 mm diameter for domestic and 18 mm or 25 mm for large sinks in institutions. Stainless steel and vitreous-enamelled sinks employ either separate 12 mm diameter pillar taps for hot and cold or mixer taps with swivel arms. Kitchen sinks are supplied with cold water direct from the main and where mixer taps are used they must be of the bi-flow pattern which does not allow the hot and cold water to mingle in the tap. Some special fittings to assist in washing-up such as flexible hand-sprays are available in addition to conventional taps. Wastes are 32 mm diameter for domestic sinks and 38 mm diameter for large sinks. Control is by plug and chain.

Scales of provision of sanitary appliances

Code of Practice 3, Chapter 7, 1950, *Engineering and Utility Services,* gives tables recommending scales of provision of sanitary appliances.

Taps and valves

The terms tap, cock and valve are loosely employed in common building usage. They are distinct in their functions, however, and should be distinguished.

Cocks consist of a body holding a bored plug, the hole through which can be aligned with the pipe, thereby allowing water to pass. A quarter turn can fully open or fully close the cock. They are not generally permitted in water supply installations since the sharp turning off can give rise to water hammer.

Valves control flow in a pipeline. The two types most frequently used in building pipe installations are globe valves and gate valves. Diagram 175 shows both types. Globe valves are used on water supply pipework where adequate head is available and complete cut-off is essential. Gate valves offer less resistance to flow when open. Full-way types are available. They are employed when head is limited and particularly on heating pipework. 'Mains stop-cocks' are normally globe valves with washers held by loose jumpers. It is assumed that the loose jumper will act as a non-return valve if conditions for backflow ever arise.

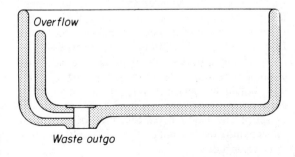

174 Section through typical 'belfast' sink

225

SANITARY APPLIANCES

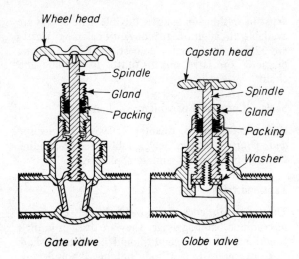

175 Valves

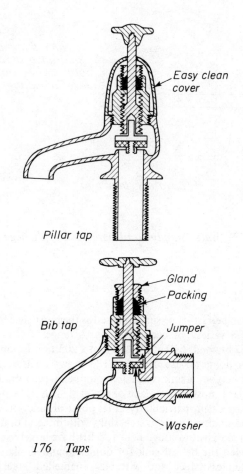

176 Taps

Taps are fittings permitting draw-off of water. Diagram 176 shows a bib tap appropriate to wall fixing over a sanitary appliance and a pillar tap which is fixed in the sanitary appliance itself.

There are many kinds of special taps for different purposes. Spring-loaded taps operated by a push button are much used in factory lavatories in an endeavour to save water. They should be of the non-concussive type, which prevents 'hammer' noise. Many water authorities do not like self-closing taps. There are many kinds of non-splash taps and some designed so that the washer can be replaced without having to turn the water supply off. Many 'combination sets' are marketed with special lavatory basins and baths wherein hot and cold supplies are valved and the outlet is combined. With such fittings the cold cannot be connected to the mains, owing to the possibility of contamination of the main supply if it is connected directly to the hot supply. Mixing valves serve the same end but the better types are now controlled by one handle, which turns from the off position through cold up to hot. This prevents the possibility of scalding and is almost essential for showers. Thermostatic mixing valves are now obtainable and are very useful in effecting economy in hot water in big institutions as they can be set at any required temperature. Most of these valves require the hot and cold water supply pressures to be balanced.

A recent development which should become a normal accepted fitting in offices, works and other

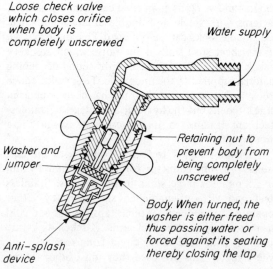

177 'Supatap' bib tap

VALVES

places where there is hand washing is a spray tap scientifically designed to give a cone of fine spray, adequate for washing but using comparatively little water. One such is marketed under the name of 'Unatap'. Under independent tests, by the Building Research Station, great economy of water used has been shown. When combined with a mixing valve, or on a hot supply of set temperature, the saving reflects in a saving of fuel.

Another type of tap development is shown in diagram 177. It is the 'Supatap' which has a check valve that cuts off the flow when the washer is removed. Washers can, therefore, be removed without turning off the water supply.

Ball valves which supply water and maintain the level in water storage cisterns and flushing cisterns should, in terms of the definitions above, be called float-operated taps. The term 'ball valve' is, however, universally applied. There are several types. Diagram 178 shows the 'Croydon' type of ball valve which was widely used and is still sometimes installed. It cannot be fitted with a silencing tube, which is a serious disadvantage. Diagram 179 shows the 'Portsmouth' type which is governed in materials and manufacture by BS 1212. The silencing tube will be noted. The Portsmouth valve is available in a range of sizes and in each case different orifice sizes allow for low-, medium- or high-pressure supplies. Trouble with slow filling of flushing cisterns is sometimes due to the inadvertent fitting of high-pressure valves in situations where there is a small head. Both the above types of ball valve are subject to trouble due to deposits of salts from the water

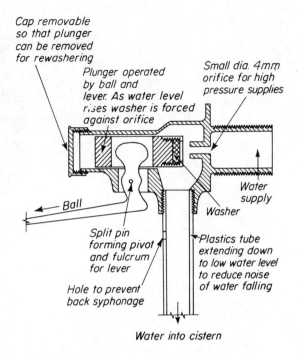

179   Portsmouth ball valve (BS1212)

impeding the operation of moving parts and wear and noise from the nature of the water flow through the valve. At the Building Research Station research into these problems led to the development of the diaphragm type of valve shown in diagram 180. This valve gives substantially quieter operation and better reliability in operation. The design of the nylon

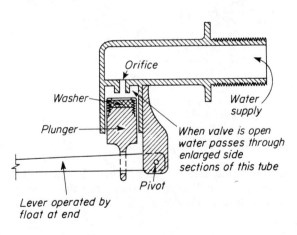

178   Croydon ball valve

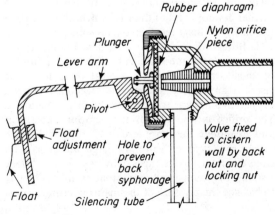

180   BRS ball valve

SANITARY APPLIANCES

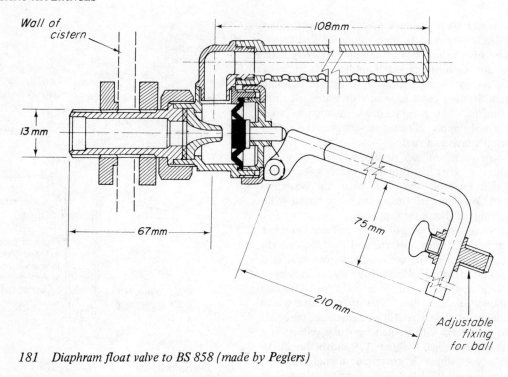

181   Diaphram float valve to BS 858 (made by Peglers)

orifice reduces noise and wear and the plunger and pivot which close the orifice are kept dry so that no deposits occur to impede operation. Although the BRS valve represented a sitnificant advance it did not meet the increasing concern felt for prevention of backsyphonage. Although the silencing tubes of ballvalves have small holes to prevent backsyphonage this does not provide a high enough standard to meet the requirements of the Water By-laws. A further development of the BRS valve which satisfies the backsyphonage requirements is shown in diagram 181. Instead of a silencing tube descending into the cistern an outlet tube at high level delivers a number of small streams of water against the side of the cistern, thus achieving complete freedom from risk of backsyphonage together with quiet operation. The orifice is flooded during operation and this is said to reduce the noise level even further. BS 858 covers this type of valve. In large buildings with large diameter service pipes delivering to the storage cisterns the buoyancy of a float is not adequate to balance the pressure of the main operating against

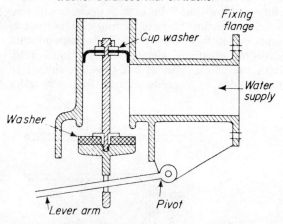

182   Equilibrium type ball valve

a substantial area of water. The equilibrium ball valve shown in diagram 182 overcomes this problem and the float is only called upon to overcome friction.

11 Pipes

Historically lead, because of its durability and ease of working, was universally employed for pipes. It has now been almost entirely superseded by other materials which are more economical, have a better appearance or are more easily installed. These are cast iron and asbestos cement for large diameter pipes, and copper, mild steel and wrought iron for smaller diameters and for heating pipes. Very recently new ranges of materials for pipework have been developed. These include plastics, thin-wall copper tube, stainless steel and coated tubes.

PROPERTIES AND APPLICATIONS OF MATERIALS

Lead and lead alloy

Lead is obtained from the mineral galena, a compound of lead and sulphur (about 86 per cent lead). After smelting the lead contains small quantities of antimony, tin, copper, gold and silver. These are removed by refining so that good commercial lead possesses a high degree of purity, over 99.99 per cent. Lead is very soft. It is ductile and malleable, and does not appreciably work harden, so does not need annealing. It is highly resistant to corrosion by the atmosphere, a protective film being formed on the surface by the action of oxygen, carbon dioxide and water vapour.

Lead is very heavy, 11325 kg/m^3, and has a low melting point, 327°C. The quality of lead is improved for many purposes by alloying with very small quantities of other metals, the silver-copper-lead alloy having increased tensile strength and much improved creep resistance.

Uses: chiefly small pipe-work, water service pipes, cold distribution pipes, overflows, warning pipes, short-length connections, waste pipes, and soil pipes where complex or where great durability is required.

Merits: easily worked, cut, bent cold, soldered, hence its popularity for short or complicated connections. Its slight flexibility makes it useful for connections where there may be slight movement as between rigid soil pipe (cast iron) and brittle sanitary fitting. It has a tendency to creep, ie expands on heating, but tends to deform on cooling rather than contract to its original form, so needs support when used horizontally or for hot liquids. It is now expensive, also, alternatives are lighter and easier to handle.

Standards and specifications: BS 602 for lead pipe; BS 1085 for silver-copper-lead alloy. BS 603 for BFN terney alloy No. 2 is being withdrawn, as this alloy is being superseded by the silver-copper-lead alloy. BS 334 (chemical lead) covers lead for waste pipes in chemical laboratories. Lead pipe is extruded. It is specified by internal diameter and weight per linear yard. Pipe up to 25 mm diameter is available in 18 m coils, up to 50 mm diameter in 11 m coils, over 50 mm diameter in straight lengths of 3 m or 3.7 m. Different requirements of pressure call for different thicknesses of wall of the pipe.

Jointing: is by soft solder (as shown in diagram 183) or by lead burning, which is gas welding in which the lead of the joint is first fused and then strengthened by the addition of molten metal from a lead filler strip. For jointing to sanitary fittings and drainware, lead pipe has to be strengthened by use of brass sleeves or brassy sockets, to which it is soldered. For connection

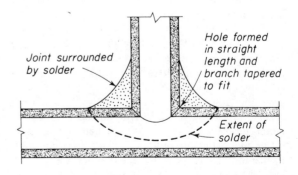

183 Wiped lead joint

to metal fittings, lead pipe is usually soldered to brass tail piece and connected to the fittings by coupling or union nuts, thus facilitating taking apart.

Fixing: by lead tacks soldered to the pipe. Tacks at 0.76 m centres are recommended for horizontal pipes and at about 1.2 m centres for vertical pipes.

Asbestos cement

A manufactured material with three main applications: pressure pipes for water mains, soil and ventilating pipes and rainwater pipes and gutters.

Merits: economical and resistant to decay. Used below ground for water mains as the material is resistant to many soil conditions and to the bacteria which attack cast iron. Flexible joints are available for underground pipes. Above ground painting is not necessary except for decorative purposes.

Standards: piping for water mains is governed by BS 486 asbestos cement pressure pipes. There are four grades capable of withstanding different pressures and diameters from 50 mm to 600 mm. Soil- and vent-pipes are covered by BS 582, 50, 62, 75, 87, 100 and 125 diameters are made.

Jointing: pressure pipes can be fitted with flexible joints. Soil- and vent-pipes have socket and spigot joints which should be sealed with caulked lead wool or asbestos jointing compound.

Fixing: when fixed above ground ring pipe-clips are used below each socket.

Copper

Uses: in light gauge for all hot and cold water pipes, including heating systems; waste and ventilating pipes; soil pipes where shop fabricated.

Merits: high strength and ductility enable tubes to be drawn with thin walls which, while light and cheap, will resist quite high pressures. Many quick methods of jointing these light-gauge copper pipes are now available. Smooth surface of copper tube gives low resistance to water flow and so allows smaller pipes to be used. The smaller amount of metal means that less heat is taken out of hot water in heating-up process, making for economy of fuel in hot water systems. Small pipes can be cold bent in a bender to quite small radii.

Standards and specifications: pipes are supplied in 'half-hard' temper. This allows for only a small amount of working (see Materials for temper and annealing). Light-gauge copper pipes are made to BS 659 for work above ground. These are of nominal bore, 9 mm to 100 mm. BS 1386 covers light-gauge tubing for use underground, which is available in coils of dead soft temper which can be bent easily by hand. BS 61 covers heavy-gauge tubing for normal screwed joints. This is not much used now owing to the high cost of copper.

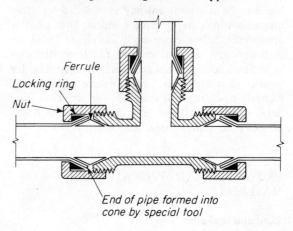

184 Manipulative compression joint for copper tube

Jointing: by welding or capillary or compression joint. The latter can be manipulative or non-manipulative. Manipulative joints are used below ground since the deformation of the pipe ends renders slipping or pulling out of the joint unlikely, Diagram 184 shows a typical manipulative joint. Non-manipulative joints, as shown in diagram 185, and are easier to form are used above ground with BS 659 tube.

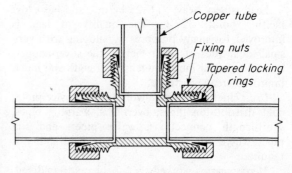

185 Non-manipulative compression joint for copper tube

PROPERTIES AND APPLICATION OF MATERIALS

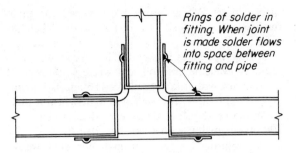

Rings of solder in fitting. When joint is made solder flows into space between fitting and pipe

186   Capillary solder joint for copper tube

Capillary solder joints, shown in diagram 186 have a very neat appearance. They are formed by introducing clean, square, fluxed pipe ends into closely fitting joints. Solder is either applied or is contained in the fitting and when heated by blowlamp spreads throughout the joint by capillary attraction.

Welding is much used for larger pipes and repeat work, such as a unit serving combined soil and waste to a bathroom or kitchen on one floor of a block of flats. Bronze welds are strong and easily made; autogenous welds can be buffed up so that they hardly show. Connections to other materials and fittings are made usually with special fittings, available in a wide range according to the type of proprietary joint chosen, or by welding.

Fixing: usually by pipe clips or holderbats. Copper pipe, being so rigid, fixings need not be as frequent as with lead. CP 310, *Water Supply*, contains a table of recommended spacings.

Mild steel and wrought iron pipe

Uses: hot water and heating installations principally, also cold water distribution and waste and ventilation pipe-work with special fittings.

Merits: cheap, very strong, both for water pressure and in resisting damage. Subject to corrosion so should always be galvanised, except for closed circuits such as heating installations. Corrosions may be excessive where water is very soft. Easily welded. Smaller diameters cold bent, but much work is forged.

Standards and specifications: mild steel pipe, referred to as 'barrel', is made to BS 1387. It is available in sizes, nominal bore up to 150 mm, and in three classes: A, with thin walls, banded brown; B, medium, banded yellow; and C, thick walls, banded green. The latter is equivalent to the old 'steam barrel', but the terms *gas, water, steam barrel* are obsolete, and should not be used now. Test pressure for all is 4.8 MN/m^2, but C quality is usually required for rising mains and work under pressure, B being used for distribution and hot water pipes where head is not too great, and A for waste, ventilating and overflow pipes. Some authorities require C quality for all work. Wrought iron pipes are the subject of BS 788 and are seldom used in general building.

No standard dimensions can be given for steel pipe work for wastes, stacks and branches as they are usually multi-branch assemblies designed to suit the job, with branches or bosses welded in to take screwed joints in the exact positions to suit fittings. This pipe can be supplied with spigot and socket joints, which are preferably caulked with a cold caulking compound. The multi-branch units and pipes should be galvanised after fabrication.

Fittings in wrought iron, used for steam work, and gas and water over 75 mm diameter, are covered by BS 143 and in malleable iron, for domestic work, by BS 2156. The latter should be galvanised.

Jointing: by screwed and socketed joints as shown in diagram 187, or by welding. Pipes, particularly the larger diameters, can also be obtained with flanges, which facilitate bolting together, to valves and to flanged fittings. Asbestos fibre washers are used in the joint. Where possible bends should be forged, or cold bent, to allow easy flow. Frictional loss and loss by elbow and other sharp fittings can restrict the flow considerably.

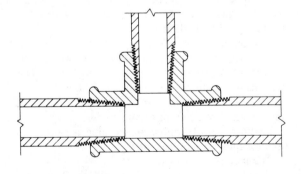

187   Screwed joint for steel tubes

Fixing: by pipe clips or holderbats. The pipe is so rigid that fixings can be at wide intervals.

Enamelled iron pipes—mostly rain-water goods and special fittings for laboratories—are made of heavy gauge plate, pressed and welded and vitreous enamelled. These have great durability and are maintenance free. Jointing is by socket and spigot with caulking, if required, in compound.

Cast iron

Uses: soil and waste and ventilating pipes; rain-water goods; larger sizes of water and gas mains.

Merits: great strength and durability; large range of fittings available (pages 260 and 269) enable plumbing installations to be built up quickly within a reasonable dimensional tolerance, without any shop fabrication: material is heavy, joints are cumbersome and ugly and system is very rigid.

Standards and specification: BS 416 covers three grades of socketed pipes and fittings—medium, heavy and extra heavy. The medium satisfies LCC requirements and Model By-law requirements for pipes up to 100 mm diameter. The extra heavy grade has the 6 mm wall thickness demanded by the Model By-laws for pipes of more than 100 mm diameter. Pipes are now supplied 1.5 m 'effective' length, ie from top of socket to top of socket when fixed, or 1.8 m overall.

BS 416 also covers the comprehensive range of fittings.

BS 460 and 1205 cover cast iron pipes, gutters and fittings for rain water.

Cast iron pipes for water, gas, and drainwork are covered by BS 78 and are used chiefly for mains. They can have joints spigot and socket, turned and bored, or flanged. Sizes are from 32 mm diameter up to 1.2 m diameter, in lengths from 1.8 m to 3.7 m according to diameter. The BS also covers a wide range of fittings.

Spun-iron pipes are also used for mains.

Cast-iron pipes and fittings are normally supplied coated with black bituminous compound which needs special priming if it is to be painted with other than bituminous paint.

Jointing and fixing: the spigot and socket joint caulked with molten lead, lead wool or with asbestos caulking compound is illustrated in diagram 192. Also the fixing with ears cast on. There are a number of proprietary ranges with special types of ear projecting to keep the pipe further off the wall, to make painting behind easier.

A new design of socket for iron pipe, which is spun, is available for soil and waste installations. It complies with the revised BS 416 and is known as FF spun pipe. The socket is much neater and incorporates the new single ear. It is available in 100 mm diameter, with a range of fittings.

For jointing copper, lead and mild steel waste pipes to cast-iron soil or soil and waste pipes, several 'quick make' joints have been devised, but they involve ordering the cast-iron pipes with special bosses in predetermined positions. The joint is then made by inserting the waste branch pipe into the boss and clamping with a back plate, which presses on to a rubber or composition ring which grips the branch. They show a great saving of time in large installations which can be well preplanned.

Plastic tubes

Two types of plastics tubing are well established in building applications: polythene, which is flexible and is widely used underground, and unplasticised PVC, which comes in straight lengths.

Polythene (polyethylene), is now made into tubes and marketed under various proprietary names such as 'Alkathene'. It is a thermo-moulded plastic produced by extrusion.

Uses: only for cold services and wastes taking hot water at low pressure; considered very good for services underground as it is not affected by acids in the soil. Long straight lengths may be laid very economically by mole plough.

Merits: very tough, very light (30 m coil of 25 mm can be handled quite easily), cheap, flexible and easy to work; non-conductor of electricity. Chief disadvantage is that it softens at low temperature, above $70°C$ (melts at $115°C$). Not approved by all water authorities yet, though preferred by some for work underground. The material can allow water standing in the pipes to be tainted by town gas where the ground is impregnated by gas from leaking mains.

Medium density polyethelene pipe is becoming used for underground gas mains.

The material is frost resistant.

Standards and specifications: BS 1972 applies. It gives two gauges: heavy gauge, which can be

PROPERTIES AND APPLICATIONS OF MATERIALS

jointed by screw threading owing to the greater wall thickness, and normal gauge, which is not capable of jointing by screw threading. Nominal bores available in normal gauge, 12 mm up to 50 mm; heavy gauge, 9 mm to 32 mm.

Jointing: the most common method is by means of a compression fitting using a metal sleeve. It is possible to joint plastic pipes by heating the open ends and pressing them together, but great skill is needed to avoid causing an obstruction in the pipe. Plastic fittings are now available and are used in conjunction with a tool which makes a flange in the tube, which gives a sufficient shoulder for the plastic back-nut to grip. Joints can be made by welding. Heavy-gauge pipes can be jointed by compression fittings to BS 864, type A, and also by screw threading mentioned, which should be to BS 21 pipe threads, with fittings to BS 143 and BS 1256. For waste pipes it is possible to make a simple joint with a length of copper tube of the same nominal internal diameter, as shown in diagram 188.

Uses: water mains, service and cold water distributing pipes, soil and waste and rainwater systems. High resistance to a wide range of chemicals but cannot be used at temperatures higher than 70°C.

Merits: light, easy to handle and install; easy to manipulate and joint; good finish and range of colours. Compares favourably in cost with cast iron for soil pipes.

Standards: BS 3505, unplasticised PVC pipe for cold water supply and BS 3506, unplasticised PVC pipe for industrial uses. Diameters range from 9 m to 150 mm.

Jointing and fixing: two main jointing systems are employed. Solvent welding is shown in diagram 189. A solvent cement is applied to a clean square

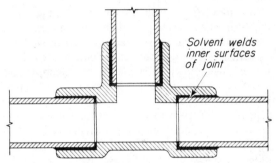

Solvent welds inner surfaces of joint

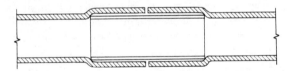

188 *Sleeved joint for polythene pipe used for wastes (no pressure)*

189 *Solvent weld joint for PVC pipe*

Bending and fixing: plastic tubing can be bent cold to a radius of eight times the outside diameter of the pipe, but it must be fixed securely, as it tends to spring back. Heating by immersion in hot water, or judicious use of a blowlamp, enables permanent bends to be made to within a radius of three times the outside diameter of the pipe. Owing to the flexible nature of the plastic, fixing should be more frequent than with metal pipe, approximately 225 mm centres for small pipe, up to 450 mm centres for 50 mm pipes for horizontal runs, and much less frequently for vertical runs.

PVC (polyvinyl chloride)

Unplasticised PVC is marketed both under its own name and by trade names. It is formed by extrusion and cut into 10 m lengths for ease of handling.

end of pipe which is introduced into a fitting of PVC. The result is a completely welded joint. Rubber ring joints as shown in diagram 190 are also used, particularly for waste and soil systems. PVC has a high coefficient of thermal expansion and installations should be designed to allow for some movement. Long straight runs of pipe should be equipped with expansion joints. Fixing is by means of normal pipe clips and spacings vary from

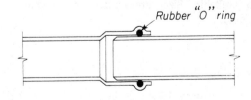

Rubber "O" ring

190 *Rubber ring joint for PVC or copper*

PIPES

1 m for horizontal 18 mm pipe to 3 m for vertical 150 mm diameter. CP 310 gives a table of spacings.

Other newly developing materials

Plastics: pipes are made in several other materials which have strength, durability and heat-resisting properties superior to polythene and PVC. They are, however, usually more expensive and consequently limited in their use in buildings. Sometimes fittings such as traps are made from these materials. The main materials are polypropylene, ABS (acrylonitrile butadiene styrene) and PTFE (polytetra fluorethylene) which has outstanding properties and is capable of conveying hot water under pressure. Expense, however, limits its use for building.

Stainless steel: aided by current shortages and the high price of copper, thin-wall stainless steel for domestic water services has recently been marketed by a number of manufacturers. It is worked and jointed in similar ways to copper and is in many other ways comparable to copper. Its use as an alternative to copper appears to depend on relative prices. At present only 12 mm, 18 mm and 25 mm diameters are available.

Thin-wall copper tubes: although the copper tubes governed by BS 659 are described as light gauge, it has proved possible to produce a thinner and more economical tube which still has an adequate performance. BS 3931 covers this type of tube, which is worked and jointed in the same way as BS 659 tube although, due to the thinner walls, bending is more difficult. 12 mm, 18 mm amd 25 mm diameters are available.

Coated tubes: Tubes are available for use on closed circuit heating installations only, consisting of a steel base coated with copper. This gives the appearance of copper and normal copper pipe fittings can be used. The material is significantly cheaper than copper. Plastic covered tubes are also available. Sizes are limited to 12 mm and 18 mm.

Pipe fixing

Pipes fixed on walls should be secured clear of the surface by a suitable clip. Diagram 191 shows several types ranging from the economical saddle

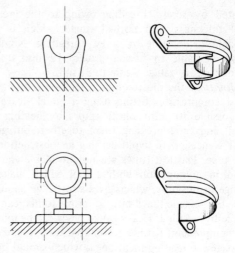

191 Pipe clips

band to the more expensive and elegant pressed and cast bronze clips. Fixing should be firm to avoid noise but not so rigid that thermal movement is prevented.

Where pipes run through walls or floors a sleeve of large diameter pipe should be provided and the entry or exit points covered by pipe clips as shown in diagram 192.

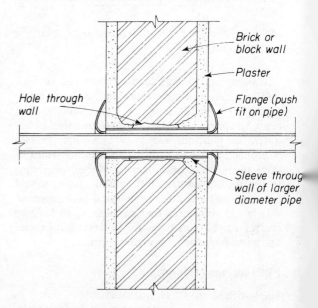

192 Detail of sleeve and flanges for pipe passing through wall or floor

ESTIMATION OF SIZES FOR COLD AND HOT WATER

For both cold and hot water supply installations it is necessary to be able to estimate the volume of water to be stored and the appropriate sizes of pipe to deliver the right flow of water to taps.

Storage for cold water supply

For dwellings the Model Byelaws for the Prevention of Waste, Misuse and Contamination of Water Supplies requires 230 litres actual capacity of storage where supply for both cold and hot is by cistern and 114 litres actual capacity for hot water supply only. Cistern sizes are usually quoted in relation to the capacity full, making no allowance for space for ball valve and warning pipe. Consequently galvanised steel cisterns of 318 litres and 182 litres nominal capacity would be required Table 60 gives the range of sizes for galvanised sheet steel cisterns according to BS 417. It is becoming increasingly usual, in domestic application, to use plastics or asbestos cement cisterns, which have a longer life than galvanised steel in most circumstances. Incases other than domestic, storage is based on the anticipated use by the population of the building. Table 61 from CP 310 water supply gives recommended storage capacities per head of population to give 24 hours reserve in a variety of building types.

Storage for hot water

Adequate storage must be provided to meet peak demands for hot water, the heat input itself being spread over a longer period (known as the recovery period). For domestic installations it is usual to use a cylinder of 114 litres actual capacity (140 nominal) for dwellings where only one bath will be taken at one time and 155 litres actual capacity (180 nominal) where two baths will be taken in quick succession. In larger buildings suitable volumes of storage in relation to population are also established. Table 62 from the IHVE *Guide to Current Practice,* gives a schedule of these. Table 63 gives sizes of hot water cylinders in accordance with BS 1565 and 1566. Cases may be encountered when these figures do not apply and there may also be restrictions of space in some buildings which make the full volume of storage difficult to achieve. In these cases it is possible to

Table 60 Sizes of cold water cisterns given in BS 417 Table 1, Dimensions of Cisterns *(galvanised mild steel)*

BSS No.	Capacity		Dimensions (m)		
	Nominal *l*	Actual to Water line *l*	Length	Width	Height
C 1	45.5	18	0.45	0.31	0.3
C 2	68	36	0.61	0.31	0.38
C 3	91	55	0.61	0.41	0.38
C 4	114	68	0.61	0.43	0.43
C 5	136	86	0.61	0.46	0.48
C 6	182	114	0.69	0.51	0.51
C 7	182	114	0.61	0.61	0.48
C 8	227	159	0.74	0.56	0.56
C 9	272	191	0.76	0.58	0.61
C 10	318	232	0.81	0.66	0.61
C 11	318	227	0.91	0.61	0.53
C 12	364	264	0.91	0.66	0.61
C 13	455	336	0.97	0.69	0.69
C 14	455	327	1.22	0.61	0.61
C 15	569	423	0.97	0.76	0.79
C 16	910	710	1.17	0.89	0.89
C 17	1137	840	1.5	0.81	0.91
C 18	1592	1228	1.5	1.14	1.14
C 19	2274	1728	1.83	1.22	1.22
C 20	2730	2138	1.83	1.22	1.22
C 21	4548	3367	2.44	1.5	1.5

Note that grade A and grade B are described in this Standard. Dimensions are identical, but grade A is made from heavier gauge steel and should generally be specified.

Table 61 Volume of cold water storage to cover 24 hours interruption of supply recommended in CP 310 : 1965 Water Supply

		litres
Dwellings	per resident	90
Hostels	per resident	90
Hotels	per resident	135
Offices	per head	37
Offices with canteens	per head	45
Restaurants	per head per meal	7
Day schools	per head	27
Boarding schools	per resident	90
Nurses homes and medical quarters	per resident	115

Table 62 *Hot water storage capacities and boiler power recommended by the Institution of Heating and Ventilating Engineers for various applications*

Type of use	Storage capacity l per person	Boiler power W per person
Dwellings (medium rental)	32	750
Schools, boarding	23	750
Schools, day	4.6	90
Offices	4.6	120
Factories	4.6	120
Hotels, first class	46	1200
Hostels	32	750
Hospitals, general	27	1500
Nurses homes	46	900

estimate the right balance of volume stored and recovery rate by the graphical method shown in diagram 193. In this method the hot water consumption estimated is entered on the top part of the chart on the basis of the figures given and totalled for each hour. Commencing after the slack period during the night when the water will have become fully heated, the amount of water consumed each hour will be deducted from the amount of cold water which can be heated (recovery rate) and if a negative answer is obtained this is deducted from the volume of hot water available and the result plotted on the chart at the foot of diagram 193. Positive answers will be added to the volume

Table 63 *Galvanised steel and copper indirect cylinders as BS 1565 and 1566*

Capacity l	Diameter m	Height m	Daimeter of connections mm
109	0.46	0.76	25
136	0.46	0.91	25
159	0.46	1.07	32
227	0.51	1.27	38
272	0.51	1.47	38
364	0.61	1.37	50
455	0.61	1.75	50

The standards deal with two grades, B and C, suitable for maximum working heads of 18 and 9 m respectively.

available up to the full capacity of the hot water cylinder or calorifier. If the curve of hot water available formed in this way is completed for the whole day any liklihood of the reserve being exhausted and hot taps running cold will be demonstrated. Since the hot water remains at the top of the cylinder and the cold water at the bottom with comparatively little mixing, particularly if the cold water inlets are properly designed, it may be assumed that the quantity calculated as being available by the method described is in fact available even though part of the cylinder is filled with cold water. If the reserve of hot water is reduced at any time to below 20–25 per cent of the total volume the system should be regarded as inadequate. Where a system is demonstrated to be inadequate by the method described two means of correction are available: the volume of storage can be increased or the heat input (recovery rate) increased and the new values checked the way described.

Pipe sizing: cold and hot water supply

Adequate rates of flow are required at taps. Table 64 shows the rates recommended in CP 310, *Water Supply*. Very small installations will be designed to allow all the taps to operate simultaneously and the figure in Table 64 can be used to give the flows required in the pipes. In large installations it would be unrealistic and uneconomic to design in this way since the probability of simultaneous use of all the taps will be very remote. The problem of arriving at a realistic

Table 64 *Rates of flow recommended for various sanitary appliances by CP 310, Water supply*

Sanitary appliance	Rate of flow, litres/sec. (same values apply to hot and cold)
WC (flushing cistern)	0.11
Lavatory basin	0.15
Lavatory basin, spray tap	0.03
Bath tap, 18 mm	0.30
Bath tap, 25 mm	0.60
Shower (umbrella spray)	0.11
Sink taps, 12 mm	0.19
18 mm	0.30
25 mm	0.40

Estimated analysis of consumption

TIME	Mdnt	1	2	3	4	5	6	7	8	9	10	11	Noon	13	14	15	16	17	18	19	20	21	22	23
Casual								2	2	2	2	2	2	2				4	4	4	4	4	4	
Hand Wash 4l								4	8	4		8	20				8		8	8			12	8
Strip Wash 6l									12								12							
Sink (Wash Up) 30l										30				30					30				30	
Sink (Prep) 30l									30			30	30					30	30					
Sink (Clean) 30l											30			30		30								
Washing Machine 40l											40		40											
Bath (Adult) 60l									60											120				60
Bath (Baby) 20l																			20					
Shower 10 l								10	10															
Total Litres								16	122	36	72	40	92	62		30	20	34	92	132	4	4	46	48

Graph of cylinder capacity — consumption + recovery rate. Shows reserve available
Consumption hatched. Quantity of hot water theoretically available shown by heavy line
600
550
500
450
400
350
300
250
200
150
100
50
6 kW
3 kW
Storage at
65 deg C
Cylinder cap, l
Litres per hour
through 55° C
Av. hourly
consump. ....
Solid line shows reserve of hot water in 155 l cylinder with 3 kW immersion heater
Broken line, as above but 6 kW immersion heater
15 l per hour per kwh

193 *Sheet for estimating storage capacity and heat input (recovery rate) for hot water supply installations*

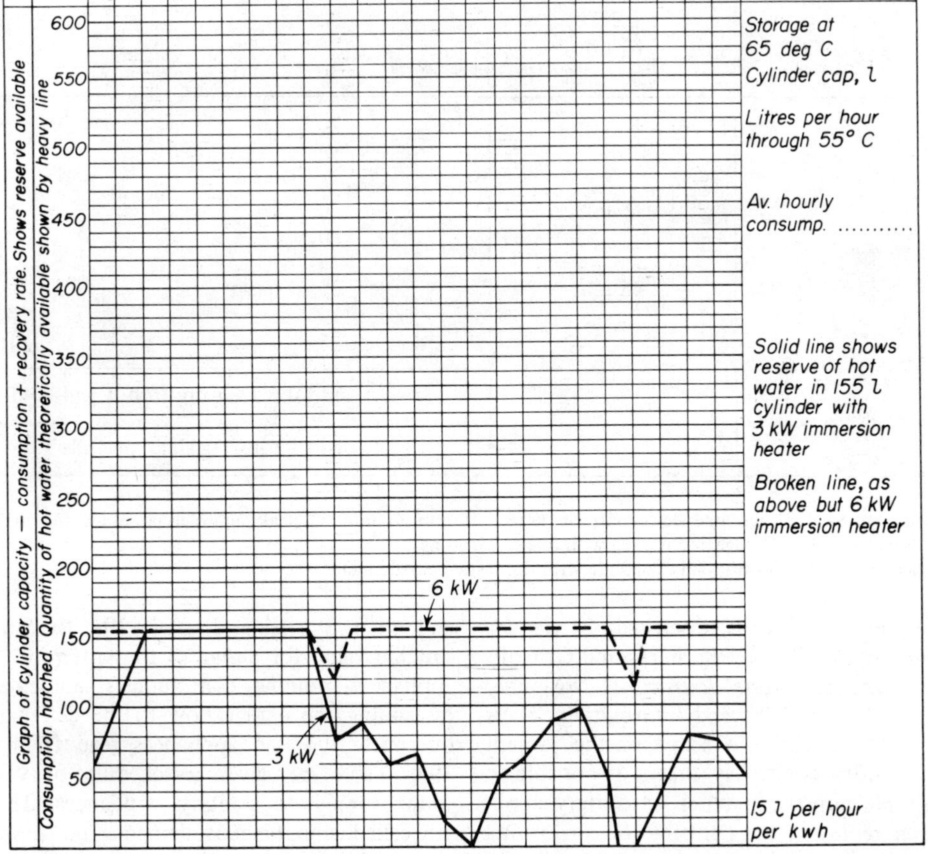

Table 65 Simultaneous demand units for cold and hot water supply pipe sizing

Sanitary appliance	Tap size	Frequency of use (minutes)	Simultaneous demand units
WC flushing cistern 14 litres	12 mm	20* 10 5	1.5 3.0 6.0
WC flushing cistern 9 litres	12 mm	20* 10 5	0.9 1.8 3.6
Lavatory basin	12 mm	20* 10 5	0.3 0.8 1.5
	Spray tap: add 0.015 l/s per tap to hot and cold		
Bath	18 mm 25 mm	75† 30 75 30	1.0 3.3 2.0 6.6
Sink	12 mm 18 mm 25 mm	20* 10 5 20 10 5 20 10 5	0.7 1.8 3.5 1.0 2.5 5.0 2.0 5.0 10.0
Showers		add 0.6 l/s to both hot and cold	
Urinals		will fill during slack periods	

*20-minute interval represents peak domestic usage; 10-minute interval peak commercial usage (eg offices, etc); 5-minute interval represents congested use (eg with queues forming).

†Peak bath usage does not normally coincide with domestic peak demand.

figure for design flow is discussed in *Service Engineering: Hydraulics*, by Burberry and Griffiths published in the *Architects' Journal*, 21 November 1962, pages 1185 to 1191, and *Information Sheets* 1163 to 1174. Within the scope of the present volume it is sufficient to say that it is possible to allocate simple numerical values to sanitary appliances which represent the contribution to loading of the system, to add the values for all the appliances served by a particular length of pipe and then by using diagram 194 to arrive at a suitable value for design of the pipework. Sanitary appliances running continuously are catered for by adding the actual flow to the design flow for intermittently used appliances read from diagram 194. Table 65 gives a schedule of values for water supply to sanitary appliances. The values are called simultaneous demands units, or simply demand units.

Pipes are sized so that the friction loss in the

ESTIMATION OF SIZES

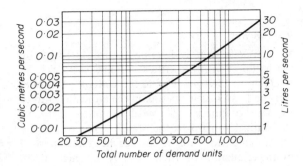

194 Design flow chart for hot and cold water supply installation pipe sizing

pipe balances the head available from cistern or main. Appliances such as showers require some head to give the spray. This is described as maintained head and, for calculation purposes, must be deducted from the head available in the system. Diagram 195 is a nomogram for sizing pipes in relation to head available, flow required, diameter and length, and table 66 gives a step-by-step method for pipe sizing. Table 67 and diagrams 196 and 197 provide an example of the method.

Table 66 Step by step method for water supply pipe sizing

1 Diagram

Prepare a line diagram of the installation showing draw-off points, lengths of pipe, values, cisterns and cylinders and heights between cistern centre and draw-off points.
Annotate draw-off points and pipe junctions by letters to identify pipe sections.
(A three-dimensional diagram will often be helpful both sizing and general use. Axonometric projections can be quickly made from ordinary plans and they enable pipe length to be scaled-off (in horizontal and vertical planes).

2 Peak loading

Decide which sanitary appliances will contribute to peak flow and what frequency of use should be catered for in design. See table 65.
(It may be necessary to consider more than one case if the peak flow condition is not clear from initial inspection.)

3 Flows required at appliances

Using table 65, mark the diagram with the following information for each draw-off point:
(a) Flows required, litres per second
(b) Demand units
(c) Maintained head required at appliance, if any.

4 Flows required in pipework

(a) On each section of pipework mark the total of demand units for all appliances supplied by that section.
(b) Using design flow and the figure for demand units calculate above, mark each section of pipe with the appropriate design flow. Steady flows for showers or any other fitting should be added.
(c) For pipe serving single appliances and in other cases where the total of demand units is not large enough for diagram 194, use the sum of the actual flows required at the appliances.

5 Equivalent pipe lengths

Allowance must be made for the loss of head through fittings and taps. The most convenient method is by adding an additional length to the pipe for calculation purposes.
A 25 per cent addition to the length for elbows, tees, etc, plus the appropriate value for the guessed size of valves and known size of taps from Table 67 should be added to the pipe length and the total marked on the diagram.

6 Establish critical run of pipework

The draw-off point having the most critical supply must be established.
Using diagram 195, link the head available and pipe run to each appliance and project the line to cut the rate of loss of head scale. The lowest value found establishes the critical run of pipework which extends from the supply point, usually the cistern to the draw-off point.
(In practice it will not be necessary to test every point since the majority can be eliminated by inspection.)

7 Established pipe diameters for critical run

Using the point found on the rate of loss of head scale (diagram 195) for the critical run, draw lines

PIPES

Table 66 cont/d...

through the various flows required along the sections of the run to cut the pipe diameter scale at the appropriate point. The next largest pipe diameter will normally be selected unless a reduced flow would be more acceptable than a larger pipe.

8 Check

Check that the diameter found does not vary from the assumption of valve size made in step 5. If it does, recalculate on the basis of the new valve size.

9 Branches

May be dealt with as described above starting from the junction with the critical run and using as head available to the total head to the sanitary appliances less the head loss in friction in the critical run up to the point of junction. The head lost in the section of the critical run up to the branch is determined by multiplying the rate of loss of head (established during calculation of critical run) by the length up to the branch, including allowance for fittings.

Table 67 Equivalent lengths of pipe to allow for friction losses in taps and valves

Type	Equivalent length metres for nominal diameters (mm)								
	12	18	25	32	38	50	62	75	87	100
Taps and globe type isolating valves	5	6	9	11	14	18	21	25	30	36
Ball valves—high pressure	75	40	40	35	21	20				
Low pressure	8									

Table 68 (based on step-by-step method shown in table 66)

1 Diagram

Diagram 196 shows a hot water supply distribution layout including diagrams of the sanitary appliances. Diagram 197 shows the related cold water supply layout containing a minimum of detail and appropriate to this type of calculation. In this very simple case elevational diagrams are appropriate. In more complex installations an axonometric diagram is likely to be helpful. Provided the diagrams are drawn to scale dimensions may be scaled off with sufficient accuracy for sizing.

All pipe junctions are labelled (eg Ⓐ, Ⓑ,) so that all runs can be identified.

2 Peak loading

In this case no reduction in flow for diversity is appropriate and the flows marked in boxes (eg 0.45) for each section of pipe represent the sum of the actual flows. In larger installations the diagram would be marked with the demand unit value for each sanitary appliance and for each pipe length the total demand units for all sanitary appliances served

In this small installation all the sanitary appliances may be in use at peak times.

3 Flows required at appliance

The flow in litres per second for each draw-off point is marked on the pipe branch leading to it (eg 0.30). See table 64 for typical rates.

4 Flow required in pipework

In this case all appliances may be used simultaneously and no diversity allowance is appropriate. Each pipe length is therefore marked with the sum of the flows of all the fittings it serves.

(In larger installations the diagram would be marked with the demand unit value for each sanitary appliance (table 65) and each pipe length marked with the total of demand units for all sanitary appliances served. The appropriate design flow can then be determined from diagram 194 and marked on the pipe length.)

ESTIMATION OF SIZES

Table 68 cont/d...

5 Equivalent pipe lengths

Scale the length of each pipe run between junctions, add allowance for elbows, tees (25 per cent assumed) and allowance for valves and taps based on assumed diameter. (See table 67).

Pipe	Length	Allowance for elbows and tees	Allowance for valves and taps	Total
Diagram 156				
AB	5	1.25	11	17.25
BC	0.5	0	0	0.5
CD	1.5	0.4	0	1.9
DE	2	0.5	9	11.5
EF	1	0.25	5	6.25
EG	1.7	0.4	5	7.1
DH	0.7	0.2	0	0.9
HI	5.3	1.3	5	11.6
HJ	7	1.7	5	13.7
Diagram 157				
KU	3	0.7	11	14.7
UL	1	0.25	0	1.25
U to WC	0.5	0	8	8.5
LM	2	0.5	9	11.5
MN	1	0.25	5	6.25
MP	1.7	0.4	5	7.1
LQ	2.5	0.6	0	3.1
Q to WC	0.5	0.13	8	8.6
QR	3	0.7	5	8.7

6 Critical run

Consider the following draw-off point in accordance with the method described in table 66:

	Pipe run	Total length	*Head available (scaled)	Rate of loss of head
Diagram 196	A–F	37.4	2	0.52
	A–G	38.25	2.3	0.65
	A–J	32.2	4.5	1.5
	A–I	34.25	4.5	1.6
Diagram 197	K–N	32.7	2	0.61
	K–P	33.5	2.3	0.71
	K–R	27.8	4.5	1.7

*Head available is the vertical height from the centre line of the cistern to the tap on draw-off point.

The lowest value found for diagram 196 is 0.52, therefore A–F is the critical run. For diagram 197 K–N is the critical run.

7 Pipe diameters

Using the nomogram (diagram 195) and the data above, determine the diameter appropriate to the various sections of AF and KN.

Pipe run	Section	Flow required	Diameter
AF	AB } BC } CD }	0.79 l/s	32 mm
	DE	0.45	25 mm
	EF	0.15	19 mm
KN	KU	0.82	32 mm
	UL	0.71	32 mm
	LM	0.45	25 mm
	MN	0.15	19 mm

8 Branches

Diagram 196. Consider pipe from D onwards. The critical run must be established. Since I and J are at the same level J is more critical since pipe length is greater.

Established head available at J and size pipe run D–J. Head available is 4.5 m less the friction loss up to D. The rate of loss of head in A–F is 0.52 and the length A–D is 20 m. Therefore the loss of head due to friction in AD (from diagram 195) is 1.1 m. Head available for flow in D–J is 4.5–1.1, ie 3.4 m. The length of DJ is 14.6 m and the rate of loss of head is 2.3. The appropriate size for DH is therefore 19 mm and for HJ 13 mm. The other branches may be similarly sized.

Note: In practice it will not be essential to set down all the figures recorded in this example. Critical runs can often be determined by inspection and pipe sections can be sized from the nomogram without tabulation. In diagram 197 the WCs have not been considered in the critical runs on the assumption that some restriction of the cistern filling rate will not be significant.

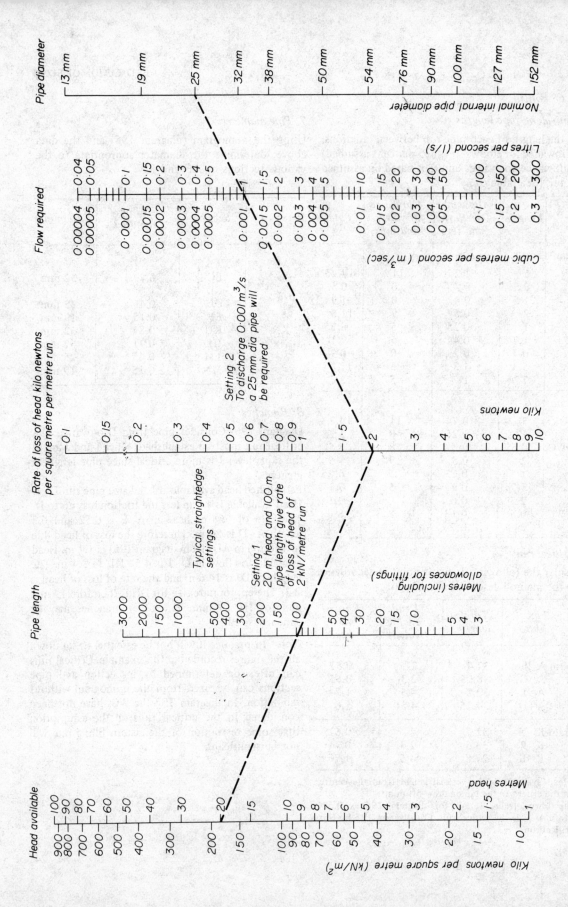

195 Nomogram for water supply pipe sizing

ESTIMATION OF SIZES

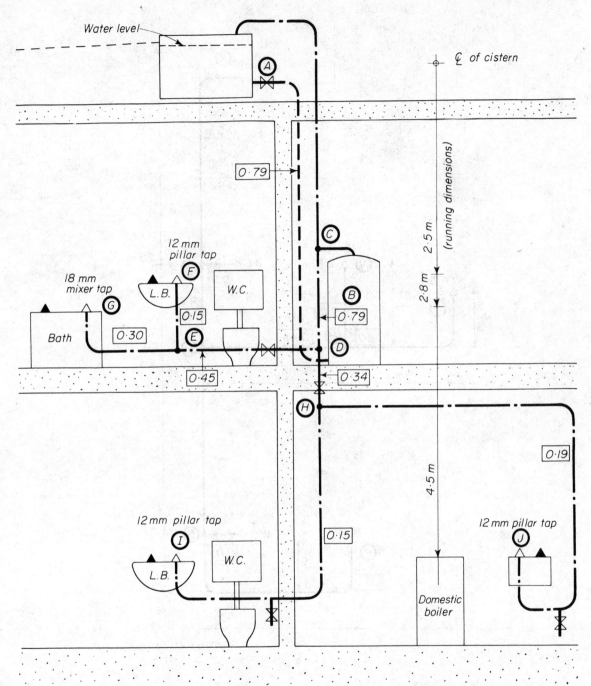

196 Pipe sizing example, see table 68

PIPES

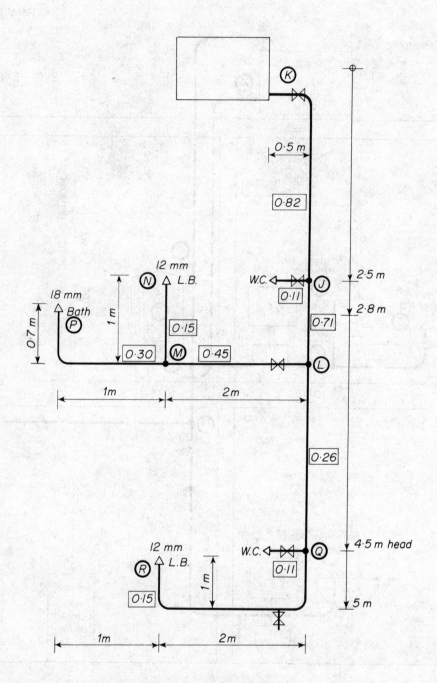

197 Pipe sizing example, see table 68

12 Drainage installations

GENERAL PRINCIPLES

A long process of development by trial and error during the late nineteenth century resulted in the establishment of a set of rules for pipe sizes and layout of sanitary installations which, when followed, ensured satisfactory performance. Drainage by-laws were framed as an expression of these rules and for nearly half a century drainage design consisted almost solely of applying these rules. It was not necessary for designers to understand the basic principles involved. Departures from established solutions, even if justified by satisfactory performance and based on sound principles, were not acceptable.

Performance standards

In recent years problems of size of installation, and particularly height of building, unprecedented in this country have had to be solved and completely internal plumbing achieved. Economy of established methods has been questioned. Fundamental studies of the hydraulic behaviour of drainage installations have been made, principally by the American National Bureau of Standards and the Building Research Station. New methods have been developed and most significantly the legislation affecting drainage has been revised so that it now requires the achievement of standards of performance rather than detailed compliance with specific precepts. New problems and methods and the freedom to design for a standard of performance in fact impose considerably more responsibility, and require from designers, if improved installations are to result, much more fundamental knowledge than was previously the case. A comparison of past and current regulations, Table 69 shows clearly the change from insistence on fixed solutions to a specification of performance requirements.

Hydraulic design

The design implications of these changes in the form of legal requirements and of research into economical systems relate mainly to the hydraulic performance of installations. It is worthwhile and not difficult to set down the hydraulic flow problem posed by removal of waste and soil flows from buildings.

Pipe systems for drainage must be able immediately to accept the flow from any sanitary appliance connected to the system. These flows must be discharged completely and without delay to sewer or treatment plant. Entry and exit points cannot be protected by valves but no water must emerge under any condition of flow and in spite of displacement by flows of water, air must at no time be allowed to escape from the pipes into the building. Solid matter is present in some of the flows and must be carried by the flow of water without becoming stranded or lodged.

Posed without precedent, the solution to this might appear to be a formidable problem. However, the principles of solution have already been developed and are implicit in the form of established drainage systems involving pipes constantly falling towards discharge points, empty except when flow is taking place, with entry points closed by water seals and within which air pressure fluctuations are limited. If it can be taken for granted that watertight, properly aligned, smoothbore pipes laid to suitable and continuous falls are part of all drainage installations then two major hydraulic flow design problems remain. They are:

1 How to estimate flows so that satisfactory but economical pipe sizes may be employed;
2 How to ensure, with economy of pipework, that excessive pressure fluctuations do not occur.

1 Estimation of flows

There are two cases to be considered: one where the pipe carries the flow from only one sanitary appliance and the other where two or more appliances can contribute to the flow.

In sanitary appliances subject to a fixed rate of flush such as WCs, urinals and showers the outgo from the appliance must be large enough to discharge the full rate of input. In other cases, however, such

DRAINAGE INSTALLATIONS

Table 69 Past and current regulations affecting drainage installations

Past	Current

London
1 If the soil pipe or waste pipe of any soil fitment shall be in connection with the waste pipe of any waste water fitment, the trap of every such soil fitment or waste water fitment shall be ventilated in the following manner:
 	
A trap ventilating pipe shall:
(a) be connected with the trap or the branch soil or waste pipe;
   (i) at a point not less than 75 mm nor more than 300 mm from the highest part of the trap;
   (ii) on that side of the water seal which is nearest to the soil pipe or waste pipe;
   (iii) in the direction of the flow
2 The branch and main trap ventilating pipes respectively shall have in all parts an internal diameter of not less then:
(a) 50 mm where connected with a soil pipe, or a waste pipe 75 mm more in internal diameter.
 	
(LCC *Drainage by-laws* 1934, clauses 9(1) A and 9(2) A)
	In order to prevent destruction of the water seal of the trap of any sanitary appliance, the trap shall be ventilated whenever necessary by a ventilating pipe positioned so as to prevent any nuisance or injury or danger to health arising from the emission of foul air from such a pipe and, where necessary, have the open end fitted with a suitable grating or other cover constructed in the manner prescribed in by-law 6 paragraph (15) *(c) (iv)* for gratings to drain ventilating pipes, or connected to the soil or the waste ventilating pipe above the highest appliance.
	
	(LCC *Drainage by-laws* 1962, clause 11)

England and Wales The drains intended for conveying foul water from a building shall be provided with at least one ventilating pipe, situated as near as practicable to the building and as far as practicable from the point at which the drain empties into the sewer or other means of disposal: Provided that a soil pipe from the water closet, or a waste pipe from a slop sink, constructed in accordance with these by-laws, may serve for the ventilating pipe of the drain, if its situation is in accordance with this by-law. *(Model by-laws,* Series IV, Buildings 1939, clause 102)	Water seals in traps. Such provision shall be made in the drainage system of a building, whether above or below the ground, as may be necessary to prevent the destruction under working conditions of the water seal in any trap in the system or in any appliance which discharges into the system. *(Building regulations* 1965, clause N3)

as baths, lavatory basins and sinks, the size of the waste outgo and associated pipework governs the rate of outflow and is itself governed only by the time considered acceptable to empty the sanitary appliance. Existing precedents for sizing are valid and well established and this aspect of sizing presents little problem.

When several sanitary appliances can contribute to flow in a length of drainage installation a problem does arise. It would be simple to add the combined flows of all the appliances but it is uneconomic to do so since the changes of all the appliances contributing to flow at the same time may be so small as to be ignored. The probability of finding a 9-litre WC in process of flushing at a time of peak domestic use is 0.004 (when 1 represents certainty of finding the WC flushing and 0 certainty of finding the WC not flushing). Even in congested use this probability will only increase to 0.02. The probability of finding two WCs flushing at the same time is the product of their individual probabilities. Therefore the probability of coincidence of flush from two domestic 9-litre WCs is 0.000016. From three it would be 0.000000064. This very low probability of finding WCs flushing together demonstrates the need, if economy of pipework is to be obtained, for taking this into account in estimating flows and consequent pipe sizes. Data for estimating flows given in *Archi-*

tects' Journal Information Sheet 1417 take this factor into account. A detailed description of the theory and method by which the design data were established can be found in *Services engineering: hydraulics* (see page 238). The use of these data will in large installations usually enable economical pipe sizes to be used. The same benefit does not arise in very small installations since the minimum size of pipe permissible from considerations of blockage (100 mm diameter in the case of foul drains or soil pipes carrying flows from more than one WC) is capable of dealing with the flows from many appliances. Even in these installations, however, unnecessary increases of drain diameter, very common at present, may be avoided.

2 Avoidance of pressure fluctuations

Modern opinion on hygiene does not regard the escape of drain air into a building with the same degree of horror that was the case in the past. No one, however, would advocate systems which would permit this—on grounds of amenity if not of health. Consequently air pressure fluctuation in the pipework must not be allowed to reach intensities which would affect the water seals of traps. The classical method of doing this is by means of vent pipes opening into waste or soil pipes near each trap outgo and leading to the external air. By this means the air pressure at each side of the trap can be kept stable at atmospheric pressure. The arrangement is expensive and unsightly and in an increasing number of suitable instances is being replaced by neater and cheaper arrangements which avoid the development of pressure fluctuations by means of more sophisticated hydraulic design rather than ugly and unnecessary trap venting.

In principle the approach depends on sizing and arranging the pipework so that full bore flow and the consequent pressure fluctuation are avoided. Even a momentary bridging of the pipe can give rise to gross pressure fluctuations. An understanding of the way in which flow in partially filled pipes behaves is vital if design is to be carried out. Diagrams 198 to 208 show some of the important cases.

Horizontal and sloping pipes flowing part full

Flow occurs in the bottom of the pipe (diagram 198). As the water level rises air is displaced from the pipework, and as flow of water takes place friction with the air in the pipe induces movement of air in the direction of flow. (It is interesting to compare this with the thermo-siphon theory of drain venting popular at one time.) Pressures induced are small.

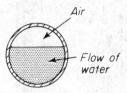

198 *Flow of water in horizontal sloping pipe*

Vertical pipes flowing part full

(a) Flow from a branch entering a vertical stack accelerates downwards because of gravity and quickly assumes the form of a sheet on the wall of the stack (diagram 199). Velocity of flow increases until fric-

199 *Flow of water in vertical pipe*

tion balances the pull of gravity. Air flow is induced down the stack and positive pressures are likely to develop at the foot of the stack due to restriction in area available for air flow as the rate of water flow is reduced on entering the horizontal drain. Paper and solid matter tend to fall in the centre of the pipe. They are discharged very rapidly.

(b) The pattern of flow described does not continue progressively with increasing flow until the pipe is flowing ufll bore. When one-quarter to one-third of the full flow capacity of the pipe is reached the whole section of the pipe is bridged at intervals by 'plugs' of water (diagram 200) which move along the pipe at the velocity of flow. Gross-pressure fluctua-

DRAINAGE INSTALLATIONS

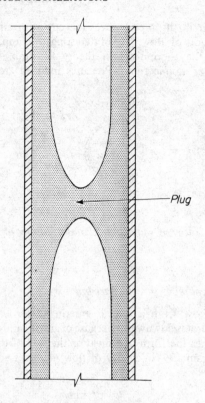

200 *Flow in vertical pipe showing 'plug' formation*

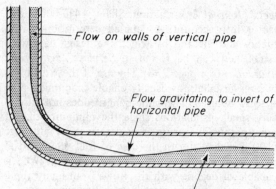

Flow at foot of stack (1)

201 *Vertical section through pipe showing change in relative position of flow from vertical to horizontal section of pipe*

tions occur as a result of the movement of the plugs and in any system where the concentration of flows could give rise to the formation of plugs complete trap venting would be required. It is essential if vent pipes are to be avoided and desirable in any case to avoid plug formation and this is one of the factors governing the number of appliances that can be connected to a stack of given size. For economical but satisfactory pipe sizing it is necessary to be able to estimate the peak flow resulting from the operation of numbers of sanitary appliances.

Transition from vertical stack to horizontal drain

(a) At the foot of a stack the flow leaves the walls and runs in the bottom of the pipe and the reduced velocity of the horizontal flow will result in greater cross-sectional area of the pipe being occupied, with corresponding reduction in air space (diagram 201). Positive pressures will develop at the foot of the stack and may be great enough to blow traps. Sharp bends at the foot make the problem worse.

(b) It is quite possible in building drainage installations to have a combination of flow rate, pipe diameter and drain fall which will give rise to a total bridging of the pipe in the drain known as hydraulic jump (diagram 202). If this situation arises gross

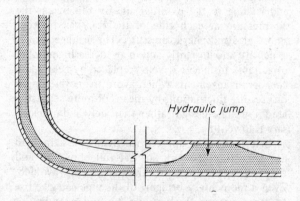

Flow at foot of stack (2)

202 *Vertical section through pipe illustrating 'hydraulic jump'*

pressure fluctuations will occur. The same result can come from a drain not subject to hydraulic jump but which has no outlet for the induced air flow as in the case of connections to sewers with no interceptor or fresh air inlet when the sewer is flowing full.

248

GENERAL PRINCIPLES

Discharge of branches into stack

The discharge of a branch into a stack will at least partially block the cross-section and if gross pressure fluctuations are to be avoided the rate at which flow can be introduced from any one branch and the

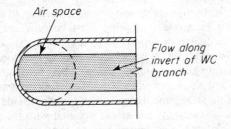

203 *Plan showing flow from WC entering stack*

grouping of branches must be controlled. Diagrams 164 and 165 show plan and section of a waste inlet to a stack, showing how the air flow in the stack is restricted by waste discharges. Diagram 163 shows the flow from a WC entering the stack. Since only the bottom part of the pipe, not the full diameter, is occupied by the flow, some space for air flow is left.

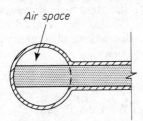

204 *Flow from waste branch entering stack. Plan*

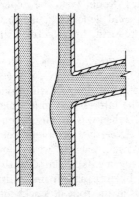

205 *Vertical section showing flow from waste branch entering stack in which flow is taking place*

Pipes for individual waste appliances

(a) In most waste appliances the theoretical free discharge capacity of the waste pipes is greater than the flow that can occur through the waste, which is partially restricted by a grating. In practice the direct connection of sanitary appliance, trap and pipe and introduction of water into the waste pipe at a sharp angle result in full bore flow at reduced velocity (diagram 206). The system operates as a siphon.

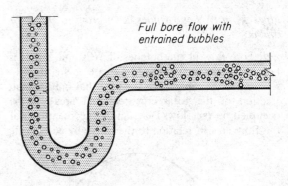

Flow in trap and waste (1)

206 *Vertical section through trap and waste showing full bore flow*

(b) Towards the end of the discharge air will pass through the trap and the full section of pipe will become a diminishing plug as the flow of water ceases (diagram 207). Provided the plug is near

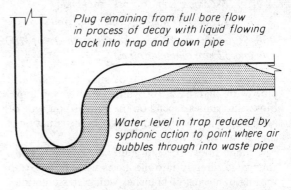

Flow in trap and waste (2)

207 *Vertical section through trap and waste showing how trap reseals when waste fall and length are appropriate*

249

DRAINAGE INSTALLATIONS

enough to the trap, water will flow back and reseal the trap as the plug decays. If the pipework is too long or too steep this resealing flow will not be possible. Sanitary appliances with flat bottoms have a long-drawn-out tailing off of flow after the siphoning has occurred, which tends to reseal the trap.

The limtis of slope and length of wastes in non-vented systems and the differences between lavatory basins and baths are the result of the phenomena described.

Pipes from WCs and gullies connecting to stacks or drains

Full bore flow in the outgoes of WCs and gullies is not likely to occur. The flow rate is so small in relation to the capacity of the pipe that only a small fraction of the total cross-sectional area will be occupied by the flow. Diagram 208 gives an idea of the flow area in relation to the whole by comparing

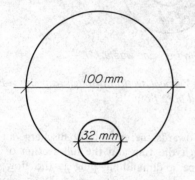

Comparison of cross-sectioned areas of 100mm and 32mm dia. pipes

208 *Comparison of cross-sectional areas of 100 mm and 32 mm diameter pipes*

the cross-sectional area of a 32 mm diameter flush pipe with that of a 100 mm diameter soil pipe. Flows into gullies depend on the number and type of appliances but clearly in normal use the situation is similar to that of WCs.

Ground floor WCs and gullies are conventionally connected directly into manholes without any trap venting by means of branches which are sometimes long and steeply sloping and may have bends at gully outgo and entry to manhole. It is clear why no pressure fluctuations develop and also that there is little risk that in normal circumstances any will arise.

Design considerations

Venting

Where pressure fluctuations cannot be kept within acceptable limits (normally, plus or minus 25 mm water gauge) by means of controlling size, length and fall of pipes with only terminal opening to atmosphere, separate vent pipes must be provided that can discharge or supply air to match the rate of air movement in the installation caused by flows of liquid. In below-ground drainage where velocities are low and pipes very rarely if ever flow full, very simple venting arrangements will serve a whole installation. In most cases the soil stacks coupled with the fresh air inlet or direct connection to the sewer will be more than enough. Inside the building small diameter pipes present a more critical problem which may be solved by venting the pipe serving each trap or, if the design of these branches themselves avoids pressure problems arising within them, by venting the main pipe runs at suitable intervals. *National plumbing code handbook,* by V.T. Manus, published by McGraw-Hill, New York, 1957, gives data on air flows required and the capacity of vents to deliver these flows. To maintain their efficiency vent pipes must be designed so that they do not become flooded in use and so that any condensation will drain to the waste or soil pipes.

Oversize pipes

It will be clear from the foregoing that the use of large diameter pipes could avoid any problems of pressure fluctuations in a drainage pipework system. If it is proposed to adopt this approach with horizontal wastes, trouble may be experienced from a build-up of solid deposits in the unflushed section of the pipe, reducing the effective diameter.

Detergents

The main effect of detergents in drainage pipework is to keep the interior very much cleaner and freer from deposits than was the case when soap was used exclusively. Excessive detergent foam may cause problems, however, where waste flows are subject to pipe arrangements which disturb the flow and produce turbulence and foam. This may arise at sharp elbows at the foot of stacks, at gullies and at hopper heads. To avoid this, waste stacks should be connected to the drain by easy bends. The so-called two-pipe system, involving trapping waste stacks from the drain by means of gullies, should not be used and nor should hopper heads.

Common misconceptions

Velocity in vertical stacks

In high buildings concern has been felt by architects about the velocity which flows down vertical pipes might reach, and in some buildings tall drainage stacks can be found to have offsets apparently intended to prevent an uninterrupted drop. In fact the forces of gravity and friction acting on the flow in a vertical stack soon balance and a 'terminal velocity' is reached. The vertical distance required for flow to reach terminal velocity varies with flow and pipe diameter. It is very likely to be no more than one storey height. Clearly, therefore, no precautions are necessary to reduce velocity in tall stacks. Offsets should be avoided since velocity of flow is reduced and plugs may be formed giving rise to gross pressure fluctuation.

Minimum falls

A velocity of flow of about 0.6 m/s is required to avoid stranding solid matter and the critical case is that of a pipe carrying the flow from one WC (2.3 l/s). For a 100 mm or 150 mm diameter drain in perfect condition a fall of 1 in 200 would give this velocity. In practice irregularities of laying, ground movement and distortion of pipes mean that sharper falls are needed. Local experience is likely to be the best guide. Local authority recommendations for minimum falls vary greatly, ranging, for 100 mm diameter pipes, from 1 in 40 to 1 in 110. One distressingly common misconception about minimum falls arises from the misuse of the old rule of thumb saying 100 mm drains, 1 in 40; 150 mm drains, 1 in 60; and so on. It is that if the fall available is only 1 in 60 then a 150 mm diameter pipe should be used even though a 100 mm diameter pipe would have adequate capacity. This is very wrong. It is quite clear that for a given fall the smallest pipe capable of discharging the flow will be the best pipe to use since the depth of flow will be greater.

Design of drainage installations

The foregoing describes some of the phenomena involved in drainage installations. It does not provide quantitative data for design. There are three ways to approach this:

1 *Following established precedents from publications and from practical observation* This is rightly the most popular method for drainage designers. The main limitation is that an existing installation which demonstrably functions satisfactorily is not necessarily an economical one. If there are doubts on this score recourse must be had to 2 and 3 below to check the performance. The main pitfall is that if the new situation, while being apparently similar to the model, does in fact differ in some significant respect, trouble can result. The analysis of 2 and 3 could be expected to bring to light differences which might be missed on more superficial inspection.

2 *Use of published hydraulic design data* In a circumstance where direct precedent is not available and where the mode of performance of the installation is understood, the necessity for venting and the size of waste, soil, drain and vent pipes may be established from published data covering flows of water and air and related pipe sizes. The references give a wide range of values but the data are not comprehensive at present.

3 *Practical experiment* Since standards for drain installations are now based on performance, observation of a specially made prototype can be of fundamental value in design. In cases where designers suspect that traditional practice is uneconomic or wish to employ some arrangement which seems theoretically sound but is as yet untried, conclusions can be reached by constructing and testing a prototype, particularly of work above ground.

A building contract provides opportunities for this and in cases where there is frequent repetition of a particular plumbing detail the procedure can be well justified. It is also possible where the designer feels that simpler than normal venting arrangements would be satisfactory to delay this part of the installation so as to give opportunity to observe the performance of the rest.

In addition to hydraulic considerations there are a number of other conditions which must be satisfied by drainage installations. Table 70 shows a summary of the general criteria for performance together with the means available to achieve them.

DRAINAGE INSTALLATIONS

Table 70 Performance criteria for drainage installations

Requirement	Achieved by means of
1 Watertight. In the case of underground drains entry of water as well as loss must be avoided and either case would encourage root penetration	Appropriate selection of pipe material and jointing system and by satisfactory workmanship
2 Non-blocking	Adequate minimum diameter Smooth internal surfaces Properly aligned joints No reduction in diameter in direction of flow
3 Adequate discharge capacity	Flows in pipe depend on diameter, gradient, internal surface and alignment of pipes
4 No nuisance. Overflowing should not occur in free-flowing installations of adequate size but nuisance may be caused by the discharge of drain air into the building	Traps at all entries to drainage pipework (surface water systems outside the building are not subject to this) Pipe installation designed to avoid air pressure fluctuations which might unseal the traps

5 Durable. Durability of drain installations is the problem. Very durable materials may not necessarily provide a long-lived drain
Factors involved include:
  Normal life of drain materials and jointing materials
  Effects of moisture, building settlement and ground movement on drain
  Deterioration of pipes and joints due to chemical nature of flow
  Deterioration of pipes or joints due to surroundings (eg soil chemicals)
  Mechanical damage or crushing
  Rust penetration
	Selection of pipe material
	Selection of jointing system and material
	Support or bedding
	Workmanship for above

6 Traceable and accessible for maintenance	Pipework arranged in straight lines between access points Access point provided so that all sections may be inspected and if necessary cleared by flexible lines and rods
7 Economic	Satisfactory siting of building in relation to disposal point for drainage Grouping of sanitary appliances to give short pipe runs

DRAINAGE ABOVE GROUND

Types of drainage system

Drainage installations for buildings have to deal not only with flows from sanitary appliances but also with rain water from roofs and paving. The established techniques for these installations are divided into work above ground (including that inside basements) described as waste and soil systems for flows from sanitary appliances and as rain water pipes for surface water. Below ground all the work is referred

SOIL AND WASTE SYSTEMS

to as drainage. Installations taking flows from sanitary appliances and those taking surface water being called foul drains and surface water drains respectively. In some cases the same pipe carries the two flows and is termed a combined drain. While the pipes remain within the boundary of the site of the particular building which they serve, the term drain is applied irrespective of the size of the pipes. When, however, they pass from the site they will be termed sewers and some different considerations of design and construction apply.

For all waste and soil and drainage systems a satisfactory standard of materials and workmanship is required. This will include watertight pipes with smooth internal bores; satisfactory junctions, which avoid risk of blockage; and properly aligned joints with no ridges to catch solid matter. Pipe materials and joints are dealt with on page 229 and the use of proper material and workmanship is assumed in the following notes which deal with the design of drainage systems.

It is important to remember that all waste and soil pipes and drains must discharge the flows that they receive immediately and that consequently all pipes must fall continuously to ensure a proper flow. Arches and dips to avoid obstructions are not acceptable (unlike water supply or other pipe systems under pressure). This point has considerable implications for planning and economy in installations both inside and outside buildings.

Considerable developments in theory and practice have taken place in recent years (see page 245, *Drainage Installations,* General Principles) and many existing buildings have installations which would not now be repeated in new work. It is important therefore to consider past as well as present methods.

WASTE AND SOIL SYSTEMS

Waste and soil systems have to convey the flows from sanitary appliances to the drains without causing any nuisance from leakage, blockage or smell and in a neat and economical manner. Soil pipes carry flows containing excrement, while waste pipes carry flows from baths, sinks, lavatory basins etc. The hydraulic problems involved have been described (page 245). In the early days of plumbing in buildings these phenomena were not so well understood and in addition it was thought necessary to exclude drain air from the building not only because smell could constitute a nuisance but because it was thought that smell was a direct menace to health. Drainage pipes were therefore taken outside the building, above or below ground, at the first opportunity and sanitary appliances were usually sited against external walls. Each sanitary appliance was provided with a trap (see diagram 213) to seal the air in the pipework from that in the room. In the case of WCs the single trap was accepted and the outgo pipe was connected directly to the drain at a manhole either by a pipe underground, in the case of ground floor appliances, or by a vertical stack for those on upper floors. In the case of waste flows, however, where, unlike soil systems, nuisance would not result from the content of a gully, an additional safeguard against the penetration of drain air into the building was provided by disconnecting the waste pipes from the drain by discharging them over a gully (see diagram 219). In addition to the air gap at the termination of the waste pipe the gully itself contains a trap to seal off the drain air. Waste appliances on first floors were discharged to a hopper at the head of a vertical stack itself discharging as before over a gully. Buildings over two storeys in height or with ranges of sanitary appliances discharging into common waste or soil pipes presented special difficulties in that the flows in the pipes could cause pressure fluctuations liable to affect the water seals in the traps. This was overcome by providing each trap with a vent pipe (full description 'trap ventilating pipe', sometimes called 'anti-siphon pipe') connected near the trap outgo and leading to the open air. This vent pipe allowed air to escape when pressure developed, or fresh air to enter to overcome suction, thereby preventing pressure fluctuation and consequent trap unsealing. The arrangement described, where waste flows are kept separate, is known as the *two-pipe system*. While many of its features are still in use the complexity and expense of pipework, and the serious disadvantages of the waste arrangements, have led to improvements in design. The main disadvantages may be tabulated:

1 The discharge of waste water over a gully grating was an unpleasant feature and caused considerable nuisance when the gully grating became blocked by leaves. This can be overcome by the use of the back inlet type of gully (diagram 219) which introduces the waste flow above the water level but below the grating. This is a considerable improvement but an open waste gully is still a

DRAINAGE INSTALLATIONS

possible source of smell and is by no means an advantage close to a building except when it is desired to pour down it liquids that could not be disposed of through the normal sanitary appliances.

2 The hopper used at first floor level in two-storey work (particularly housing) to collect the flows from waste pipes at that level constitutes a potential nuisance from smell, blockage and overflowing. The problem can only be avoided by eliminating the open hopper and connecting the wastes directly into the waste stack which would then have to be carried up above the eaves as a vent and if more than one sanitary appliance contributed to the stack then a vent pipe to protect the trap seals would also be needed.

3 The widespread use of detergents has produced a very critical problem, particularly in multi-storey buildings where the turbulence arising as the waste is discharged into the gulley causes foaming of the detergent, which can rise and spill over the gully grating, causing considerable nuisance.

These disadvantages, coupled with the extent of the pipework required, led to a desire for economy and improvement. More confidence in plumbing design, observation of American practice and less fear of drain air as a health risk, led to the use of the *one-pipe system.* In this system waste flows are connected into soil stacks and thereby to the drain, without the trapped gullies usually provided. By this means the gully and hopper are eliminated and the duplication of pipework substantially reduced. Each trap must, however, in this system, be provided with a vent pipe. In domestic work where, by the use of hoppers, common wastes requiring venting could be avoided, this meant that the savings in waste stacks and pipes might well be more than offset by the additional vents needed and consequently the two-pipe system continued to be employed.

The desire for economy and efficiency in the extensive post-war building programmes directed the attention of the Building Research Station to this problem and as a result of studies and laboratory work carried out there it became apparent that undesirable air pressure fluctuations in drainage pipework could, in a certain range of circumstances, be eliminated by the observation of a number of simple rules without the necessity for trap ventilating pipes. The method of design resulting was termed the *single-stack system.* Initially it was limited to 5 storeys in its application; after some years modifications were proposed which enabled buildings up to about 10 floors to be served, while recently installations up to 20 floors high, using 150 mm diameter stack, have been considered feasbile. The single-stack system depends for its performance in retaining trap seals on:

1 Preventing pressure fluctuations arising from the operation of other sanitary appliances by
 (a) connecting each appliance separately to the stack

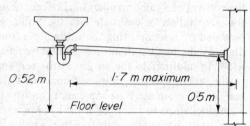

32 mm dia. waste for lavatory basin or small sink

If L is less than 0·69 m slope may be not more than 5° 1 in 12
For length between 0·69 and 2·3 m (maximum) slope not more than $2\frac{1}{2}°$ 1 in 24

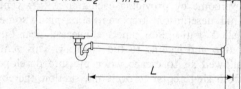

38 mm dia. waste for large sink

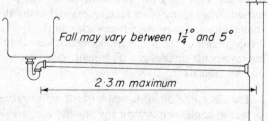

38 mm dia. waste for bath

Note: In all cases slow bends (but not elbows) may be used in the horizontal plane

209 Limits of fall and length for single stack system wastes for lavatory basins, sinks and baths

WASTE AND SOIL SYSTEMS

(b) limiting the flows in the stack to values at which plugs will not form (see pages 247 and 261)

(c) maintaining stack straight to avoid plug formation (see page 247).

2 Preventing self-syphonage arising from the operation of sanitary appliance. This is achieved by careful control of the diameter, length, fall and connection to the stack of the waste pipe servicing the appliance (see page 254 and diagrams 207 and 209).

Another important development in waste and soil installations, intended to avoid freezing, is the requirement in the Building Regulations 1965 for all waste and soil pipes (except wastes to back inlet gullies) to be inside the external walls.

The single-stack system giving neat and economical pipe installations is clearly the most desirable system to employ wherever possible. It lends itself mainly to dwellings. The need for a straight stack and limited length and fall to the individual wastes means that if advantage is to be taken of this system it is important to bear its requirements in mind at the planning stage of the building design and in the layout of sanitary appliances. Diagram 209 shows the requirements for single-stack wastes and diagram 210 shows some problems which may have to be over-

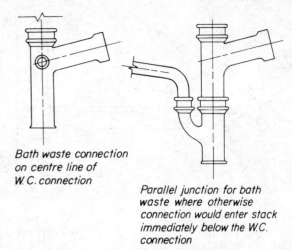

Bath waste connection on centre line of W.C. connection

Parallel junction for bath waste where otherwise connection would enter stack immediately below the W.C. connection

210 *Acceptable arrangements for juxtaposition of bath waste and WC outgoes in single stack systems*

come when branch connections from WC and bath meet the stack itself. The difficulties arise because the levels at which the bath waste and WC outgo meet the stack are inevitably very close. If the bath waste centre is not lower than the centre of the WC outgo a correction can be made, as shown in diagram 210. At points lower than this, however, connection is not satisfactory because of the possibility of the

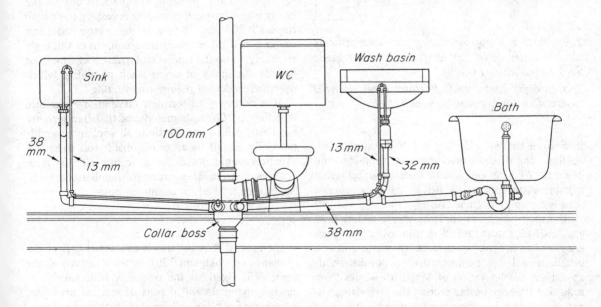

211 *Marley collar boss for neat and economical single stack plumbing*

DRAINAGE INSTALLATIONS

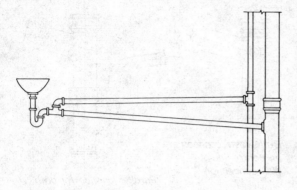

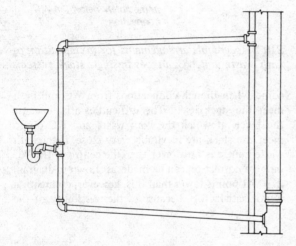

212 Waste and trap ventilating pipe arrangements for a lavatory basin too far from stack for single stack arrangements to apply.
Top: Straight run to stack. Bottom: Pipes arranged to miss doorway between basin and stack

flow from the WC affecting the bath outgo. Connection cannot, therefore, be made until 200 mm below the WC junction. A suitable detail for achieving a bath waste connection below the unacceptable zone is shown in diagram 210.

The single stack system is now very widely used and has had a major effect in improving the efficiency, economy and appearance of waste and soil installations. It has, however, three major disadvantages. First the limitation of length of wastes from sinks and lavatory basins, second the very flat gradient appropriate for these wastes which is likely to conflict with other appliances, and third the prob-lem of bath waste connections described above. Development work at Marley Plumbing has resulted in the Marley Collar Boss fitting (for the Marley PVC system). This fitting consists of a gallery with bozzes for pipe connections running round the 100 mm soil stack. Flows from the branches are introduced into the stack without risk of interference and small vents can also be connected to the gallery. This fitting enables the layout shown in diagram 211 to be achieved. All three disadvantages of the single stack system are overcome. The bath connection can be made in the previously unacceptable zone below the WC connection. The sink and lavatory basin wastes can drop vertically since they are provided with 13 mm diameter vents to overcome syphonage and the same vent makes it possible to connect the lavatory basin waste into the bath waste, instead of having a separate connection to the stack.

Where it is not possible to plan the location of sanitary appliances within the limits described for the single stack system, or the stack cannot be kept straight, trap ventilation pipes will have to be used to ensure that trap seals are not destroyed. Diagram 212 shows arrangements which may have to be made when a lavatory basin is so situated that normal single stack principles cannot be applied and the same layouts would be appropriate for other sanitary appliances similarly situated. Where the stack cannot be kept straight, pressure fluctuations due to the bends may occur and it may be necessary to vent all traps. CP 304 gives details for sizing large waste and soil and vent pipes, and future editions of this code are likely to contain details of a range of situations outside the limits of single stack plumbing, which nevertheless do not require trap venting.

It is necessary to test new installations to ensure that they perform adequately and that trap seals are not destroyed in use. Flushing all appliances simultaneously would be an unreasonable test since this situation would not occur in reality. CP 304 gives tables which enable appropriate numbers of appliances to be flushed for testing purposes.

Traps

In the case of WCs the trap is an integral part of the appliance (see diagram 120); in other sanitary appliances it is fixed to the outgo. A wide variety of designs, materials and depths of seal are available. Appearance may be an important feature, or ease of access for cleaning, flexibility of outgo position or

WASTE AND SOIL SYSTEMS

a variety of other features. Diagram 213 shows basic trap types; diagram 214 a bottle trap, which can be chromium plated for neat appearance; diagram 215, a three-jointed trap which can give a wide range of outgo positions and the bend of which can be removed bodily in the case of blockage (this makes clearing easier than the conventional clearing eye and also avoids the internal ridges associated with the clearing eye which can give rise to blockages themselves); and (diagram 214) a trap, the body of which is formed by a glass bottle which can retain kitchen sediment and give visual indication of the need for cleaning. These considerable variations mean that, as with sanitary appliances, it is usual to select and specify a particular type from a particular maker for each application.

In addition to the features described above the depth of seal must be specified. It is normal at present for a minimum of 25 mm of seal to be retained under all conditions of operation. Traps were used with seals of 75 mm, 38 mm and of nominal depth of a few mm. WC traps have 50 mm seals. The nominal depth traps were used in situation where in past regulations very short lengths of waste pipe discharging into the fresh air were not required to be trapped but were provided with traps to avoid cold draughts and the entry of insects. This arrangement was particularly popular for bath traps, since space under the bath was restricted and the waste was often very short, discharging into a hopper on the wall immediately outside the bathroom. 38 mm seal traps normally give satisfactory performance when provided with trap ventilating pipes, although when the one-pipe system was introduced 75 mm seals were used because of the fear of health risk from soil pipe air. In the single-stack system 75 mm seals are used, partly to overcome the slight seal losses occurring in normal operation of this system and partly to give an additional safety factor. The Building Regulations 1965 require 50 mm minimum depth of seal for traps discharging to drains, and the use of 75 mm seal traps is becoming usual.

The diameters of traps given in chapter 9, *Sanitary appliances,* have no special theoretical basis. They are sized that have been found to give accepted times of discharge for normal use.

A quite different approach to the problem of maintaining trap seals, from that described in the paragraphs on waste and soil installations, lies in the design of special traps which will allow air to pass through while retaining sufficient water to re-

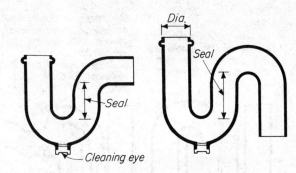

213 P and S traps, standard and deep seal

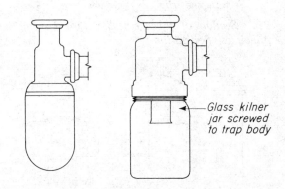

214 Left: chromium plated resealing bottle trap
Right: Econa kilner jar trap

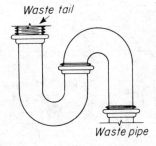

Note that the additional union obviates the need for a cleaning eye by enabling part of the trap to be removed and allows a wide range of waste pipe positions as the two parts can be swivelled

215 Three jointed trap

257

DRAINAGE INSTALLATIONS

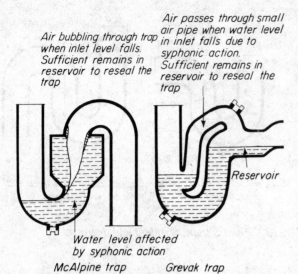

216 Sections of typical resealing traps showing the principles of operation

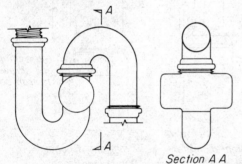

217 Econa resealing trap

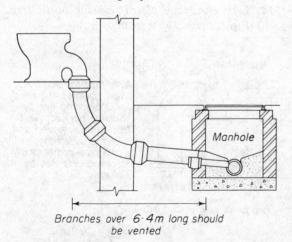

218 Ground floor WC connected directly to drain

establish the seal. Diagrams 216 and 217 show a number of traps of this type where a special reservoir of water, coupled in one case with an air by-pass, enables performance of this sort. Opinions differ on traps of this sort and they are not universally accepted for use in buildings. It is possible by their use to make substantial pipework economies but many people take the view that deposits inside the trap over a period could render its operation less effective. When called upon to function they have a serious disadvantage, particularly in domestic applications; this is the gurgling noise of air passing through the water.

Practical installation details

1 Entry to the drain

Flows from waste and soil installations have to be introduced into the drain without any discharge of drain air. This is accomplished in two ways:

(a) Stack Wastes and WC outgoes are discharged into a vertical stack rising to an open vent at above eaves level and leading down to a bend below ground level which delivers the flow into the drain itself. The junction is almost invariably made at a manhole.

(b) Ground floor WC It is well-established practice to connect ground floor WCs, usually of S-trap types, directly to the drain via an underground bend. The junction is usually made in a manhole. The problem of self-syphonage of the trap does not arise because the rate of flow from the flush only fills a small part of the pipe (see diagram 208). Induced syphonage does not arise since the drain is itself ventilated and usually not subject to marked pressure fluctuations. Diagram 218 shows a typical arrangement.

(c) Ground floor gully The connection is similar to that of the ground floor WC and the traps of gullys are not subject to syphonage for the same reasons as WC traps. Gullies are therefore used to introduce waste flows into the drain. Diagram 218 shows a typical detail. The waste connection to this type of gulley is the only waste or soil pipe which is now permitted outside external walls. (A drop inside the wall, leading to a horizontal back inlet, gives additional frost protection but is more complicated to install and maintain.) Diagram 220 shows caulking ferrules which can be used to join the small diameter waste pipe to the 200 mm gully inlet.

These limits on drain entry must be borne in mind when planning buildings. Sanitary appliances on the ground floor may be freely disposed, the only

WASTE AND SOIL SYSTEMS

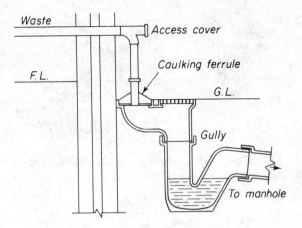

219 Waste connected to drain via gully

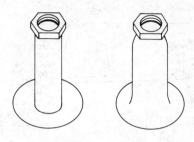

220 Caulking ferrules for joining small diameter (32 mm and 38 mm) waste pipes to gully inlets or to 100 mm diameter pipes

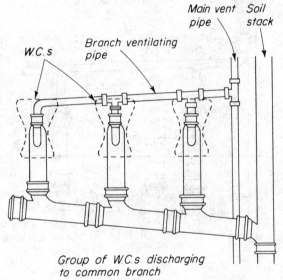

221 Group of WCs discharging to a common ranch with vented traps

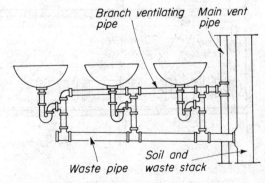

222 Row of lavatory basins with common waste and vented traps

limit being the need to construct drains to lead to them. It is recommended that branches to WCs and gullies should not be longer than 6 m unless they themselves are provided with a vent pipe. On upper floors, however, unless the sanitary appliance is within reach of a stack, a special one will have to be provided running both down to the ground to the drain and up to the roof as vent. This means that isolated sanitary appliances on upper floors should be avoided as much as possible and appliances should be grouped on each floor and that groups on different floors should be vertically over one another.

2 Groups of sanitary appliances

Several appliances discharging into a common waste or soil branch is an arrangement frequently required in practice, particularly in non-domestic buildings. The conditions for single-stack design do not apply and it has in the past been thought necessary for every trap in such a layout to be individually vented. Diagram 221 shows such an arrangement for a group of WCs. It is, however, now becoming standard practice for ranges of up to 5 WCs to be connected to a common branch without individual trap ventilation. Larger numbers of bends in the common branch would necessitate ventilation but vent pipes leading to selected traps or to the common branch are now more usual than venting every trap and adequate performance is achieved.

Ranges of lavatory basins have often been individually vented as shown in diagram 222. More

DRAINAGE INSTALLATIONS

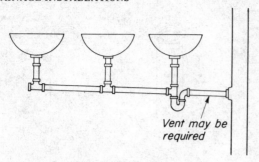

223 Row of lavatory basins with common waste and trap. (Maximum length of waste 5 m)

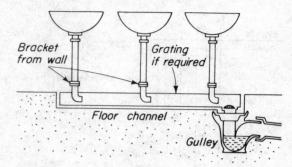

224 Group of appliances discharging into floor channel

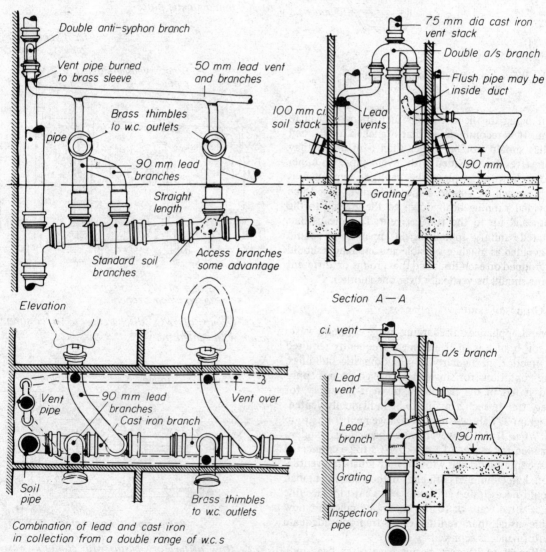

225 Plumbing duct between two lavatories

economical and neater possibilities are shown in diagrams 223 and 224. The channel and gully make hosing down and mopping the floor easier. The arrangement is a particularly convenient one on the ground floor since there is no need for a vent stack.

Where rows of WCs are required in adjacent male and female lavatories the plumbing arrangements can be made very much simpler and neater if a duct is used. A space approximately 1 m wide can accommodate and give access to all the plumbing for the WCs. Flushing cisterns can also be fixed inside the duct operated by a lever fitting passing through the wall. Diagram 225 shows a duct of this type. While the duct illustrated runs through several floors, it is possible to have one on a single floor. It will be noted from the diagram, however, that the floor level of the duct in this case will have to be somewhat lower than the floors of the lavatories.

Sizes for waste and soil pipes

Chapter 10, Sanitary appliances, gives the diameters of the waste or soil outgoes of the most common sanitary appliances. The first section of pipe is normally of this diameter and if a single stack system is used, all branches will be sized in this way. The discharge capacity of stacks can be determined using the discharge unit method described in pages 281 - 284 and the units given in table 75 may be used in conjunction with table 71 to determine appropriate sizes for vertical stacks. CP 304:1968 (soon to be revised) gives details of sizing for more complex installations and for large trap venting pipes. The step by step method for foul drains (see page 283) and the associated data, together with table 71 for capacity of vertical pipes may be used for pipe sizing in large installations.

Table 71 *Discharge capacity of vertical waste and soil stacks*

Nominal stack diameter, mm	Maximum number of discharge units
50	20
76	200 (not more than one WC)
100	850
125	2700
150	6500

RAIN WATER PIPES AND GUTTERS

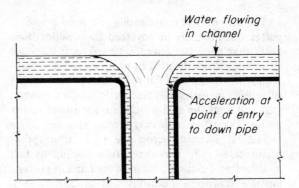

226 *Shows why the water flowing in a gutter or a roof channel is not likely to give full bore flow in a rain water down pipe. (Wire balloon to prevent leaves entering pipe not shown)*

RAIN WATER PIPES AND GUTTERS

Pitched roofs are normally provided with eaves gutters to catch the rain running off and with downpipes to carry the water to the ground. Flat roofs are usually provided with a slight fall to carry rain water directly to a roof outlet, or to a channel formed in the roof leading to an outlet or to an eaves gutter. Where eaves gutters are not used, or are used only on part of the perimeter of a flat roof, strong winds may overcome the slight falls provided and blow the rain water over the edge of the roof. A slight upstand will prevent this. If it would be unsightly at the eaves ti may be placed a short distance back. The small quantity of rain falling between eaves and upstand will not normally constitute a problem.

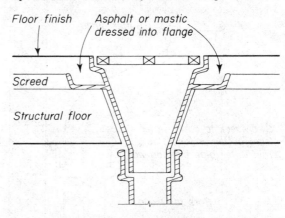

227 *Roof outlet for surface water. Note the tapering from the diameter of the outlet to the smallest diameter rain water pipe connection*

DRAINAGE INSTALLATIONS

The area of roof contributing to a given length of gutter will normally be governed by considerations other than rainwater run-off. Pitched roofs will have spans and pitches dictated by planning and structural requirements while on flat roofs the cost and weight of screeds or firring required to form the necessary falls will limit the area which can contribute to the gutter. Large flat roofs may require channels to be formed at intervals across the roof or in appropriate cases they may have falls leading to bell mouthed roof outlets instead of channels or gutters. Possible downpipe positions will be closely constrained by planning, structural and aesthetic considerations and gutters and downpipes will often need to be sized to suit these limitations. Gutters should be provided with a very slight fall, 1 in 360 is recommended. Falls which are too steep cause difficulties as gaps between the gutter and roof finish. Diagram 228 shows a typical eaves gutter.

In the early design stages it is often sensible to consider the layout of gutters and downpipes in relation to economy of drain layout. It will often be possible to save some drain runs and even manholes by judicious gutter layout. In the case of industrial buildings there is much to be said for using pipes at high level to convey surface water rather than bring a large number of rainwater pipes down to ground level and provide an underground drain along the same route at considerably greater expense. The high level pipe would be provided with falls and calculated in the same way as a drain.

Roof channels may be provided with bell mouthed outlets, as shown in diagram 227, but if it is possible to form a box type receiver, or sump, at the location of the downpipe a greater depth of water can be provided to give a closer approach to full bore flow in the downpipe.

Rainwater pipes may run either externally or internally. In both cases the joints should be watertight so that no leakage will take place in conditions of full bore flow. Internal rainwater pipes do not run entirely silently. Care must be taken that pipes are enclosed behind walls with adequate weight to prevetn noise penetrating to habitable spaces. There is always a possibility that unusually heavy rain, or a blockage in a pipe, may cause gutters to overflow. In every building this should be considered and the designer satisfy himself that, in the case of overflow, the water flows down the facade and does not make its way into the building where damage would result. If necessary special provisions for overflow, such as spouts, should be provided. Northlight factory roofs, where gutter ends are sometimes stopped by structural beams, sometimes give trouble if water can overflow under the glazing before finding a better way out of the gutter. Access for rodding should be provided for all rainwater pipes. If bends would prevent rodding from the top, access covers should be provided, especially in the case of internal pipes. CP 308:1974 deals with all aspects of roof drainage.

Estimation of gutter and downpipe sizes

Studies of rainfall intensities and gutter and downpipe performance have led to the establishment of 75 mm per hour as a suitable rate of rainfall for gutter design. For horizontal surfaces this figure can be used direct. The wind blows rain against vertical surfaces and a suitable estimate can be made by halving the area of the vertical surface and taking the

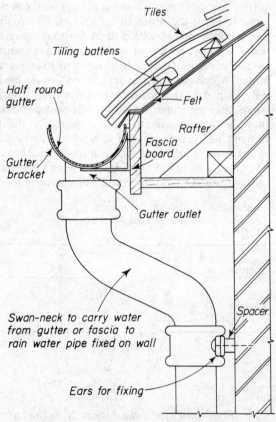

228 Typical eaves detail showing half round gutter, gutter outlet, swan-neck and rain water pipe

full rate of rainfall on this area. Pitched roofs also receive more rainwater than their plan area would indicate. An appropriate estimate of the effective area for a pitched roof can be made by taking the length, parallel to the gutter, and multiplying by the plan width, taken at right angles to the gutter plus one half of the vertical height of the roof from gutter to ridge. On this basis the volume of run off from any roof can be calculated and an appropriate gutter size selected from table 72. Where a gutter is provided with more than one downpipe, it should be calculated in sections, taking into account appropriate areas of roof. Table 73 gives sizes for downpipes which have discharge capacities appropriate to the gutter sizes shown in the table.

Bends reduce the flow in gutters so that reduction should be made as follows: Within 2 m of outlet:
sharp-cornered bends, 20 per cent
round-cornered bends, 10 per cent.
For valley and parapet gutters the formula:

flow capacity (litres per sec) = area of cross section mm^2 x 0.000156 may give some guidance as to suitable sizes, provided the proportions of the cross-section of the gutter remain somewhat similar to the half-round or ogee section, is ratio of width to depth is two to one.

Outlets with sharp corners will restrict the flow in the gutter and downpipe and should be avoided if possible.

CP 308:1974 and BRS Digest 34, give further data on sizing.

Materials for gutters and rain water pipes

Cast iron was, for many years, the most usual material for gutters and downpipes. It was, however, heavy, liable to fracture and required maintenance. Plastics are now replacing cast iron in popularity. Pipes and gutters made of plastics (usually uPVC) are light, easily fixed and require no painting. In general the plastics systems are similar to the cast iron that they replace, but special jointing methods and support and gutter sections designed for better flow are used and manufacturers' catalogues should be consulted for specific details.

Other materials are also used.

Asbestos cement will not corrode but is not so strong as cast iron. Cast aluminium is usually in the shapes of cast iron. It is quite strong, rustless and very light. Long lengths of pipe are available, thus reducing the number of joints. Extruded aluminium is used for gutters. BS 1430 covers both types of aluminium rain water goods. Galvanised mild steel, 1.3 mm for gutters, 1.0 mm for pipes, are made to BS 1091. They are light and cheap but subject to corrosion if galvanising is damaged. Shapes are somewhat similar to cast iron.

Table 72 Flow capacities in litres per second for level gutters

(a) Half-round gutters

Gutter size mm	True half-round gutters* flow l/s	Nominal half-round gutters† flow l/s
75	0.43	0.32
100	0.84	0.67
112	1.14	0.84
125	1.52	1.06
150	2.46	1.82

(b) Ogee gutters

Gutter size mm	Asbestos cement To BS 569 flow l/s	Metal gutters Aluminium or cast iron† flow l/s	Metal gutters Pressed steel* flow l/s
100	1.06	0.49	0.91
112	1.52	0.52	1.37
125	1.82	0.84	1.75
150	2.88	—	2.66

*Pressed steel to BS 1091; Asbestos cement to BS 569.
†Aluminium to BS 2997; cast iron to BS 460.

Table 73 Recommended down pipe sizes

Half-round gutter size dia mm	Sharp (SC) or round-cornered (RC) outlet	Outlet at one end of gutter dia mm	Outlet not at one end of gutter dia mm
75	SC	50	50
75	RC	50	50
100	SC	62	62
100	RC	50	50
112	SC	62	75
112	RC	50	62
125	SC	75	87
125	RC	62	75
150	SC	87 } provisional	100 } provisional
150	RC	75	100

DRAINAGE INSTALLATIONS

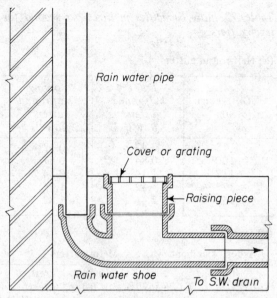

229 Rain water down pipe delivery to untrapped rain water shoe. This arrangement is only possible with separate drainage systems

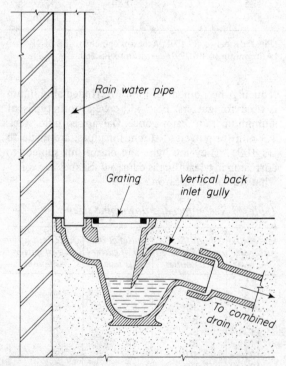

230 Rain water down pipe delivery to trapped gully. This arrangement is essential if the rain water pipe is delivering to a combined drain

Wrought copper and zinc pipes and gutters are covered by BS 1431. They resist corrosion, but are not so robust as cast iron or aluminium, and therefore have a different shape, the edges of gutters being beaded. Method of fixing the OG gutter is through the tubular stays. The sockets on the pipes are slimmer and neater than with cast iron.

Outlets for flat roofs are made in cast iron as shown in diagram 186. These are for use with asphalt, which is turned into a special flange. The outlet with junction is useful for balconies in flats.

Enamelled steel gutter and pipe, marketed under the name 'Vitreflex', produce very neat sections with an internal socket and simple bar stay for fixing.

Large gutters for industrial buildings are available in asbestos cement, in rectangular sections, BS 569; widths up to 600 mm; depths, usual 150 mm; joints, spigot; and socket, external. These are very strong with all thicknesses of 9 mm or 12 mm. Similar profiles are used for heavy pressed steel gutters in a range of sizes from 300 mm wide up to nearly 1 m, the thickness being in 12 BG for the smallest, up to 8 mm plate for the largest. Standard length, excluding socket, is 3 m. Standard fittings for both asbestos cement and steel are stopped ends and outlets, bends are not normally needed. These gutters are normally used with the roof sheeting, asbestos cement corrugated, metal deck, etc, or roof glazing clipped to both edges. For this reason a fall is difficult to arrange and they are usually fixed level. This necessitates rain water pipes at frequent intervals. Details of factory roofs using these gutters are shown in *MBC: Structure and Fabric*. Joints are made in red lead, or more often bituminous mastic, and bolted. Support is best at 1.5 m intervals, usually on steel brackets as part of the roof framing, but sometimes on timber framing, which is easier for site adjustment.

Connections from rainwater pipes to drains.

There are two distinct cases to consider in connecting an RWP to the drains. If the drainage system is separate and the RWP is delivering to a surface water drain no trap is necessary and a direct connection between pipe and drain may be made. Diagram 229 shows a typical arrangement. Where the drainage is combined, however, it is necessary to provide a trap between the surface water pipework and the combined system. This may be done by collecting a number of RWPs into a drain and making a trapped connection to the combined drain or by providing a trap at the foot of each RWP as shown in diagram 230. In most new buildings with combined drainage systems both systems will be used to achieve economy of pipework and while some RWPs will be trapped and deliver to the combined drain others may contribute to areas of separate surface water drains only trapped at the point of junction with the combined drains. (See page 252 for definitions of drainage systems).

Drainage of paved areas

Paved areas must be laid to falls towards channels or gullies in the same way as flat roofs. This is very frequently overlooked in modern buildings which, in wet weather, force everyone approaching the entrance to splash through puddles formed on the paving. As a general rule all paved surfaces over $6\,m^2$ in area should be drained and should be given a minimum fall of 1 in 60 which will discharge rainwater swiftly and avoid puddles. Falls should never run up to the external walls but preferably away from the building. Paved areas may discharge to channels in site roads, or to specially formed channels forming part of the paving. Channels, which are more carefully laid and levelled than paving and where some pounding would not matter, because of the narrow defined shape, can be laid to substantially lower gradients down to 1 in 200. Often the slope of the building site will give greater falls than this in any case. Gullies, either untrapped or trapped, depending on whether the drainage system is separate or combined, will discharge the surface water flows to the drains.

Snow

Snow is not a serious problem in buildings in most of the country. Drifting across openings, or excessive weight on roofs rarely occur and the rate of run-off from melting snow is invariably lower than the 75 mm per hour standard for gutter and downpipe design. Large valley gutters may be provided with snowboards to ensure a free passage for water underneath the snowboards particularly if overflowing could be a problem. Where sharply pitched roofs could spill snow directly on to glazing below or outside an entrance vertical snow guards are sometimes provided at roof eaves.

DRAINAGE BELOW GROUND

The general principles of good drainage described on pages 245 to 252 and Table 69 apply to underground drains. In addition a number of special considerations arise from the planning, constructional, organisational and hydraulical points of view.

Planning

Drains carry foul and surface water away from buildings to points of disposal. In many cases disposal from the site will be by means of discharge to a sewer. Public sewers are maintained by local authorities and usually sited below roads. In some cases building owners may combine to have flows from several properties passing across the different sites in a private sewer leading to a public sewer or other means of disposal. Where discharge to a sewer is not possible means of disposal on the site itself will have to be arranged. Whichever is the case, in designing a new building, thought must be devoted to the ways in which siting and planning may be influenced by the need for feasible and economic drain layout. In normal situations it will be desirable to site the building so that it is at a level from which drains can be laid to fall towards the disposal point so that liquids discharged into the drain will flow by gravity. It will clearly be economical to keep the length of the drain involved to a minimum. Within the building itself the grouping together of sanitary appliances, and particularly their grouping vertically, reduces the number of entries to the drain, branch drains and manholes, while the positioning of these groups and that of rain water pipes can reduce the length of drain running round the building.

While it is possible by means of long runs, deep excavations and pumping to drain most conceivable building situations it is desirable to avoid these features unless they make possible some important and otherwise unattainable planning requirement.

Construction

Table 69 sums up the main performance requirements which must be met by drain installations. Drains must be watertight; not affected by the flows they carry or the soil chemicals; able to resist root penetration; able to withstand earth pressures, the effects of earth movement and building settlement;

DRAINAGE INSTALLATIONS

and capable of withstanding thermal and moisture movement. The materials available for use as drains are in themselves watertight and durable. The main problems for drain construction arise because of the various types of movement to which drains are subjected and to crushing. This situation has not been understood for very long. Its realisation is giving rise to the use of new materials, jointing and bedding techniques. In the past drains were constructed to be as rigid as possible. Joints allowed no movement and in cases where specially high performance was required concrete beds were provided under the drains. These precautions were not able to resist the forces involved in moisture, thermal and ground movement and consequently drains laid in this way were liable to damage from fracture. The present tendency is to replace rigid drain construction with flexible. In fact the construction of rigid drains with rigid joints must now be regarded as bad practice in almost every circumstance. Either material which are themselves flexible or joints capable of accommodating elongation or contraction (draw) or angular movement (slew) are increasingly used in drain construction. Joints capable of adjusting to both types of movement are known as flexible/telescopic.

Materials

The main materials available for drain use are:
Vitrified clay (glazed or unglazed) until recently described as salt-glazed ware or stoneware.
This is the traditional material for drains. It is a very durable material although brittle. It is widely used and a considerable range of special junctions, bends and fittings is made and stocked by builders' merchants. The traditional method of jointing (diagram 231 (1)) is by placing the plain (spigot) end of one pipe into the socket forming the end of the next pipe. Tarred hemp is wrapped around the spigot and compressed into the space between spigot and socket, thereby centering the spigot in the socket and preventing the cement mortar which is used to complete the filling of the socket from penetrating into the bore of the pipe. The joint resulting is not only rigid but is liable to fracture even without ground movement, because of the differing thermal and moisture movements of the cement joint and VC pipe. To overcome this a number of flexible telescopic joints have been developed. Diagram 231 (2) shows one type employing plastic bedding rings fixed to socket and spigot and a rubber sealing ring.

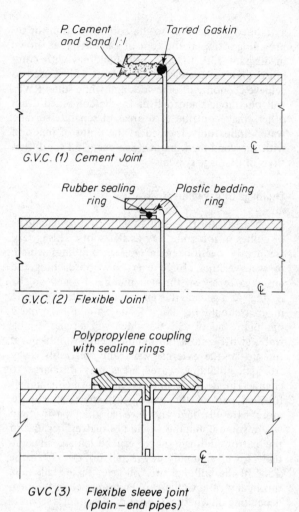

231 *Joints for VC drains*

Joints of this type can withstand up to 5° in bending and 18 mm extension or shortening. Using this type of joint with proper bedding VC pipes make excellent and durable drains. They are available in a wide range of sizes (only some of which are stocked by merchants) between 75 mm and 750 mm diameter. Until new metric lengths become available standard lengths based on the imperial sizes are 610 and 914 mm.

Very recently a new type of joint for VC drains has been introduced. It uses plain pipes without sockets, an arrangement that has a number of advantages in terms of economy, ease of cutting and hand-

DRAINAGE BELOW GROUND

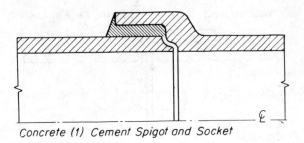

Concrete (1) Cement Spigot and Socket

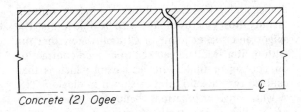

Concrete (2) Ogee

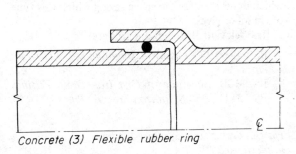

Concrete (3) Flexible rubber ring

232 *Methods of jointing concrete pipes*

competitive in price with VC except where heavy crushing loads have to be met. Concrete pipes can be constructed to give special resistance to crushing. The methods of jointing are similar to those for stoneware. Flexible rubber ring joints have been available for concrete pipes for many years but their use has normally been for large civil engineering works. Several forms are available for building drains and their use is much to be preferred to the traditional cement joint. Diagram 232 shows some of the jointing methods which are employed with concrete pipes. The relevant British Standard is BS 556, *Concrete Cylindrical Pipes and Fittings.*

Asbestos cement

Asbestos cement pressure pipes have been available for many years and can be used for underground drainage. Recently however, a special asbestos cement drain-pipe has been developed. It is supplied in 4 m lengths, thereby reducing the number of joints to

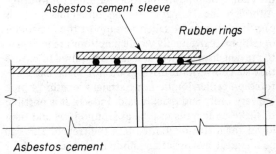

Asbestos cement

233 *Rubber ring joint for asbestos cement drains*

ling, linked by a plastic sleeve which will allow both elongation and angular movement. Diagram 231 (3) shows this type of joint. The relevant British Standard is BS 65: 1971 *Clay Drain and Sewer Pipes.*

Concrete

Concrete drain pipes are notable for the availability of very large diameters. Many manufactures list diameters from 100 mm to 1.83 m and special egg-shaped pipes are available to give improved depth of flow at times of slack discharge. In the smaller diameters concrete pipes are not usually considered

be made. The method of jointing which is shown in diagram 233 is inherently flexible and telescopic. Diameters from 100 mm to 600 mm are available.

The relevant British Standard is BS 3656: 1963 *Asbestos Cement Pipes Fittings for Sewerage and Drainage.* (Class 3 is appropriate for most building drainage appliaction.)

Pitch fibre

Pitch fibre pipes are formed by impregnating wood-fibre with pitch. The resulting pipe has adequate durability for normal drain use (although not suitable

267

DRAINAGE INSTALLATIONS

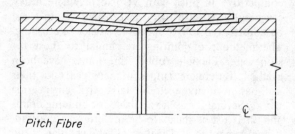
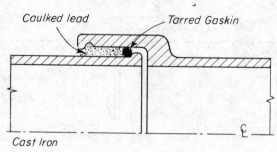

234 Tapered sleeve joint for pitch fibre drains   *235 Caulked lead joint for cast iron drains*

for continuous very hot discharges or some types of chemical flow). It is available in 2.5 m lengths, is very light in weight in relation to other drain materials and has a very simple jointing technique. Diagram 234 shows the type of joint. A taper is formed on the end of the pipe and this is driven into a tapered socket on a sleeve junction piece. A sledge hammer and a wooden dolly to prevent damage to the end of the pipe are the tools required for jointing. Although the formation of an individual joint is easy, care has to be taken to support the remainder of teh pipe, particularly where junctions are involved, to prevent displacement and care is required to ensure that successive joint drivings do not place excessive force on earlier joints. Long straight lengths of pipe are very easily and quickly laid. Indeed, it is possible to make up lengths of pipe at ground level and then lower them already jointed into the trench. This can be of great assistance in minimising excavation or with difficult trench conditions. Junctions and fittings with appropriate joints are available for manufacturers. Bends are cut from lengths of pipe of appropriate radius and the tapered ends formed on site by the use of a special tool. Lengths of straight pipe are cut in the same way. Open channels for manholes are formed by cutting standard pipe. Although the jointing does not act telescopically the pipes are flexible and will accommodate themselves to a considerable degree of ground movement.

Diameters from 75 mm to 200 mm are available and the range may be extended.

The relevant British Standard is BS 2760: 1956 *Pitch-impregnated Fibre Drain and Sewer Pipes.*

Cast iron

Traditionally used where ground movement might occur and where leakage must be prevented (eg drains under buildings) although modern thought favours short lengths of pipe with flexible joints. The spigot and socket joints of CI drains are either run with molten lead and caulked to take up contraction on cooling or filled with lead wool which is then caulked into a solid joint as shown in diagram 235. A wide range of junctions, bends and other fittings are available. Diagram 236 shows some of these, including a 236 typical cast iron inspection chamber which demonstrates the space saving which this type of chamber gives.

The relevant British Standards are:

BS 437: 1933 *Cast Iron Spigot and Socket Drain Pipes*
BS 1130: *Schedule of Cast Iron Drain Fittings*
BS 1211: 1958 *Spun Iron Pressure Pipes.*

Plastics drains (manufactured from unplasticised PVC)

The potential advantages of plastics drains in terms of lightness, ease and speed of assembly and smooth internal surfaces are apparent. Designers have, however, felt doubts about the viability of the material for drains and about the cost. In 1968 an Agrèment Certificate (see *MBC : Materials, Components and Finishes* for details of the Agrèment Board) was issued for 100 mm drains, and in 1971 a British Standard for Unplasticised PVC Underground Drain Pipes and Fittings followed. This standard also covers only 110 mm diameter pipes and fittings (the outside diameter is used to define plastics pipes rather than the inside diameter normally used for other pipes). Since 1968 there has been rapidly increasing use of plastics drains and several Agrèment certificates have been issued to a number of companies including some certificates for 160 mm diameter pipes. The

DRAINAGE BELOW GROUND

range of sizes available varies between different manufacturers. It is possible to obtain 82.4 mm (for surface water only), 110 mm, 160 mm, 200 and 250 mm diameter pipes. Imported pipes can be obtained up to 630 mm diameter. Pipes can be obtained up to 6 m in length. Although solvent welded joints may be used, they are difficult to make in the difficult conditions found in drain tranches and push joints with rubber sealing rings are normally used. Diagram 190 shows a joint of this type.

uPVC pipes are light in weight, relative to other drainage materials, easy to handle and joint and the drains may be tested as soon as they are laid which enables trenches to be backfilled quickly, a significant advantage for progress of work on the site.

While uPVC drains may be laid in conventional layouts with normal types of brick manholes manufacturers working in this field have identified manholes as a source of some drainage troubles and as an important element in the cost of drainage systems.

Most manufactures include, in their ranges of pipes and fittings, special manholes designed to minimise cost, or in one case a system which replaces a proportion of manholes with rodding points. Agrèment certificates have been awarded to a number of systems of this sort. Designers wishing to use uPVC drain systems must therefore consult manufacturers' literature and should clearly satisfy themselves that Agrèment certificates cover the system that they propose to use.

Two systems are described on page 278.

Drain construction

Proper bedding for drain pipes is essential for drain durability. Traditionally, and most existing drains will follow this pattern, tranches were dug and their bottoms levelled to the falls required. Pipes were laid by scooping grooves for the sockets or packing to support the barrels to the final level. Where a par-

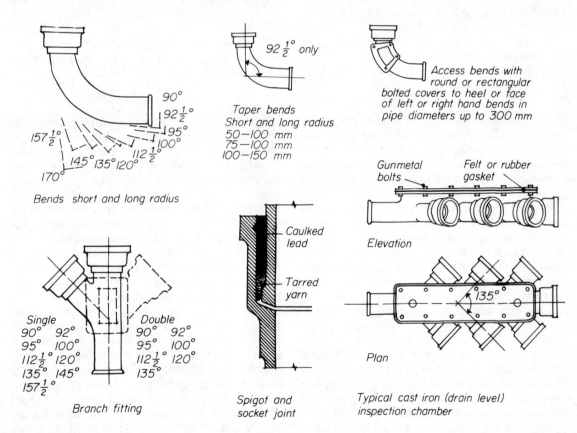

236 Cast iron drain fittings

DRAINAGE INSTALLATIONS

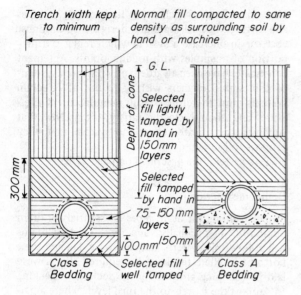

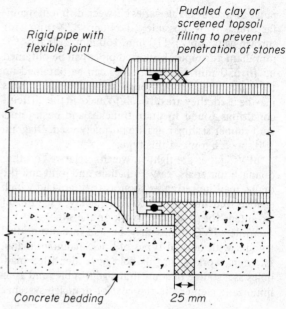

237 Drain beddings recommended by BRS
Left: class B, granular bedding
Right: class A, where pipe is supported against crushing by concrete bed
(See diagram 195 for joints in the bedding)

238 Concrete bedding for drain pipe showing gap cut in bedding at the joint to prevent fracture and allow flexibility

ticularly sound job was required, as in the case of foul drains, a continuous strip of concrete was provided for pipe bedding. This is not now regarded as satisfactory. The concrete strip will not be capable of resisting the pressures of earth movement, fracture will occur and serious drain failure is likely to result at the point of breakage. In trenches without concrete bedding local hard spots could cause pipe fracture.

At present the Building Research Station recommends the use of grandular bedding, which is used under as well as at the sides and over the pipe and is laid and compacted in thin layers. This type of material provides a uniform bed for drain pipes, avoiding the problems of irregular support, and materially contributing to the crushing strength of drain pipes. The effects of ground movement are distributed rather than concentrated by granular bedding. Suitable graded sand, gravel and broken stone may be employed. Diagram 237 shows how a pipe is bedded in this way. In cases where the pipe is not strong enough to resist the crushing forces additional strength can be provided by providing in addition to the granular bedding a concrete base for the pipes.

Diagram 237 (right) shows this arrangement. It is important, however, that the base should not be continuous but should support each pipe separately. If laid as a continuous strip a gap should be cut out to correspond with each joint. Diagram 238 shows a section through a typical joint and bedding. It will be noted that the gap in the concrete and in the socket of the drain pipe are filled with clay to prevent stone penetration and consequent loss of flexibility. Where more strength is required than can be provided by this type of bedding complete concrete encasement can be used, provided the casing is provided with gaps at each joint.

BRS Digests 124 and 125 (first series) *Small Underground Drains and Sewers,* 1 and 2, describe these principles and give details to enable pipe strengths and loads from the ground to be estimated, together with references to more detailed studies. It is worth noting that the load on drain pipes is determined by the width of the drain trench at the level of the crown of the pipe, above this the trench can be any convenient width. The width at the crown should clearly be kept to a minimum. BRS recommendations for width at crown are:

Drain diameter	Width of trench
100 mm	530 mm
150 mm	610 mm
230 mm	690 mm
305 mm	760 mm

It is also recommended that the minimum cover over drains should be 1 m under fields and gardens and 1.3 m under roads. Shallower drains than this may require a protective concrete slab placed over the pipe, but not in contact. Under roads this slab should be reinforced and wide enough to transmit the loads from above to the undisturbed ground beyond the sides of the trench.

Where pipes pass through walls special precautions should be taken to ensure that differential settlement does not cause drain failure by crushing. Diagram 239 shows a suitable detail for a drain passing through a wall. Note that not only is the wall kept clear of the drain but also the soil is kept out by asbestos cement sheets.

Inspection chambers

For ease of tracing and maintenance, drains are laid in straight lines between access points which should allow drain rods to be introduced for clearing the drain if necessary. Suitable access points may be made by providing rodding eyes, access gullies or inspection chambers. Rodding eyes and access gullies can only be used at the heads of branch drains. At intermediate points and junctions access, if required, must be provided by inspection chambers. Rodding eyes are formed by bringing a drain bend up to ground level and sealing with a cover plate. Access gullies have a removable cover, usually on the outgo pipe, which allows access to the drain. Inspection chambers are pits formed below ground through which the drain runs in an open channel. They are required at all bends or changes of gradient and at junctions when either the main drain or the branches cannot be rodded from other points. Traditionally inspection chambers were provided at all junctions on soil drains. Present thought allows soil junctions without inspection chambers provided that both main drain and branch can be completely rodded. A small number of manholes will enable main runs of drain to be rodded but branches from stacks or ground floor WCs cannot easily be rodded except from manholes at the junction with the main run. Gullies can be provided with special access points and it is easy to connect branches from them to runs of drain, without manholes at the junction, while ensuring that each length can be rodded. Inspection chambers are also provided at intervals along straight lengths of drain without jucntions. This is done to ensure that the whole length of drain can be satisfactorily rodded. Opinions vary about the satisfactory spacing of inspection chambers for this purpose. A survey of local authorities has shown that their requirements for manhole spacing varied between 23 m and 90 m. The Building Regulations 1965 require that no part of a drain shall be more than 45 m from an inspection chamber and that no junction, including junctions with private and public sewers, if not itself provided with an inspection chamber, shall be more than 12 m away from one.

Inspection chambers must be of adequate size to permit a man to work and also to allow the branches to enter sufficiently far apart not to impair the stability of the construction. Where manholes are very deep a working chamber of suitable dimensions is constructed, roofed by a concrete slab or brick arch, and this chamber is approached by an access shaft leading down from the surface. Table 74 shows minimum working sizes for rectangular manholes and gives details of sizes to allow for branches. Different considerations apply to manholes formed

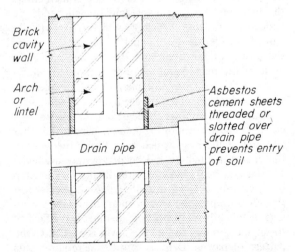

239 *Detail of provisions for drain passing through a wall (some uPVC drain systems have special fittings for passing through walls)*

DRAINAGE INSTALLATIONS

Table 74 Minimum sizes for rectangular manholes

	Depth m (to invert)	Minimum internal dimensions m			Thickness of walls mm	Thickness of slab mm
		Length	Width	Height above benching		
Shallow manholes (no access shaft)	Up to 0.6	0.75	0.7	—	115	100
	0.6 to 0.9	1.2	0.75	—	230	150
	0.9 to 2.4	1.2	0.75	—	230	230
Deep manholes (working chamber and access shaft)*	3.3 and over	1.2	0.75	2†	230	230–450

* Access shafts for deep manholes should be not less than 0.7 m x 0.7 m (corner step) or 0.85 m x 0.7 m (ladder). Manholes over 6 m deep should have a shaft provided with a rest platform every 6 m.

† Minimum cover opening size 0.6 m x 0.6 m.

Rules for sizing manholes with branches:

Length: Established by side with most branches on the basis of 300 mm per 100 mm diameter pipe or 380 mm diameter pipe plus an allowance for the angle of entry if this is not 90.
Width: Branch bends are designed for a 300 mm gap between wall and channel.

from circular precast concrete rings and manufacturers should eb consulted.

Inspection chambers are usually constructed of either brick walls with concrete base and slab or of precast concrete rings. They must be able to withstand soil pressures and be watertight both to prevent leakage of liquid into the surrounding soil and to prevent seepage of ground water into the drain. Table 74 gives the thickness of walls and slabs often employed in brick inspection chambers and diagram 240 shows the general arrangements in a shallow manhole. Class B Engineering Bricks in English bond with joints fully filled with cement mortar treated with waterproofer form a suitable wall. Internal rendering should be avoided. Frost causes the rendering to break away from the walls near the top of the manhole, falling into the channel and perhaps blocking the drain. Step irons as shown in diagram 240 must be provided if the inspection chamber is more than some 610 mm deep. They should be set at about 300 mm centres and alternately placed to right and left to suit the movement of feet of workmen climbing into the inspection chamber they may be bedded directly on to the top of the brickwork. Where, as is more usual, the cover is smaller than the manhole, a concrete slab should be used to support the cover. This gives a more permanent and satisfactory result than corbelled brickwork.

The main run of drain passes through the bottom of the manhole in an open half-round GVC channel, either straight through the manhole or as a bend. Branch drains shown in diagrams 240 and 241 are discharged into the main drain by means of open curved members known as branch bends, which rest on the edge of the channel and turn the flow so that it joins the main flow in the direction of its travel, irrespective of the angle at which the branch enters the manhole. Both channels and branch bends are bedded in concrete 'benching' which rises vertically from the edges of the channel to above the level of the crown of the drain and then slopes up to the walls of the manhole at a rise of 1 in 6. This allows a reasonable platform for men to stand while providing no lodgement for sewage in cases of surcharge.

Diagram 242 shows a precast concrete inspection chamber base incorporating the channels and branches and one of the rings which would form the walls of the chamber. Shallow and deep types are available. The base is heavy and requires lifting tackle. Step irons are ready fixed into the rings. A special top ring gives support for a rectangular cover. One advantage of this type of inspection chamber is the facility to set the top ring so that the cover is parallel to the pattern of paving. Delivery and lifting requirements mean that inspection chambers of this sort are most likely to be used on large contracts.

Diagram 243 shows a GRP inspection chamber

DRAINAGE BELOW GROUND

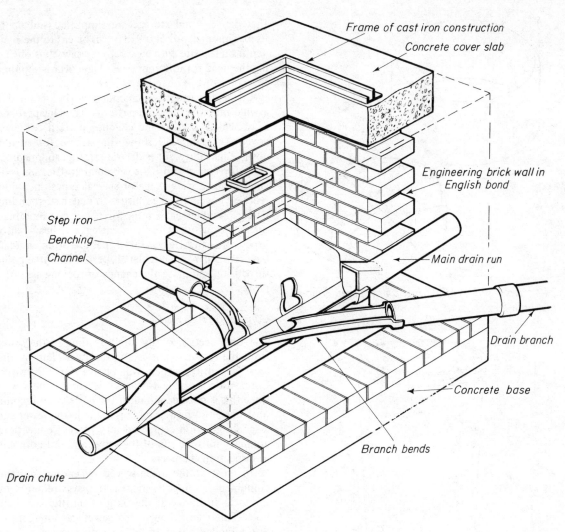

240 Cut away view of typical construction for a shallow brick manhole

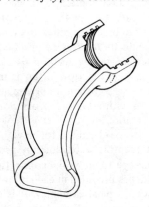

241 Branch bend for use in manhole to achieve junction between branch and main drain

273

DRAINAGE INSTALLATIONS

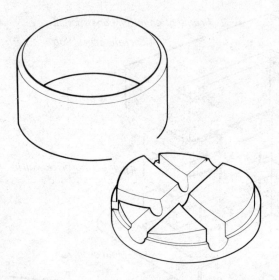

242 Base and typical ring for pre-cast concrete manhole

243 GRP (glass reinforced plastic) preformed inspection chamber for use with uPVC drains. (There are five inlets and one outlet for 110 mm dia drains. Inlets not required are fitted with stoppers)

and cover specially developed for use with uPVC drains. The inspection chamber is light and easily installed and the connection of drains is simple since the inlets and outlet of the chamber are equiped with rubber ring joints similar to those in the run of drains. The chamber is supplied in stacking form so that delivery and storage are simplified and up to its maximum depth of 0.9 m it can be cut to the right depth on the site with a carpenter's saw, thus obviating the need for different sizes. The cover is supplied specially for this chamber.

Covers for inspection chambers can be obtained in a wide range of sizes and patterns. The cheapest and most widely used is the chequer pattern cast iron cover. Diagram 240 shows the frame for such a cover. Light duty, for pedestrian traffic, and medium duty for trucks and light vehicular traffic, are available as are a wide range of special types, including recessed covers to take finishes to match surrounding paving, double seal covers for use inside buildings, open grating types for ventilating and heavy duty types for roads. Heavy duty types are often circular, but triangular or circular types with three-point suspension minimise the chance of rocking as traffic passes.

Connection to sewers

The final section of the drain leading to the sewer from the last manhole is almost always laid by the local authority, which will charge for this work. Connection is made towards the top of the sewer, since this is most economical and gives the minimum obstruction to flow. A hole is made in the sewer and a saddle junction cemented into place. Connections are made into the run of the sewer and not normally at inspection chambers.

In most of the areas served by earlier foul and combined sewerage systems a trap was required by the local authority at the outgo from the last manhole. This trap, known as a sewer gas interceptor, was intended not only to prevent sewer gases from entering the house drains but also to prevent the passage of rats. In fact the interceptor was a major cause of drain blockage, since any heavy solid matter would remain there. Present practice in areas served by new sewers, taking account of this problem and recognising that the flow of air is more frequently from drain to sewer than otherwise, is to make direct connections without interceptors. Some local authorities still require interceptors in new work and their use may also be desirable for new buildings in areas where interceptors were generally provided.

A cleaning eye to allow the connection to the sewer to be rodded is provided in each interceptor (diagram 244 shows a typical example). It is desirable to use the type of interceptor which has a cleaning eye

DRAINAGE BELOW GROUND

cover opened by a chain, the far end of which is secured near the cover. This arrangement enables the liquid to be drained easily from the inspection chamber rather than dipped out in the case of blockage. Where the interceptors are used a special air vent, not less than 75 mm in diameter, is required to the final manhole.

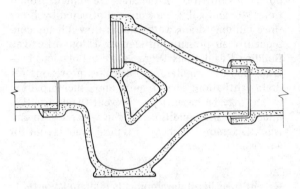

244   Sewer gas interceptor

Systems

It will normally be economical and desirable to discharge building drains to a sewer if one is available. Local authorities may require this if a sewer is within 30 m of the site or if they are prepared to bear the cost of any additional length of connection. In early sewage systems both foul and surface water drainage was discharged into the single 'combined' sewer. Purification plants capable of dealing with the foul flow cannot, even if it were economic to do so, deal with the flow in wet weather. When six times the dry weather flow was reached untreated sewage would be discharged direct. This is not a very satisfactory arrangement and more modern systems separate the foul flows, which are all treated, and surface water which is discharged untreated into watercourses. In country districts with suitable subsoil it may be convenient to use soakaways on the site to dispose of surface water and many local authorities will insist on this to keep down the load on existing combined sewers. The sewage system clearly affects the drainage layout. If it is a separate system then separate drains will be required for surface water and foul flows. If not the flows can be mingled in one set of combined drains. In some cases local authorities will have proposals for providing a new surface water sewer to relieve an existing combined sewer which will subsequently be used only for foul flows. They may then ask that the building drains are constructed as separate foul and surface water systems combined initially into a common connection to the sewer until the new sewer is built.

Hydraulic considerations

It is necessary, in the case of foul drains, to be able to introduce flows into the drain without nuisance arising from the escape of drain air, and for flows to take place without causing air pressure fluctuations which would destroy gully seals. In both foul and surface water drains the discharge capacity of the drain must be adequate to deal with the flows likely to take place.

Entry to drain:

Waste and soil stacks are connected to the drains normally at manholes directly by means of a bend below ground. The stack must be carried up to such a level that no nuisance can arise (usually above the eaves) from the escape of drain air.

Ground level WCs can be connected directly to the drain as shown in diagram 218. As described on page 250, there is no risk of siphonage of the WC trap.

Ground floor gullies and urinal traps can also be connected directly to the drain since the same considerations apply as in the case of WCs. Diagram 219 shows an external waste gully with waste pipe connection.

Rain water pipes discharging to combined drains must deliver into a trapped gully. In conditions of heavy rain the flow may be substantially greater than the foul flows previously described. The restrictions and bends, however, will not normally allow sufficient water to pass to give full bore flow in the 100 mm diameter outgo from a gully and consequently siphonage of the gully trap is unlikely. In any case rainfall flows do not end abruptly and a final period of limited flow which will not give rise to siphonage is certain. Where rain water pipes discharge to surface water drains, direct connection, or connection to an untrapped gully and thence to the drain, is feasible and normal, since no nuisance problems arise from the escape of air.

DRAINAGE INSTALLATIONS

Drain ventilation

Apart from the avoidance of air pressure fluctuations, drain ventilation is advocated as preventing the build-up of smells within the drain and as causing the drying and subsequent flaking off of deposits on the drain walls. It is questionable how critical the last two items are. In the case of air pressure fluctuations it is clear that the nearly horizontal drain pipes do not prevent such a critical problem as the vertical and horizontal ones of waste and soil installations. Traditional practice has been to provide a low-level vent at the last manhole (called the fresh air inlet) and a vent-pipe running up the building at the head of the drain. A soil pipe could be employed to act as this vent. Branches longer than 6.4 m were often not separately ventilated but special ventilation for such branches was considered good practice. Diagram 245 shows a simple system of this sort. The arrangement was based on the concept that warm drain air would rise up the high stack and be replaced by cold air entering the fresh air inlet. This would give a circulation of air through the drain. The fresh air inlet was provided with a mica flap intended to allow air to pass only in the desired direction. In operation this flap does not normally serve its purpose and the overriding influences on the movement of air are

1 Filling of the drain which forces air out at all vents, and

2 Entrainment of air in the direction of flow which causes an opposite movement to that originally envisaged

When interceptors are not employed the drain will be ventilated directly into the sewer where, under most conditions, a current of air is drawn in the direction of the flow. Although the number of points of ventilation of the drain are limited, ground floor WC and gully traps are not affected by flows since building drains very seldom flow full and consequently air pressure differences seldom arise. The Building Regulations 1965, unlike the previous bye-laws, do not stipulate any specific provisions for drain ventilation. They require only 'such provision . . . as may be necessary to prevent the destruction under working conditions of the water seal in any trap'. Diagram 246 shows a typical drain layout for a housing scheme.

Recent specialised developments in drain layout

The two systems described below were the result of extensive studies of the problems of drainage undertaken by manufacturers of uPVC pipes. Both systems are based upon reducing the cost of manholes which represent a substantial proportion of the total cost of drains. In both cases the principles employed could be used with drain materials other than uPVC.

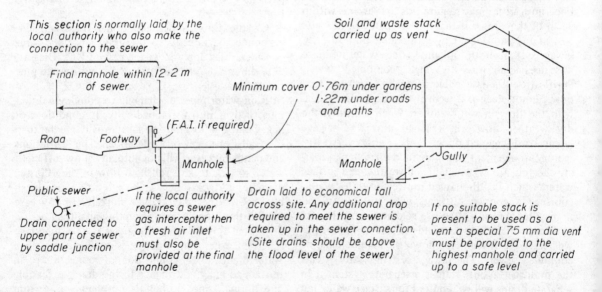

245 *Principles of simple drain layout and drain ventilation*

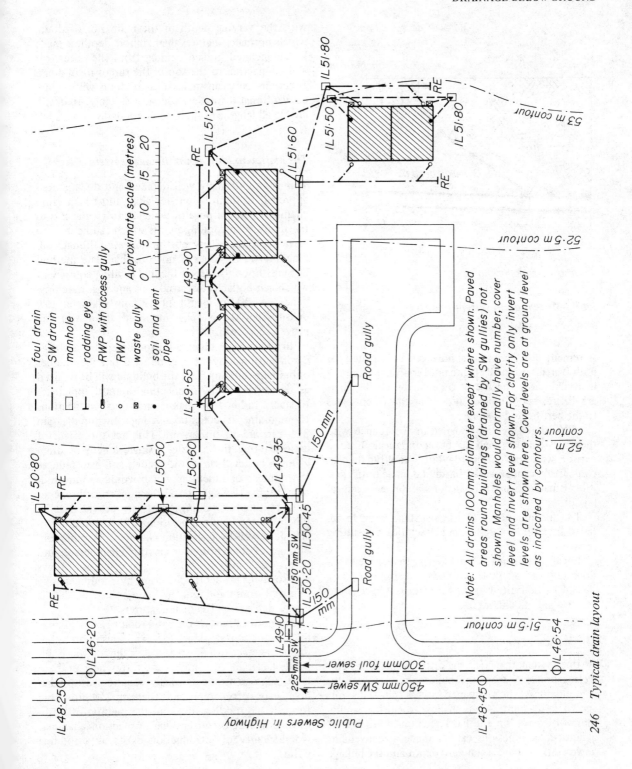

Typical drain layout

DRAINAGE INSTALLATIONS

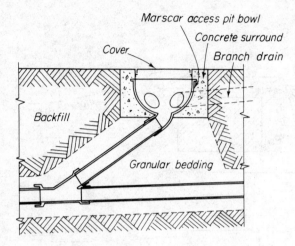

Building construction fig B

247 Marscar drain system

Agrèment certificates have, however, been issued for both systems using uPVC and until equivalent support is available for the use of other materials designers are likely, very judiciously, to restrict themselves to the certificated systems.

uPVC drain layouts designed in accordance with these systems are said to be economical in comparison with conventional layouts. An ultimate cost comparison can, however, hardly be made until traditional materials are used in comparably economical layouts.

The manufacturers of these systems provide detailed literature on design and a technical consulting service.

Local authorities often have particular views on drain falls and manhole positions and spacings. They should be consulted at an early stage if specialised systems are to be used.

The Marscar system

In this system the great majority of manholes collecting branches from the building are replaced by a standardised bowl shaped access pit. This pit, shown in diagram 247, is shallow, requiring a minimum of excavation and has knock out holes so that a number of junctions can be made at convenient angles. All pits have a standard depth and the linkage with the varying depth of the main run of drain, which normally dictates the manhole depth, is made by an inclined branch coming from the access pit and connecting to the top of the run of main drain. Junctions between main runs of drain will require conventional manholes, which, if suitably sited, will enable all lengths of drain to be rodded.

The Marley rodding point drainage system

This is an extremely well detailed and documented approach to drain layout replacing most inspection chambers, the purpose of which is to give access to the drain, with rodding points which can be used to clear the drain of any obstruction even though rodding in straight lines is not possible. The method depends upon the use of flexible rodding equipment which can both clear obstructions and negotiate slow bends. A 30 m length of 13 mm diameter low density polythene tube mounted in a reel has proved successful.

In preparing a scheme for rodding point drainage a conventional layout may be used but the number of branches coming from the building will be reduced as far as possible by connecting sanitary appliances to stacks rather than making individual connections to the drain. The collar boss fitting shown in diagram 211 can aid in the layout. This has the effect of concentrating flows along a single length of drain giving a better flush in the branch and also reduces the number of junctions which would require rodding points. Having reduced the number of branches to a minimum, rodding points are provided at all junctions and bends in positions which would normally be served by inspection chambers. A minimum number of conventional inspection chambers should be provided sited to allow visual inspection of flows and so that any debris dislodged by rodding can be removed from the drain. Very comprehensive details are available from the manufacturers.

Diagram 248 shows the basin rodding point. Bends in the horizontal plane and junctions with other drains are made just downstream of the rodding point bend. Diagram 249 shows how differences in drain level, such as would normally require a drop manhole, may be accomplished. Diagram 250 shows the form of the rodding tube and the method of introducing this at a conventional inspection chamber. No difficulty of introduction exists at a rodding point.

DRAINAGE BELOW GROUND

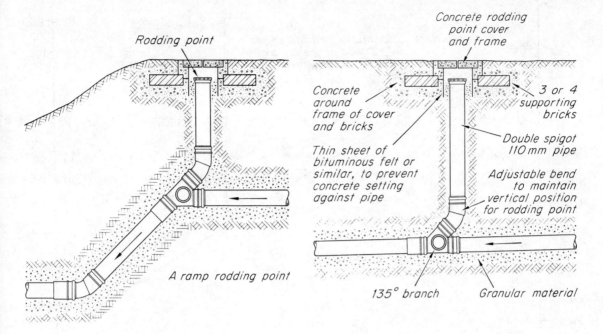

248 Marley rodding point

249 Ramp rodding point

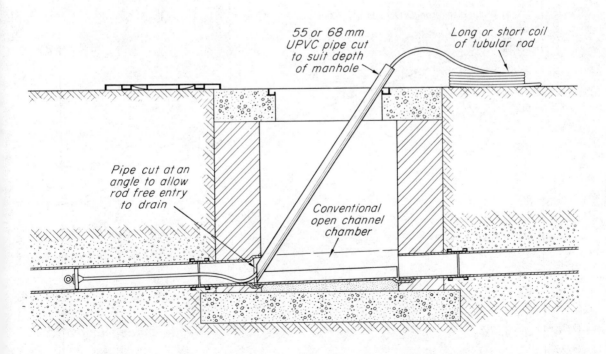

250 Rodding tube being fed into drain at a manhole in the Marley Rodding Point system

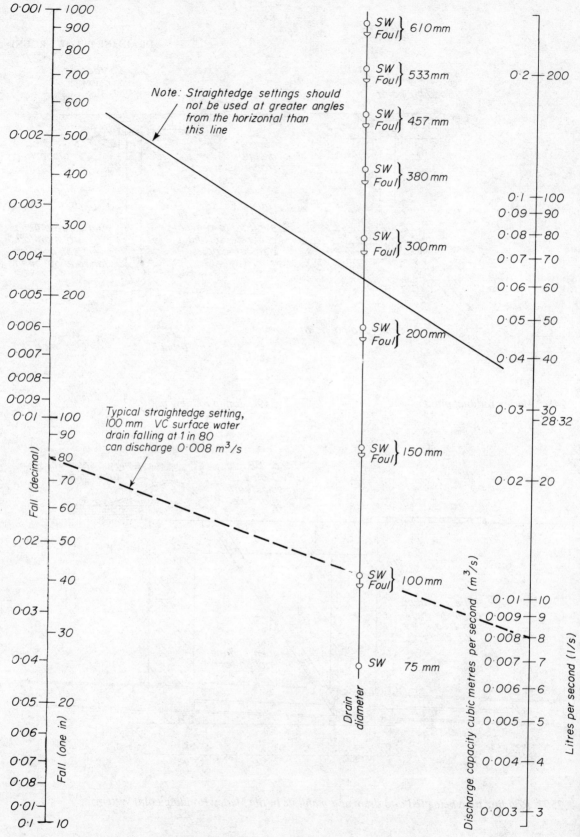

251 Nomogram for estimation of drain sizes

Discharge capacity and fall

The discharge capacity of drains depends on the diameter of the pipe, the fall or gradient at which it is laid and the nature of the inner surface and accuracy of alignment of the joints. Formulae which give actual discharge capacities sufficiently accurately to be useful for drainage design have been available for some 200 years. Diagram 251 is a nomogram, based on the Hydraulic Research Station's Research Paper No. 2, *Charts for the Hydrualic Design of Channels and Pipes,* which relates fall, pipe material and diameter and flow and thus enables the size of pipe for a given fall and flow, which is the information normally needed, to be established. It will be noted that SW and foul pipes are distinguished in the chart. This arises since the calculations for surface water drains assume that the pipes flow full, whereas in the chase of foul and combined drains the design standards is only 0.75 depth of flow, which allows some factor of safety and helps to prevent air pressure fluctuations. If the flows are known it is therefore a relatively simple matter to size the drains. The total fall available is usually established by the configuration of the site and relative levels of building and disposal points and the appropriate diameter of drain may be read from the chart. The minimum velocity of flow in drains recommended by BRE is 610 mm per second. This means that a minimum gradient is required for each diameter of pipe. The guide line on the chart will indicate when gradients are hydraulically satisfactory for drains in good condition. Local authorities, however, have their own requirements for minimum gradient, which vary considerably from place to place. A survey showed that the variation is from a minimum fall of 1 in 40 to one of 1 in 110 for 100 mm diameter drains. Larger diameters are laid to slacker gradients. This consideration may mean that the fall of a run of drain will vary as the diameter increases to take the flows from branch drains. Until recently it was thought that the maximum fall of drains should be limited to avoid stranding of solid matter and scouring of the pipe. This led to the provision of drop manholes on sloping sites. Views have changed on these points and up-to-date practice is to allow drains to follow the fall of the land where possible, thereby keeping the amount of excavation to a minimum. Drop manholes in these circumstances would only be employed to minimise drain excavation where branches have to be connected which are much shallower than the main drain. Diagram 252 shows a typical drop manhole. Care must be taken at the foot of a slope that the flatter section of drain is capable of discharging the full flow likely to come down the steeper section.

Table 75 Discharge units for estimation of flows in foul drains

Appliance	Frequency of use, minutes				
	5†	10‡	20 §	30	75
WC 14 litres	40	20	10		
9 litres	28	14	7		
4.5 litres*	12	6	3		
Lavatory basin					
with plug	6	3	1		
with spray tap and no plug	0.5	0.25	0.15		
Sink 38 mm waste	27	14	6		
Small sinks as LB					
Bath ‖				18	7
Shower ‖	Add 7 litres per minute per spray to the design flow				
Uring (per stall)			0.3		

*Where dual flush is used (see page 219).
†Corresponds to congested use with queueing.
‡Corresponds to normal commercial or office use.
§Represents the peak rate of domestic use (morning).
‖ Use does not contribute to morning domestic peak.

Estimation of flow

For surface water drains this presents no difficulty. The Building Research Station recommend that design should be based on a rate of rainfall of 50 mm per hour. To estimate the flow all that is required is to find the area of site and building which is contributing to the flow in the drain, multiply by 0.05 m to give the flow per hour in cubic metres and to divide by 3600 to give the flow per second for use in diagram 251. In sewer calculations more factors have to be taken into account. The intensity of rainfalls fluctuate over a period and it is possible that by the time an area some way up the sewer is contributing at its full design rate, nearer areas are producing a very much reduced rate

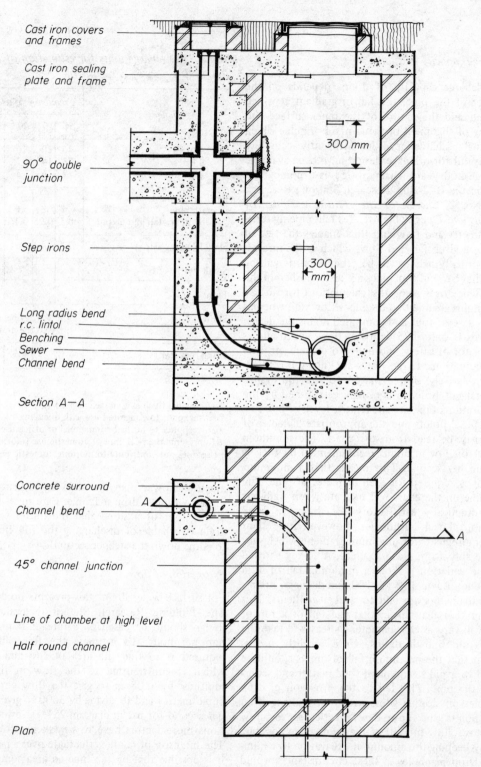

252 Construction of a drop manhole to allow shallow branch at high level to joint a main drain at lower level

DRAINAGE BELOW GROUND

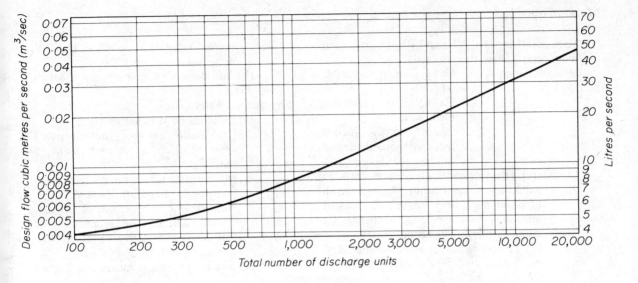

253 *Design flow for foul drains*

of run-off due to falling off of the rainfall intensity. This problem is very rare on building sites which are not, in the majority of cases, large enough for this phenomenon to have any significance. Similarly, it is found that rain falling on grass and other non-paved areas does contribute a small proportion of its volume to the drain. The flow is unlikely to coincide with peak flows in building surface water drains and these can most satisfactorily be based on the assumption that there is 100 per cent from roofs and paved areas.

Flows in foul drains present a more complex problem. Sanitary appliances operate intermittently and deliver varying quantities at varying rates of flow into the drainage system. It is difficult to decide what the peak flow will actually be. Table 75 and diagram 253 provide a means of estimating flows based on the application of theory of probability (see also *Water Supply* page 236). Each sanitary appliance has a number called a discharge unit which represents the rate of discharge, its capacity and its frequency of use. The flow in any section of the drain may be determined by summing all the discharge units of the sanitary appliances contributing to the drain and consulting diagram 253 to find the flow appropriate to the sum. Any continuous flows are added at their actual value.

Combined drains require separate calculation for surface water and foul flows. The drain size can be based on whichever flow is greater. The chance of peak foul and peak surface water flow coinciding is so remote that it is inappropriate as a design standard.

The foul drain flows given by this method assume that the flush from individual appliances is maintained as it passes down the drain. This is not the case and in fact the duration of the flow increases while the rate decreases as the flush passes further along the drain. The effect of this may be to increase the number of sanitary appliances which can be served by given sizes and falls of drain but this is not firmly established and no practicable method exists for taking decay of the flush into account. Until such a method is developed it is sensible to use the one described below.

Table 76 Step by step procedure for hydraulic design of drains

1 Prepare drain layout plan for foul and surface water drains showing all points of entry (stacks, ground floor WCs and gullies) drain runs, junctions, manholes and rodding eyes.

2 Mark cover levels at MHs and REs.

DRAINAGE INSTALLATIONS

Table 76 cont/d...

3 Mark invert levels at MHs and REs to give minimum cover. (Drain size assumed)
Minimum drain cover: Under gardens and fields 0.6m.
Under lightly loaded site roads 1.2m.
Under heavily loaded roads special calculation will be needed.

Check: (a) Drain cover is maintained between manholes. If cover is not adequate depth of drain must be increased. Resiting of manholes to coincide with changes of ground fall is usually economical.
(b) Surface water and foul drains clear each other. In case of need reassess levels. SW drains are normally laid to minimum falls and foul drains dropped to clear.
(c) Drain falls are not below minimum acceptable.(hydraulically or to satisfaction of local authority). If not reassess levels.
(d) Drains are clear of service runs and any other underground features.

4 Size foul drains:
(a) Establish flows in each section:
(i) Establish the type and number of sanitary appliances and the frequency of their use.
Consider which appliances will contribute to the peak flow. In some cases, more than one combination of appliances may have to be tested to establish which is critical (See table 15 for discharge units and frequency of use in various building types.)
(ii) Mark against each point of entry the total of discharge units for sanitary appliances contributing to peak flow in drain.
(iii) Mark cumulative totals of discharge units carried by each section of drain.
(iv) From diagram 253 establish flows appropriate to each section of drain and mark on layout.
(v) If there are any continuous flows (eg from manufacturing processes) add these to the flow in all sections of drain which will carry them.
(b) Size drain using nomogram in diagram 251.

5 Size Surface Water drains
(a) Establish flows in each section
(i) Measure areas of roof or paved areas contributing to each point of entry.
(ii) Calculate rate of run-off from areas contributing to points of entry.
Normally at rainfall rate of 50 mm per hour and mark on layout.
(iii) Mark cumulative totals of flow carried by each section of drain.
(b) Size drains using nomogram in diagram 251.

6 Size Combined drains: Peak SW and foul drain flows do not usually coincide and in most cases the combined peak may be assumed to be the same as the greatest individual flow rate. Care should be taken, however, to add any continuous foul flows to the surface water flow.

7 Check: That adequate cover has been maintained (See 3a) above).
That adequate falls have been maintained (See 3c) above).
That revised levels do not result in clashes between foul and surface water drains.

Note

For housing schemes CP 301:1971 suggests that to avoid inconvenience from blockages not more than 20 houses should be connected to a 100 (or 110 mm for uPVC) mm diameter drain and not more than 150 houses to a 150 mm diameter pipe. The statistics which lead to these figures are not made clear but it would be injudicious for designers not to take them into account in case blockages occur and the design is blamed. This consideration will influence both layout and sizing.

DRAINAGE BELOW GROUND

Drainage of basements

Provided a sewer at adequate depth is available no special problem exists in the drainage of basements. It is important to remember, however, that many sewers become surcharged at times of peak flow and water levels may rise in manholes. This could result in flows of sewage into a basement from the drain. Special gullies and traps exist to overcome this but they cannot be regarded as completely effective and if there is any risk of backflow it would be better not to have sanitary appliances in the basement. If their presence is unavoidable the flow can be collected in a sewage ejector and raised to ground level before being put into the drain. This arrangement will be inevitable if the basement is below sewer level.

Even where no sanitary appliances are present in a basement it is necessary to deal with seepages of water and water from leaks or draining down of pipework. These flows are catered for by draining ducts and boiler rooms to a sump in which, usually duplicate, sump pumps are sited. The sump pumps raise the water from the sump to a gulley at ground level. Diagram 254 shows a typical installation.

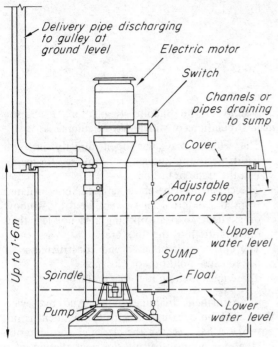

254 Sump pump installation

13 Sewage disposal

FOUL

Where public sewers exist the simplest and best method for disposal of flows in foul drains is directly into the public sewer and local authorities can, if there is a sewer within 30 metres of the site boundary, insist upon connection being made. If sewers exist but are further away than this, connection can still be required but the local authority may have to bear the additional cost involved.

Where there are no public sewers available arrangements must be made to dispose of foul sewage. The methods available are:

1 *Dilution* The sewage is discharged to a sufficiently large body of water.
2 *Conservancy* The sewage is retained on the site and periodically removed.
3 *Treatment* A plant on the site renders the effluent sufficiently innocuous to be discharged to a stream or to be allowed to percolate into the soil by means of surface and sub-surface irrigation.

Dilution

If the volume of water receiving a flow of sewage is sufficiently large, oxidation of the organic matter can be achieved by the oxygen dissolved in the water. In England it is not now considered appropriate to use this method with rivers since the population of the catchment areas would make necessary a much greater flow than is available. Many local authorities do discharge sewage, often after mechanical disintegration of the solid matter directly into tidal waters. In recent years this method has been increasingly questioned. It is in any case often not practicable for individual buildings, since even if a seashore forms part of the site boundary the outfall arrangements required would involve considerable planning and expenditure.

Conservancy

In temporary buildings chemical closets may be used. Their construction and siting is governed by the provisions of the Building Regulations 1965, the main stipulations of which may be summarised:

The closet may not open directly into an habitable room, kitchen or room used for manufacture, trade or business.
The closet must be either entered and ventilated from the external air or ventilated by window or skylight with an opening area not less than one twentieth of the floor area.
The closet must be sited and constructed so as not to cause pollution or nuisance.
There must be no outlet to a drain.

In permanent buildings not served by sewers there may be occasions when site treatment is not possible either because the size and configuration of the site do not permit the construction of a treatment plant and the disposal of effluent, or where the amount of flow is inadequate in volume and too intermittent to be treated successfully. In these cases a cesspool is appropriate. This consists of a watertight but ventilated underground container which can receive the flows from the drains and be pumped out at intervals. Local authorities normally provide an emptying service for which there is usually a charge. In order to reduce the cost involved cesspools are often allowed to overflow, or in suitable soils, even have their bases deliberately broken to allow seepage to take place. This practice is both contrary to the regulations and extremely undesirable.

Cesspools must be impervious not only to liquids inside but also to surface and subsoil water from outside. They must not be sited so as to give rise to risks of pollution of water supplies or nuisance or danger to health. In the building byelaws which preceeded the present building regulations arbitrary figures for appropriate separation distances of cesspools from houses, wells, etc, were specified. Under the present regulations, however, cases can be considered on their own merits. It is unlikely that distances smaller than those previously required will be acceptable to local authorities. Cesspools will normally be sited at a lower level than the building so that drains can be laid to fall and pumping will not be required. There must be ready access for cleaning and emptying without passing through any residential or trading or business premises which normally implies a road suitable for cesspool-emptying vehicle to a point close to the cesspool. Some vehicles carry sufficient suction hose to run for some distance but if this is to be depended upon in the siting of the cesspool the authority should be consulted to see what maximum run and lift is possible with their vehicles. Diagram 256 is of

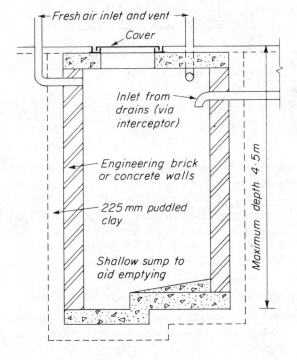

256 *Typical cesspool*

a typical cesspool, showing statutoty and other desirable features.

The capacity of a cesspool is often based on the estimate of 45 days flow. This is considered to give a reasonable frequency of emptying. For dwellings 0.11 m³ to 0.14 m³ per person per day is used to determine the input. The minimum capacity should not be less than 18 m³ and it is unusual to have cesspools larger than 45 m³. In fixing an actual size the capacity of the cesspool-emptying vehicle is often taken into account so that an exact number of loads will give the most economical emptying costs. Capacities of vehicles vary — 3.5 m³ is very usual. The local authority should be consulted.

Treatment

Although various chemical and electrical means of sewage treatment have been developed, virtually all sewage treatment plants in England, both large local authority and small private, are based on the system of first removing suspended solid matter by means of sedimentation, settling or septic tanks; then oxidising the organic matter still contained in the liquid by means of biological agencies and finally discharging the final effluent to a watercourse, or if this is not possible, by surface or sub-surface irrigation. The sludge accumulating during the first stage is removed periodically (6 months is the usual interval) and either dumped or used directly or in dried form as fertiliser. Large installations with constant supervision can have many refinements such as screening to arrest foreign objects, maceration of solids, chemical treatment to encourage flocculation and precipitation of solids, sludge-drying beds or digestion plant, facilities for recirculating effluent if desired standards have not been achieved and more sophisticated procedures and design, but the general principles of operation are essentially similar. The following notes are mainly concerned with small plants such as might be used for a single building or small group of buildings.

Septic tank

The first stage of treatment in small plants is traditionally termed the septic tank. Sewage is allowed to stand in the septic tank, which will usually have a capacity of between 16 and 48 hours flow. Sludge will settle to the bottom and scum form on the top and a clear liquid called liquor will overflow as new flows come in. To some degree digestion may take place. This process is a breaking down of the organic content by means of the anaerobic bacteria which can thrive under the conditions of a septic tank. This process reduces the quantity of sludge and renders the odours less offensive. At ambient temperatures this process can occupy a period of two months or more, consequently it can only be partially effective in a septic tank. In large sewage works special digestion facilities can be provided which have not only the advantages mentioned above but also can result in the production of useful

SEWAGE DISPOSAL

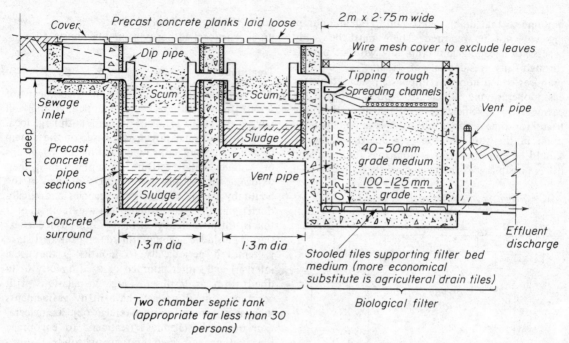

255 Typical small private sewage treatment installation. Dimensions are based on a population served of 10 people. The text gives details of sizes required for other populations

quantities of sludge gas which can be used for power and heating.

The size of septic tanks is governed by a number of factors, they must be sufficiently large to ensure that the contents are not noticeably disturbed by flows entering. This will fix the minimum size at about 3.5 m³. The tank must be large enough to allow the sewage to remain in it for an adequate time. This is usually taken to be between 16 and 48 hours. The tank must also be large enough to contain the accumulation of sludge which will take place between emptying without restricting the necessary capacity. 0.8 litres per person per day is the volume of sludge used for design. CP 302 : 100, *Small Domestic Sewage Treatment Works,* gives a formula for sizing septic tanks which are to be emptied at six-month intervals. In SI units the formula is capacity (m³) = number of persons in full-time residence × 0.14 + 1.8.

Septic tanks must be designed to allow flows to enter and leave without being affected by scum and to allow a gentle passage of liquid without short circuiting. Diagram 255 shows a septic tank for 10 people. The features of this installation are appropriate up to populations of about 30 people.

It is important to bear in mind that in the building regulations septic tanks are treated as cesspools so far as siting, emptying, and construction are concerned.

Biological filters (often called percolating filters)

Although referred to by tradition as filters, in fact biological filters achieve oxidation of organic compounds, not filtration. The purpose of the material contained in the bed is to present a large surface area over which the liquid will spread and be exposed both to the air and to the action of bacteria which form the main agency for oxidation. To function properly therefore a biological filter must have an adequate volume in relation to the flow so that the liquid will be adequately distributed; the bed must be well ventilated and, to ensure that the liquid remains in the bed for sufficiently long, the depth must not be reduced below 1.2 m and preferably not below 1.8 m. A spreading device is required to distribute the liquid over the surface of the bed. This usually takes the form of a series of channels with notched edges fed by a trough which tips when the liquid in it reaches the critical level, or by a rotating arm distributor which is operated by the waterflow but requires a higher head.

For convenience biological filters are usually sited close to the septic tanks which serve them, although this is not essential. Unlike the septic tank, the outgo from which is at almost the same level as the input, a considerable loss of height occurs in the passage of liquid through a filter bed (nearly 2 m even in the shallowest cases). From both the points of view of ventilation and disposal of effluent it is desirable that the base of the filter should be as near ground level as possible. On sloping sites this is easily achieved. On flat sites it may be necessary to raise the flow to the filter by pump.

The materials forming the filter bed must be durable, strong enough to resist crushing, and frost-resistant. Hard-burnt clinker, blast furnace slag, gravel or crushed rock are commonly used, clinker and slag giving the best results because of the greater surface areas presented. Two sizes of medium are usually employed. At the bottom of the bed over and around the drain tiles, medium of 100-125 mm grade is desirable for about 0.2 m, while above this the rest of the bed can be formed of 40-50 mm grade.

The volume of material depends on its surface area, the strength of the sewage, etc. For small domestic plants from 0.75 m^3 per person for up to 10 persons, reducing to 0.5 m^3 for larger installations, is recommended by CP 302 : 100, with a further reduction to 0.4 m^3 per person for part-time occupation.

It is neither necessary nor desirable to cover filter beds except by a wire mesh to exclude leaves.

Diagram 255 shows a biological filter using a tipper and channels. In diagram 255 advantage is taken of a sloping site. In some cases the sewage may have to be lifted by pump. The means of ventilating the filter beds will be noted.

In large installations many refinements to the oxidation process are found. The most notable, and one which is widely used is the activated sludge process. In this method, instead of exposing a thin film of liquid to air over bacteria-laden filter medium, the bacteria are introduced to a volume of fresh liquid by means of sludge retained from the last treatment. Air is introduced either by blowing compressed air through the liquid or by means of paddles which agitate the surface and encourage absorption of oxygen. The method requires continuous attention and has a considerably energy consumption. On the other hand the process can be continued until a satisfactory effluent is achieved. A degree of control not possible with the normal filter bed where, if the first passage does not give satisfactory results, the effluent must be reprocessed. Considerable space is saved, the plant is not liable to harbour flies and the loss of head through an activated sludge works will be significantly less than a filter bed type which may save pumping in some cases.

Small packaged plant is beginning to come into private use.

Humus chamber

The effluent from a biological filter can be greatly improved if it is given a final treatment to remove the humus which is a by-product of the bacterial action in the filter bed. For small installations without regular supervision the best method is to distribute the flow over an area of rough grass or scrub. 1 to 3.5 m^2 per person may be needed, depending on the nature of the soil.

Provided regular weekly maintenance is possible a better method is to provide a chamber similar to the septic tank but of about one-quarter the capacity. Special care is desirable to achieve smoothly flowing, low velocity input and outgo. Humus chambers should be cleared once a week. The humus can be used as a fertiliser, being very much better suited to this purpose than the sludge from the septic tank.

Final disposal

If a watercourse is available and the approval of the appropriate authority can be obtained, this is a simple and effective method. Diagram 257 shows a method of discharging an outfall drain to a watercourse.

Where no watercourse is available effluent may be disposed of below ground provided the soil is reasonably permeable and the water table does not approach too close to the surface (1.5 m is probably a minimum). A soakaway may be used but a system of agricultural drain tiles similar to that described under land drainage is considered superior. The total length of drain and area of land is difficult to estimate and local experience should be taken into account. Areas required can vary between 1 and 4 m^2 per head of population.

In siting sub-surface irrigation disposal systems care must be taken to avoid pollution of water

SEWAGE DISPOSAL

supplies which, in some circumstances, can take place over considerable distances. Water supply undertakings should be consulted when siting this type of installation.

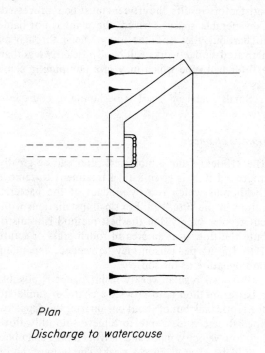

Plan
Discharge to watercouse

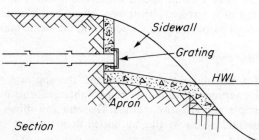

Section

257 *Details of outgo from sewage disposal system to a watercourse*

Local authority standard details

In many areas local authorities have prepared standard details of sewage disposal plants which they consider particularly appropriate for use in their areas. It is clearly desirable to make enquiries of the local authority before embarking on the design of sewage disposal plant.

SURFACE WATER

Public sewers represent the easiest method of surface water disposal but in cases where they do not exist or where connection is impracticable, soakaways may be employed. In some cases combined sewers may be overloaded due to extensive building development. Local authorities may, as a consequence, call for site disposal of surface water to minimise further overloading.

Permeable soil and a water-table some way below the surface are needed for soakaways to be successful. In many cases a pit is dug, filled with rubble and the drain arranged to deliver water into the rubble, after which the whole is backfilled. This system, although it often enables site disposal of debris, cannot be recommended for soakaways. Silting up is possible but cleaning is not and the reservoir capacity is reduced. A more satisfactory method is to construct a chamber lined at the sides with perforated, or open-jointed walling. In some types of soil no base is needed but a cover slab with access for inspection and cleaning should be provided over the soakaway. Where the soil is not satisfactory, or where cleaning may have to take place, a concrete base is desirable. If no base is provided, some cushion to receive the inflows of water should be incorporated to prevent erosion.

The proper size of soakaways depends on the permeability of the soil, the area to be drained and the intensity of rainfall anticipated. Permeability can be tested but the capability of the ground to absorb moisture can vary from place to place and the varying level of the water-table with the season of the year can affect the rate of percolation. Many successful soakaway installations have been based on a capacity equal to 12 mm depth of water over the whole area being drained. It is clearly sensible to site soakaways so that any overflow will not affect the building. Many authorities consider that several small soakaways are better than fewer larger ones.

BRE Digest 151 Soakaways deals with the construction and sizing of soakaways, describes a method of testing soil for rate of percolation and gives a chart for sizing soakaways.

Land drainage

On some building sites it may be desirable to reduce the level (called the water table) of the subsoil water or to improve the drainage of surface water. The need may arise because of the planting or horticultural use of the site, because it is important to reduce surface flooding or because for health (damp penetration of cellars) or structural reasons it is necessary to keep down the level of ground water near the building.

Before embarking on land drainage works the actual ground water levels should be investigated by means of trial holes. The water table is usually at its highest in the spring.

Sub-soil drains are usually laid 0.6 to 1 m deep and the pipe layout will be arranged to follow the falls of the land. The spacing between drains will vary between 10 m for clay soils to 50 m for sand.

Diagram 258 shows how subsoil drains lower than the water table and demonstrates how drains may be more widely spaced in soils which allow water to permeate freely.

Sometimes pipes are laid round buildings in what is described as a moat system, intended to intercept the flow of water towards the foundations for more extension areas. However, it is usual to employ a 'herringbone' pattern of pipes leading to main runs. The length of the subsidiary runs is normally limited to 30 m. It is not possible in most cases to make an estimate of likely flow. Pipe sizes are determined by connection. For building application 75 mm diameter for the subsidiary drains and 100 mm for mains are usual sizes.

Sub-soil drains have traditionally been formed from but-jointed and usually porous pipes, laid in narrow trenches. The tops of the pipes protected by turf or building paper to prevent soil penetration and rubble, or broken stone laid around the pipe to aid in water collection. In cases where it is desired to catch water running on the surface, the rubble

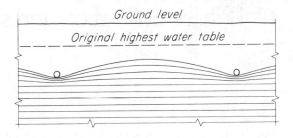

258 Reduction of water table by means of sub-soil drains

fill should be carried up nearly to the top of the trench, the top spit only being filled with top soil. Polythene and PVC pipes are now being used extensively for subsoil drains. They are provided with holes on the underside, which allow water to enter while minimising the penetration of soil which is a problem with normal drain tiles.

The outflow for laid drainage systems will normally be discharged to ditches or watercourses.

Sometimes the pipes are dispersed within land drainage work. A trench filled with rubble or broken stone (called a french drain) will give passage for water and is particularly effective at dealing with flows on the surface. In clay sub-soils 'mole drains' can be used. They are formed by a plough drawing a cylindrical cartridge through the soil. A tube is formed in the soil which can remain operative for many years. This form of drainage does not have a life comparable to buildings and its use is more appropriate to agriculture.

BSCP 303, *Surface Water and Sub-soil Drainage*, contains a table of sub-soil drain depths and spacings for various types of soil.

14 Refuse collection and storage

Bins

The traditional, and still the cheapest, method of refuse storage in buildings is the dustbin. If adequate capacity is allowed, the placing of the bin, arranged satisfactorily both for the user and the dustman, protection from the rays of the sun provided, adequate ventilation ensured and an impermeable and easily swept standing place provided, dustbins form an acceptable means of refuse disposal for many types of building. The noise inevitably associated with galvanised steel bins is largely overcome by the use of rubber and plastic bins. Small blocks of flats, provided they have lifts, can be served acceptably by dustbins placed conveniently at ground level. To simplify collection and make mechanical handling possible larger communal containers are sometimes used in these circumstances. Special collecting vehicles are required to handle and empty the containers which are very much larger than dustbins. About 0.08 m^3 of storage per dwelling is required for a weekly collection. British standard dustbins occupy a space about 460 mm in diameter and 610 mm high and hold 0.09 m^3. It is worthy of note that open fires and solid fuel domestic boilers can be used for some refuse disposal and their absence in many new dwellings add to the problem.

Paladins

The paladin, shown in diagram 259b, is a very large cylindrical bin on wheels. It has a capacity of almost 1 m^3, the equivalent of 10 dustbins. A special collection vehicle is required but most local authorities are equipped to collect paladins. Their use is in commercial and industrial premises but they are also used in blocks of flats.

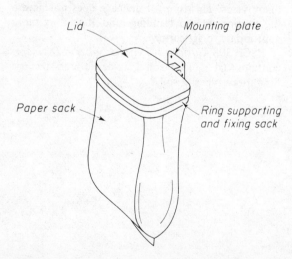
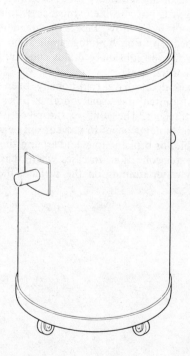

259a Typical paper sack for refuse collection and mounting

259b Paladin type refuse container

REFUSE CHUTES

Refuse chutes

In high residential buildings it is not practicable to require that all rubbish is carried by the occupants to containers at ground level. One way of overcoming this problem is to provide a chute running down the building which will take refuse from the upper floors and deliver it to a container at ground level. Diagram 260 shows a typical refuse chute. Refuse, preferably wrapped in paper, is introduced by means of hopper flaps at each floor level and falls into the container which is housed in an enclosure at ground level. The chute is carried up to roof level and vented. Refuse chutes have an influence on planning. They must rise vertically through the building and dwellings will often be arranged in handed pairs to make maximum use of chutes. Delivery points are often arranged on communal landings. This sometimes results in nuisance from careless handling of refuse. Delivery points within the area of the dwelling encourage careful use and maintenance. The kitchen, although convenient, is not acceptable from the point of view of hygiene. Delivery points on a private balcony have much to recommend them. The ground level containers must have access by a special refuse collection vehicle, being larger than the chute itself they may affect the ground floor planning. Complaints of smell often arise from dwellings near the containers.

Chutes are constructed of large-diameter pipes. 380 mm has been usual but 450 mm is the present trend, which reduces the risk of blockage. GVC, concrete or asbestos cement pipes are used, set into a builders' work duct with specially designed hopper units for delivery. Provision should be made for hosing down the container enclosure but not for washing the chute itself since this appears to make things worse rather than better. The possibility of a fire being started in the refuse must be borne in mind and also the noise which is inevitable when solid items fall down the chute and perhaps break in the container. 0.14 m^3 of container capacity per dwelling for weekly collection is thought appropriate to this method. Any type of refuse can be dealt with provided it will go into the hopper. Larger items will have to be carried down by the occupants to a special communal container.

Chutes are used mainly for domestic buildings. Offices present special problems in relation to

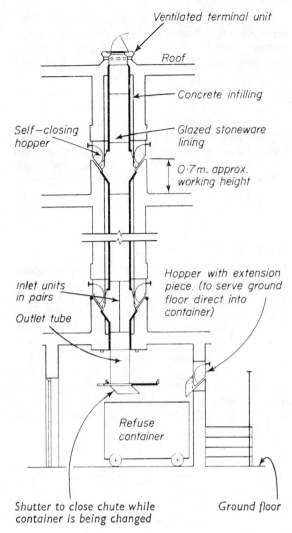

260 *Typical refuse chute installation serving a block of flats*

Paper sacks

The labour involved in collecting refuse can be substantially reduced by using paper sacks rather then bins and since the sacks are not emptied, collection is considerably more hygienic and simpler vehicles can be used. The cost is at present high and hot ashes must not be put in. Increasing use is, however, being made of this method. Diagram 259 shows a typical paper sack and mounting. The planning and hygienic considerations for bins apply equally to paper sacks.

REFUSE COLLECTION AND STORAGE

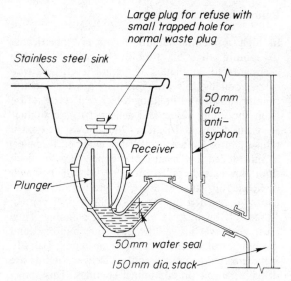

261 Section through the sink unit of a Garchey refuse disposal system. Note the receiver for refuse, the plunger for disposal and the trapped outgo leading to the stack. (The stack must take waste flows only)

their use, having large volumes of paper waste with a high fire risk but little putrefiable matter.

Garchey system

This system was developed in France but British rights are held by a British firm. In this system refuse is introduced into a receiver under the kitchen sink through a large plug in the sink. A smaller plug inset in the larger one deals with ordinary sink waste discharge. When sufficient refuse and water have accumulated in the receiver a plunger is raised and the contents of the receiver are discharged through 150 mm or 180 mm diameter cast iron drain-pipes running down the building. Waste appliances, but not soil ones, may also be discharged into this stack. Diagram 261 shows a sink unit. In the original system the refuse was collected in pits near the foot of each stack and then drawn by suction to a central plant for dehydration and burning in a furnace intended to produce heat for laundry use. The central plant is expensive, limiting the use of the system to large installations, and the utilisation of the heat from burning not always very effective. The British patentees have therefore developed a collecting vehicle which replaces the central plant and itself draws the refuse from the collecting pits. The vehicle is very much cheaper than the plant and can serve a number of buildings. Provided the vehicle is available quite small buildings can be economically served by this method.

Any type of refuse can be handled, including bottles and tins, but the size is limited. Large items must be carried down and deposited in a communal container. The system is very much more expensive than chutes or bins but is much more hygienic and gives a high proportion of satisfied users.

Grinders

Diagram 262 shows a grinding unit fitted under a sink which can reduce organic waste to fine particles suspended in water which are carried away in the normal drainage pipework. Tins and bottles cannot be dealt with and normal bins must be provided, but all refuse which could putrefy can be disposed of in this way. In individual dwellings or small buildings where chutes or Garchey systems would be impracticable grinders are the only way of improving on the service given by bins. Care must be taken when siting the

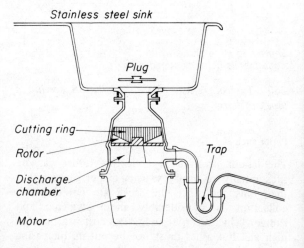

262 Sink grinding unit for refuse disposal

grinder and when mounting and connecting it to minimise the risk of trouble from noise.

Incinerators

Small incinerators for particular purposes such as hospitals and women's lavatories are widely used and are sometimes installed in blocks of flats without solid fuel heating appliances. Larger incinerators capable of dealing with all the refuse are often employed in America in domestic buildings, frequently in conjunction with refuse chutes which automatically feed the incinerator. In this country at present on site incineration of refuse is largely confined to hospitals and industrial premises.

Compactors

One of the problems of modern refuse is the low density and large volume. Small compactors, similar in size to a domestic refrigerator, are now available. Refuse is powerfully compacted into solid blocks contained in plastic bags. Small compactors are electrically driven.

Non-domestic premises

Trade refuse presents a special problem to be considered in each case on its merits. Apart from this—unless incinerators are used—refuse disposal from buildings other than dwellings is usually by bin or container. The general considerations involved are identical with those for domestic premises and have been described. The main problem with non-domestic use is in establishing a proper volume of storage to suit the accumulation of refuse and the frequency of removal. It is not possible to lay down general rules and careful thought should be given to the number of bins that will be required and to the possibility of increasing them if necessary. Many buildings are rendered unsightly and unhygienic by inappropriate and inadequate spatial provision for bins or containers.

BRS Digest 40 (Second Series), *Refuse Disposal in Blocks of Flats,* describes a study of user experience of several systems of refuse disposal and gives useful tables of capital and running costs.

15 Electricity

GENERAL PRINCIPLES

Electricity for supply to buildings and industry is generated in power stations sited conveniently for fuel supplies and cooling and distributed by means of the national grid which serves the whole country by means of overhead conductors. It is economical to distribute electricity at high voltage and the grid operates at 132 000 volts. For distribution in a particular area transformer stations convert the current to lower voltages which are supplied direct to some industrial users and to some buildings having their own transformers. 6600-volt distribution is used in many areas but is being superseded by 11 000 volts, which gives more economical distribution. For the majority of buildings, however, the supply voltage is further reduced by local transformer substations each serving an immediate locality with a supply described as three-phase, four-wire, 415/240-volt, 50 cycles per second. In this system an electrical current whose voltage alternates from positive to negative and back 50 times each second is distributed by three wires, described as the phase wires, and a neutral wire which is earthed at the substation. The voltage between any two of the phase wires is 415 volts and between any phase wire and the neutral, 240 volts. Diagram 263 shows the voltage relationship between the wires.

Small buildings are supplied with electricity by two wires, one phase wire and the neutral. This is known as a single-phase supply and gives a voltage for the premises of 240 volts. The loading of the supply wiring is balanced between the phases by using the phases in rotation so that each one serves every third building. In larger buildings it is more satisfactory to bring all four wires into the building. This is known as a three-phase supply. The balancing of the load on the phases is then achieved by serving different areas of the building by different phases. Electric motors, except for very small loads, are usually designed for three-phase operation. The standard of insulation required for 240 volts is not the same as that for 415, and in buildings with three-phase supplies care has to be taken to keep the

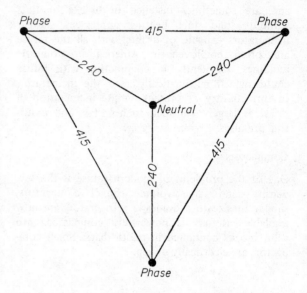

263 Voltage relationship between wires in a three phase four wire 415/240 volt electricity supply

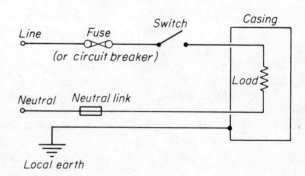

264 Schematic diagram showing the main features and safety precautions in a basic electric circuit

areas supplied by each phase distinct as otherwise special safety precautions become necessary.

The risks of electrocution and fire must be guarded against in all electrical installations. The precautions which are established practice are shown in principle in diagram 264. They are:

Insulation

All conductors are covered with insulating material or supported on insulators within an earthed casing with a clear air gap round each conductor. Standards of insulation vary with voltage. If a number of wires carrying different voltages are enclosed in a trunking they must all be insulated to the standard of the highest voltage. To avoid this, systems at different voltages are usually run in separate trunkings (eg GPO wiring and electricity supply wiring).

Fusing

Each section of wiring must be protected by having in the circuit a fuse wire which will melt if a current passes higher than that which is safe for the wiring. This prevents overheating of wiring with the possible risk of fire. Fuses may be of the traditional type where a fuse wire is stretched between terminals in a ceramic holder, or of the cartridge type where the wire is held in a small ceramic tube with metal ends. The cartridge types of fuse are much easier and quicker to replace. It is also possible to use circuit-breakers instead of fuses. These operate by thermal or magnetic means and switch off the circuit immediately an overload occurs. They can be reset immediately by a switch. They are more expensive than fuses but have the advantage that they can also be used as switches to control the circuits they serve. They are particularly valuable in industrial uses where circuits may become overloaded in normal operation and the circuit-breaker will switch off but may be switched on again immediately the overload is removed. In domestic circumstances the blowing of a fuse or operation of a circuit-breaker usually indicates a fault.

In early electrical installations both ends of each circuit were provided with fuses. This practice is now no longer employed, although many existing buildings are equipped with it, single-pole fusing is used. The difficulty with double-pole fusing was that if the fuse in the neutral side blew, through a fault in itself or because of long operation, while the fuse in the phase side remained intact, electrical apparatus would not operate, although it would remain live. Anyone attempting to dismantle or inspect the apparatus might then receive an electric shock. Modern practice is therefore to provide a fuse, or circuit-breaker, at the phase end of the circuit (called the line) and a simple link at the neutral end.

Switch polarity

The position of the switch has the same effect upon safety as that of the fuse. If the switch is fitted on the neutral side of the apparatus this will always be live even when the switch is turned off. Switches are therefore always fitted on the phase side of the apparatus they control.

Earthing

Any metalwork directly associated with electrical wiring could become live if insulation frayed or if wires became displaced. Anyone touching such a piece of apparatus would run the risk of serious electric shock. This is avoided by earthing the metalwork so that a heavy current flows to earth and the fuse is blown immediately the fault occurs. Although the neutral wire is earthed it will not serve for this purpose and a separate set of conductors for earthing are provided in almost all electrical installations. The earth connection itself is made locally in the building. A water pipe supplying the building is often used, although some water companies do not approve, particularly where corrosive subsoil conditions exist, and care must be taken that the water pipes are of metal. Plastics and asbestos-cement pipes are increasingly used as service pipes and water mains. In the absence of a satisfactory pipe the electricity supply authorities' cable sheath may be used or copper plates or rods buried in the ground. In cases where satisfactory earthing is difficult an earth leakage circuit-breaker may be used. This cuts off the supply to the circuit as the result of a small flow of current to earth which would not be sufficient to blow the fuse and so overcomes the problem.

Bathrooms

In bathrooms and similar situations, where water is present to ensure good electrical contact, and metal fittings or wet concrete floors provide a good passage to earth, special safety precautions are called for. No socket outlets may be provided in the area at risk and no switches must be within reach. They must be sited outside the area, or on the ceiling, operated by non-conducting pulls. Where any electrical apparatus is present, all metal, including not only the casing of electrical apparatus but also pipework, baths, etc, must be bonded together electrically and earthed.

ELECTRICITY

BASIC WIRING SYSTEMS

Wiring in buildings is run either on the surface or concealed in the construction. Surface wiring is cheaper but its appearance limits its use to industrial conditions or alteration work where cost is a prime consideration. The types of wiring system available for use in building are:

Sheathed

Two or more wires consisting of metal conductors each having its own insulation, are enclosed in a protective sheath. Traditionally the insulation and sheathing was rubber and the system referred to as TRS (tough rubber sheathed). Recently PVC has been used as insulation and sheathing, giving a slightly smoother and neater cable. This type of wiring is well suited to surface use and can be concealed in timber floors and joinery. It must be protected by conduit or metal channels, where it is to be covered by plaster or screed. Lead-sheathed cables are available where the sheathing can be used as an earth conductor. It is not now widely used for small buildings.

Conduit

This is the system most widely used for electrical distribution in buildings. A system of tubing is laid to the points where electricity is required and cables insulated with either rubber or PVC are subsequently drawn through. Conduit systems may be used either on the surface or concealed in the construction. One of the principal advantages of the conduit system is the ease with which modifications and rewiring may be carried out by drawing in new cables. Conduit is available in a variety of materials. Steel is the most widely used but aluminium may be employed and some plastics systems are available. Plastics conduit are usually flexible and easy to fix and airtight, which protects against condensation, but an additional cable is required for earthing, which in metal conduit is achieved by the conduit itself. The best metal conduit is of drawn tube with screwed joints; this gives good earth continuity and makes a waterproof system. Lighter tubes made from strip metal with welded longitudinal joints, or even open seams, are available with grip rather than screwed joints. These types give less satisfactory earth continuity and less protection from damp.

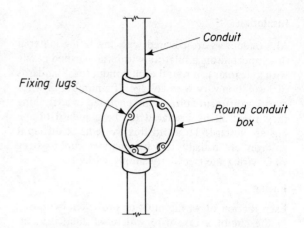

265 Electrical conduit showing box

Condensation may occur in the conduit. Ranges of bends and fittings are available for most conduit systems, which include special boxes to take switches, ceiling roses, socket outlets, etc. Diagram 265 shows a conduit box.

Conduit is normally laid on the surface of the slab in concrete floors and covered by the floor screed. In walls the thickness of plaster is not usually enough to contain the conduit and vertical chases in the bricks or blocks have to be made so

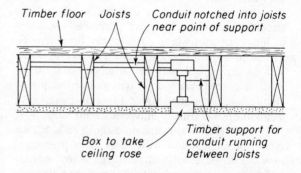

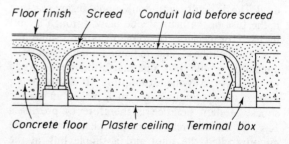

266 Conduit installed in timber and concrete floors

DUCTS FOR ELECTRICAL DISTRIBUTION

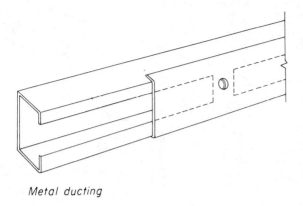

Metal ducting

267   Typical metal cable trunking

that there will be sufficient plaster depth over the conduit. In timber construction, floor joists and wall nogging pieces are notched to take the conduit. Diagram 266 shows conduit installed in a timber and in a concrete floor.

Conduit gives a good mechanical protection to wiring and can itself be arranged to span from point to point, whereas the other basic wiring systems require frequent support.

MICC cables (mineral insulated copper covered)

These cables consist of one or more copper conductors embedded in powdered magnesium oxide and sheathed with copper. The magnesium oxide must be protected from damp and cable ends are sealed with 'pots' containing an insulating sealer. MICC cables can give very high quality installations, they are heat resisting and consequently can be used with underfloor heating. They can be buried in concrete and since they are small in diameter compared with other types of cable, they can often be run in joints of brickwork, or in plaster, without charring the wall. Mineral insulated cables are also available in aluminium.

WIRING SYSTEMS FOR LARGER INSTALLATIONS

Cable trunking

Round conduit is manufactured in sizes up to 62 mm diameter, but the expense of bending and threading the tube and providing junction boxes is such that above 25 mm diameter a rectangular metal duct with continuous removable cover may well give a more economical installation. The cables are simply placed into the duct and the cover fixed on. Outlets from the trunking are made very easily since a hole may be drilled at any point to take a branch conduit. Metal trunking is available in sizes 35 × 35 mm to 100 × 230 mm. Diagram 219 shows a typical metal cable trunking.

Cable tap system

Where frequent and perhaps additional future connections to a large main cable are needed it may be economical to employ a cable-tap system where, in a standard metal trunking, a large capacity three-phase four-wire supply is supported, each cable separately, on clips. Where connections are required the insulation is stripped from the cable and a connection made by means of a special clamp and lead to a special fuse-box fixed to the trunking from which the appropriate sub-circuits may be run.

Busbars

For heavy loads and particularly where flexibility of connection is desired, the bushbar system is appropriate. This employs the standard cable trunking, but instead of insulated cables, the current is conveyed by bare copper rods or bars, supported at intervals on insulating spacing panels. Cables can be connected by clamps as in the cable-tap system. When flexibility is required, however, it is possible to obtain connection units and fuseboards which can be fixed without turning off the current. Busbar systems are available for loads of from 100-600 amps.

Bushbar and cable-tap systems are appropriate in industrial and commercial buildings but they are also well adapted to multi-storey buildings of all types, where they can form a main vertical distribution rising through the building, serving distribution boards for each floor, and in large buildings to serve local distribution boards. Diagram 268 shows a cable-tap system and diagram 269 a busbar system.

DUCTS FOR ELECTRICAL DISTRIBUTION

In additional to the wiring systems there are a number of ducts available, specially designed to

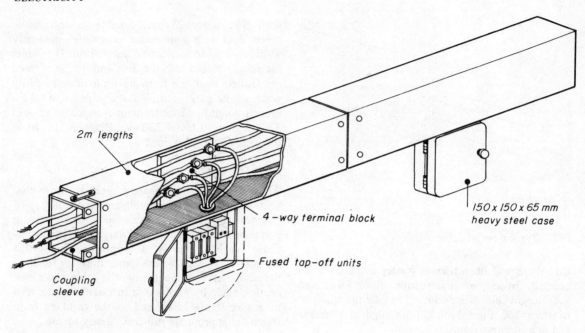

268 Cable-tap system of electrical distribution

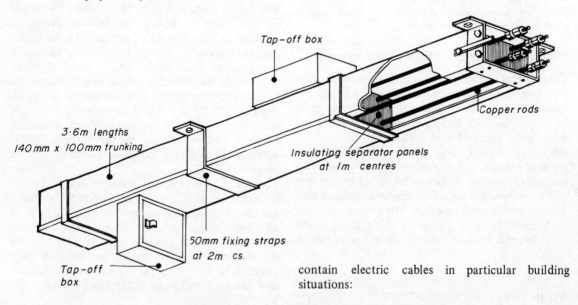

269 Busbar system of electrical distribution

contain electric cables in particular building situations:

Ductube

Consists of inflatable rubber tube which is placed in concrete formwork before pouring the concrete. After the concrete has set the ductube is deflated, withdrawn from the concrete, leaving a duct for electric wiring, or other purposes.

DUCTS FOR ELECTRICAL DISTRIBUTION

Skirting trunking

One of the most convenient positions for electrical outlets is in or above the skirting. The convenience of a trunking system which would form a skirting as well as a cableway is apparent. Diagram 270 shows a metal skirting trunking showing the electrical continuity link and a plywood one. It is very usual to run these trunkings right round the perimeter walls of an office building, using conventional skirting for partitions. In larger or more complex buildings all the skirtings may be in the form of trunking.

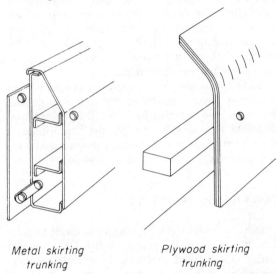

Metal skirting trunking   Plywood skirting trunking

270 *Types of skirting trunking for electrical cable distribution*

Floor trunking

In large offices where desks must be placed remote from walls and other buildings with similar planning problems, cables must be run under the floor. Where there are comparatively few isolated points, the positions of which are known, conduit can be laid in the floor screed. Where flexibility for future re-planning must be achieved, a grid of underfloor ducts will have to be employed. The spacing will be governed by planning requirements but considerations of economy will call for the widest reasonable spacing. There are two main types: pitch-fibre ducts of approximately half-round section from 35 mm to 55 mm deep overall, and steel ducts from 30 mm to 35 mm deep. To give proper cover to ducts of this sort a screed substantially thicker than normal is required. The pitch-fibre ducts can be broken into, even after laying, anywhere in their length and provided with outlet fittings. It is desirable to have the ends of the duct clearly identified by outlets so that location of the duct is simplified. In the case of steel ducts, outlets are provided by boxes with removable covers. In both cases ducts can be run in pairs to cater for electrical and telecommunications wiring and special junctions are available which enable the two sets of wires to cross without touching. In rooms with a great deal of underfloor wiring (eg Computer rooms) a suspended floor is often used giving complete wiring flexibility.

Overhead distribution

Overhead distribution systems are clearly more economical and more flexible than underfloor ones. They are used mainly in industrial situations when the pendant connections to apparatus would not be considered unsightly. Some overhead systems which allow flexibility in the layout of lighting fittings are available. They employ a special trunking system which distributes the wiring and supports the lights and are mainly intended for use with suspended ceilings. Diagram 271 shows a trunking intended to be embedded in a concrete floor.

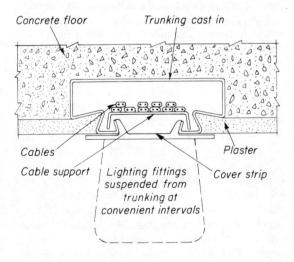

271 *Overhead electrical trunking cast in to concrete floor*

ELECTRICITY

ENTRY TO THE BUILDING

In urban areas electrical cables are usually underground and are brought up to an entry point at ground level or into the basement. Service cables cannot be bent to small radii and this should be borne in mind when considering points of entry (750 mm radius should be satisfactory in most cases). In small and particularly domestic buildings the cable run is kept as short as possible, terminating in a distribution board at the first convenient position. In these buildings the distribution board will be fitted with a sealing box to prevent moisture from entering the insulation of the service cable; a main fuse for the premises in a box sealed by the supply authority; a meter belonging to the supply authority and the consumer's unit or other switch and fuse gear belonging to the building. Where an off-peak tariff is being applied to particular items of electrical equipment (eg thermal storage space or water heaters) a time switch, a contactor and a separate meter will be needed for switching on and recording the electricity consumed. The position chosen for the distribution board should be readily accessible both for meter reading and for replacing fuses. In some cases special glass panels are provided so that meters can be read without entering the premises.

In country areas the supply is often overhead. It is usual in this case to carry the cable to the building at high level, run it down the wall to about first floor level and then turn the cable up into a conduit running through the wall. Turning the cable up before entry to the building prevents rainwater from penetrating. Apart from this the requirements for small buildings are similar to those in urban areas.

In somewhat larger buildings the entry arrangements will be the same but the distribution board will become a main board serving a series of other boards located throughout the building. The boards will be linked by large-capacity cables and the final sub-circuits run from local distribution boards.

In buildings of substantial size or having a heavy loading (100-1000 kva) a condition of supply may be that space is allocated for a transformer station which will take high voltage current and deliver a supply of normal three-phase 415/240-volt electricity to the building and perhaps to other premises nearby. Transformers are heavy, require considerable air circulation for cooling, constitute a fire risk and must be accessible to the electricity board. They may be sited in the open air, provided a security fence is provided. It is usual, therefore, to find transformers at the edge of the site near the access road. If for visual reasons it is not acceptable to have the transformers in the open air they may be enclosed in a building provided adequate air circulation is provided. If they are installed in the body of the main building, fire-resisting construction must be employed, adequate ventilation provided and easy access for delivery and for electricity board officials. Heavy cables will run from the transformer station to the building and sharp bends must be avoided. A duct about 750 mm square will be satisfactory in most cases, or if the transformer station is remote from the building the cable may be laid underground. In buildings of this sort the service cable will terminate in a switch room rather than at a small distribution board.

Industrial consumers sometimes purchase their electricity from the board at high voltage and use their own transformers. In these cases it is desirable for economy to site the transformer at the centre of gravity of the load it serves and keep the low voltage cables as short as possible.

DISTRIBUTION CIRCUITS

In most buildings electricity supply is divided finally into three types of circuit: 1 lighting, 2 socket outlets and 3 fixed apparatus. Lighting is normally carried out in circuits with 5-amp wiring and fuses although 15-amp circuits for lighting are permitted and used mainly in industrial situations and large buildings. Each circuit consists of a cable with two conductors, live and neutral, which link a number of ceiling roses or wall lighting points. The lamp and switch are connected in the way shown in diagram 224, thereby allowing independent control of the lamps although they are part of a single circuit. Although diagram 272 shows only one lamp operated by the switch, it is clearly possible to have several lamps controlled by the one switch. The converse is also the case and in many instances it is desirable to have one or more lamps controlled from more than one switch. Diagram 273 shows a method of achieving this from two switch positions, while diagram 274 shows a system whereby any number of switches can be used. The switches for this arrangement are known as intermediate switches and have four terminals which are connected by

DUCTS FOR ELECTRICAL DISTRIBUTION

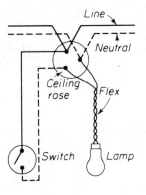

272 One way switch control of lighting

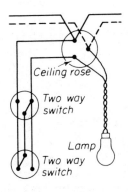

273 Two way switch control of lighting

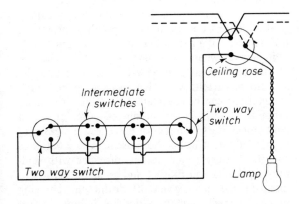

274 Intermediate switch control of lighting

the switch in pairs (horizontally or vertically in the diagram). The operation of any single switch in this system will switch the lamp on if it is off, or off if it is on.

Socket outlets

Socket outlets into which portable electrical apparatus can be plugged are a vital part of almost all installations in buildings. At one time each socket outlet was served by an individual circuit leading from a fuse at the distribution board (15 amp round-pin type). Separate outlets had to be provided for small apparatus not having 15 amp wiring, giving rise to 5 amp and 2 amp sockets. There are very many installations of this sort still in operation. The economy and flexibility of the ring main circuit has, however, become apparent and this is now an established method of supplying socket outlets. In the ring main system a heavy cable with two conductors and an earth runs in a circuit round part of the building, starting and finishing at the distribution board where the line conductor is served by a 30 amp fuse. Sockets are provided on the ring and apparatus is connected to them by means of plugs, each of which is fused. The fusing of the plugs can be varied (normally 13, 7 or 3 amps) to suit the particular piece of apparatus controlled, thus enabling complete flexibility in the movement of apparatus and maximum utilisation of sockets. A special type of plug with rectangular pins, fuse, and capable of carrying a 13-amp current, has been developed for use with ring mains. When a fault develops in a piece of apparatus the fuse in the plug blows, leaving the main circuit unaffected. The ring circuit is for most applications very much more economical of wiring than individual connections to the distribution board. The installation of ring circuits is governed by a number of rules mainly concerned with electrical efficiency. The main point of general planning interest is that the maximum number of 13-amp socket outlets allowed on one ring main is 10, except in dwellings where there must be a separate ring for every 100 m^2, or part of 100 m^2, but where an unlimited number of socket outlets may be provided. Normal domestic operation in these conditions does not overload ring mains but the economical provision of an adequate number of socket outlets is of very great convenience to inhabitants of buildings so provided. Spurs may be provided from the ring, thereby economising

ELECTRICITY

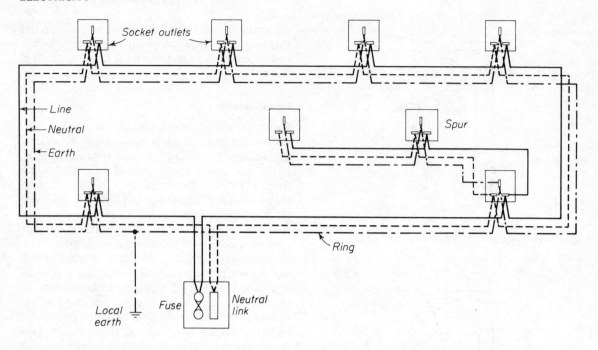

275   13 amp ring circuit

wiring, provided there are not more than two sockets per spur and that no more than 50 per cent pf the sockets on the ring are served by spurs. Diagram 275 shows the principles of wiring to a ring main.

Fixed apparatus

Pieces of fixed equipment such as cookers, water heaters, etc, usually have their own individual fuses and circuits. There should be an isolating switch adjacent to the piece of apparatus in each case so that work on it may be carried out in safety.

Diagram 276 shows a number of lighting, ring main and fixed circuits served from a consumer's unit.

Bathroom

Because of the presence of water there is increased risk of electrocution in bathroom and special precautions are needed. All metal in the bathroom, whether it is part of electrical apparatus or not, must be bonded together with cable and effectively earthed. No socket outlets, except for shaver sockets conforming with BS 3052: 1958, may be used. Fixed apparatus is permitted but switches for apparatus and lights should be of the ceiling pull pattern.

EMERGENCY SUPPLIES

In a number of building types, particularly where public assembly is involved, emergency lighting in public spaces and escape routes is required in case the normal lighting is put out of action by failure of the public electricity supply. Emergency lighting is usually provided by low-voltage lamps, sometimes separately fixed and sometimes incorporated in the lignting fittings. Electricity is provided from a battery installation, kept charged from the mains supply while this is in operation and supplying current for the lamps, either continuously or automatically switched on when the mains supply fails. The size of the battery installation varies according to the amount of current which has to be supplied. Small installations can be completely contained in a metal cabinet, while in larger buildings a complete room containing a large number of batteries on racks may be needed. Most installations use lead/acid batteries and the battery room must be provided with acid-resisting floors and finishes and good ventilation to dissipate the gases given off during the charging process. Nickel/alkali batteries are lighter, have a longer life and do not give rise to such acutely

DUCTS FOR ELECTRICAL DISTRIBUTION

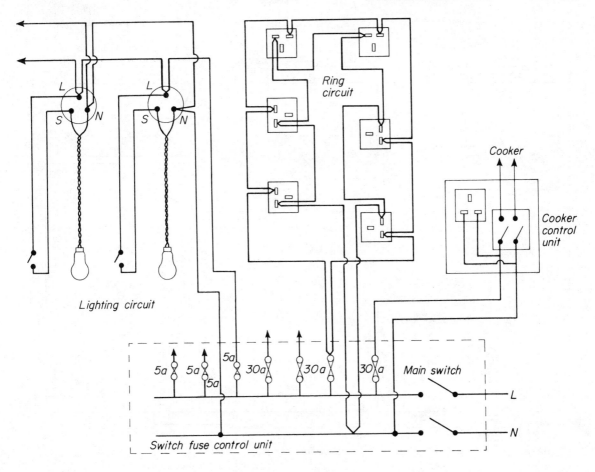

276 *Principles of distribution for domestic electrical installation*

corrosive conditions as lead/acid. Delivery and storage of carboys of acid and distilled water must be borne in mind in siting and access.

In very large buildings, or where heavy electrical loads such as operating theatres or fire-fighting pumps have to be served, stand-by generators have to be provided. These are often diesel powered and automatically brought into operation on power failure.

Electrical layout drawings, and schedules

In small buildings the extent of the electrical installation does not justify specialised design of the installation. The architect will prepare layout drawings showing where electrical appliances are to be fitted, a schedule giving details of the appliances themselves, and a specification defining contractural obligations and quality of materials and workmanship. The specification will say that the work must be carried out in accordance with the Institution of Electrical Engineers Wiring Regulations (current edition) which is intended to be specified in order to define workmanship and electrical design and safety. For work carried out in accordance with the IEE regulations, the contractor is called upon to issue a certificate undertaking that the work does comply with the standards defined in the regulations. No detailed wiring plans are prepared. Electricians work out runs on the site. Few problems normally arise but if architects wish to have all wiring concealed (eg in screeds and plaster) they must ensure that practicable wiring routes exist to all the pieces of electrical apparatus. Diagram 277 shows a typical

ELECTRICITY

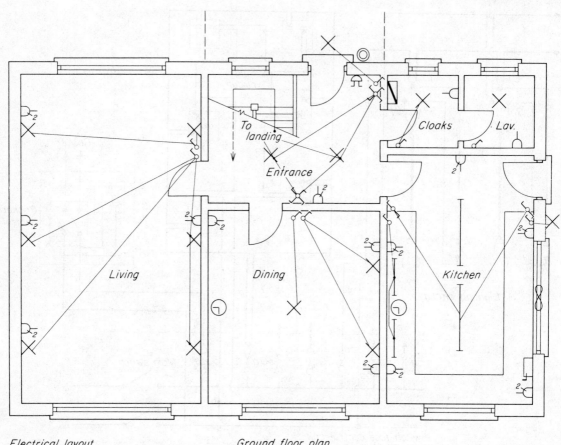

Electrical layout   Ground floor plan

Key to electrical layout

✕	Ceiling lighting point	⏚	Electric bell
⊠	Wall lighting point	⊃–	Socket outlet
⊢—⊣	Fluorescent light	⊃–₂	Two-gang socket outlet
✗	Single pole, one-way switch	▨	Intake and main control
✗	Two-way switch	⏁	Clock point
✳	Intermediate switch	∞	Fan
⊚	Push button	▭	Cooker control panel

277 Electrical layout

306

LOCATION			LIGHTING FITTINGS							SOCKET OUTLETS				SPECIAL FITTINGS
BLOCK	FLOOR	ROOM	Unit No	Description	Cat Ref	Lamps	Height	Switch	Sw Ht	Unit No	Type	Cat Ref	Height	

278

electrical layout and typical symbols, while diagram 278 gives a typical format for an electrical schedule.

TELEPHONES AND OTHER TELE-COMMUNICATION SYSTEMS

These include:

Telephones—public service (GPO)
Telephones—private
Tape-machines and teleprinters
Broadcast-reception apparatus (sound and television)
Bell and similar systems
Sound distribution systems
Staff location systems
Alarm systems
Controlled-time services

All these services are 'specialist' services and their design and installation is usually done by specialists. In most cases the architect or the engineering consultant will give the user requirements and the facilities which will be available for installation, the specialists will prepare a detailed specification based on their own equipment. The specification and design can then be submitted for approval of the engineering consultant. (BS Code of Practice 327.101 applies.)

The important aspect is the provision made in the building for the installation, alteration or removal of the necessary wiring and apparatus. As with the electric lighting and power installation, new developments are available within a few years and user requirements change equally frequently, so that ease of alteration is one of the most vital requirements. This particularly applies to GPO telephones in an office building let in separate offices; tenancy changes take place constantly and lines are often discontinued and new ones connected in different positions in the building. Provision should also be made for much greater use of these services. Detailed provision for this flexibility is gone into later.

All these services operate at low voltage so there is usually little danger. On the other hand, because of the reduced insulation proper precautions are not taken to separate telecommunication wiring from other types.

Separation of telecommunication circuits from electric light and power systems is essential. They must not run 'in the same tube, groove or section of conduit' (IEE Regulations). Where an electric conduit crosses a telephone cable there should be physical separation by a screen or partition. This does not prevent electric light cables from being in one side of twin ducts or being accessible in combined junction boxes, provided there are proper partitions in the boxes to ensure separation and there is continuity to earth in the casing.

Telephones

GPO telephones

These are usually installed after the builder has handed over, or just before, and too often, if provision has not been made for this service, inconvenience and damage to the building can result.

The position and the arrangement for the entry of the GPO cable should be agreed with the GPO engineers before building commences. In a large town building this will be a cable buried in the pavement, which will need a sleeve through a basement wall or duct such as a stoneware pipe carried up through the surface concrete to near the position where the switchboard will be. Where the intake is from overhead wires entry can be through an air-brick or ventilator. Drilling through a wood sill can do much damage, though too often it is the only simple way possible when prior provision has not been made. The tacking of telephone cable to lead flashings, gutters, parapets, stonework, etc, has in the past often led to damage to the structure, giving rise to much expense.

In a large building a switch-room will be needed to accommodate the distribution frame and the switchboard operator, though sometimes the gear can be in a separate space not needing natural light. Good ventilation and dry conditions are essential.

GPO switchboard sizes are as follows:

2.1 m high x 865 mm wide x 610 mm deep.
2.1 m high x 1.14 m wide x 610 mm deep.
1.4 m high x 675 mm wide x 850 mm deep.

The last is by far the most commonly used.

When a GPO installation is being planned it is essential that allowance be made for extra extensions and extra 'lines', as the substitution of a larger switchboard later is costly not only in money but in inconvenience.

Telephones operate on 50 volts. They are usually powered by battery which is automatically recharged from the main supply to the autocharger.

Internal telephones (separate from GPO system)

These give automatic inter-connection to any number of instruments and can incorporate a number of additional services, such as call systems, amplifiers, fire alarms, etc. All the equipment for an internal installation can be bought outright or can be hired. Hiring usually includes servicing.

The switch-gear consists of relays through which the impulse is passed to a uniselector, then to two-way selector and then out again to relays, the whole being comprised in a metal case, several of which, depending on the size of the installation, are mounted on a steel frame standing free or in an enclosed glass case. The size of the board or frame or the number of cases depends on the number of lines or instruments. For instance, a private automatic installation with 100 instruments enabling ten persons on different instruments to speak at once would require two boards with accommodation for 50 extensions on each, measuring together 1.8 m long x 1.8 m high and 640 mm wide. The battery may be 1.8 m long x 460 mm x 460 mm. The space taken by the autocharger is quite small, about 616 mm x 300 mm x 300 mm. It is usual to supply an instrument on the internal system for the use of the GPO switchboard operator and it is also used by the GPO installation engineers when installing their own system.

Internal wiring for both GPO and private systems should be provided for by ducting wherever possible, and it is advisable, in order to allow for change to be made later to ensure that all parts of the installation, particularly cabling and wiring, be readily accessible and not buried behind plaster or hidden above false ceilings or below tongued-and-grooved floors.

Each instrument requires a pair of wires taken back to the switchboard, which is usually sited as near as possible to the incoming cable from the telephone exchange. For a 100-pair cable, lead covered, the diameter may be about 35 mm. This is connected to the main distribution frame which, in turn, is connected with the switchboard. The cables are taken back to the distribution frame and from there the various pairs, in groups of 10, 15, 20, 25, 30, 50, etc, are distributed to serve every instrument in the building.

For each of the various groups, therefore, provision has to be made to accommodate them on or in the structure, according to their respective diameters. PVC cables are usual, obtainable in grey or cream, and a 20-pair and a 25-pair can be accommodated in 18 mm, and a 50-pair in 25 mm. As with other electric cables, there must be no sharp bends in the conduit. The minimum radius must not be less than five times the diameter.

The groups of pairs are branched off the main large cable, which may contain 50 or 100 pairs, by means of junction boxes which are situated usually one or more to a floor, depending on the number of vertical risers that are installed. Each group, in turn, is branched off by smaller junction boxes until the final branch to the instrument is made by the smallest junction box.

In buildings with an extensive telephone system, such as a block of offices, vertical runs are usually in *rising ducts,* with a distribution case at each floor and serving the instruments on that floor. The limit of horizontal cable from the distribution case is 10 m, so in a building with a large plan several rising ducts may be needed. These rising ducts may be part of a general service duct, but is preferably partitioned off. In either case a minimum space of 450 mm wide x 150 mm deep is needed to take the distribution case if it is to be in the duct, and in large installations it may have to be much larger. The aperture in the structural floor need only be 15 mm x 75 mm. Access to each riser should be easy: a door of normal height is best.

Horizontal cable runs present the greatest problems, as the ideal aim is to be able to connect an instrument at any desk at any position on the floor. In fact, it is only normally possible to provide a system of ducts serving a grid over the floor. The effectiveness of this depends on the type of duct and that often depends on the structural floor system used. Many floor systems, such as concrete hollow beams, T-beams, steel decking, etc, have large voids which can take conduit, or in some cases, where the voids are small and smooth as with steel floors, can take cables direct. Continuity from bay to bay has to be provided in the design of the beams; if the floor deck rests on and passes over the beams this presents no difficulty. With reinforced concrete beams holes can be formed with cardboard tubes in the shuttering; with steel beams holes can be cut in the shop, but this calls for a degree of preplanning, which is unusual. Where suspended ceilings are provided these give a convenient space for conduits.

Duct space in the structure can in any case form only part of the provision for horizontal runs. Most of the small horizontal runs to the outlets have to be in ducts in the floor screed or under the floorboards.

There are many proprietary systems of ducts for telephone and other cables. *Fibre ducts* and steel ducts are described earlier in this chapter. Fibre ducts are particularly suitable for telephones, as an outlet can be formed at any time later by chipping away a little concrete and cutting a hole in the fibre with a special tool and screening in an outlet fitting. The precise position of the ducts must be clearly marked or recorded on a plan. Fibre ducts are from 35 mm to 55 mm deep overall. Steel ducts of rectangular section, BS 774, are from 30 mm to 35 mm deep. Both of these can be run in pairs, to give proper separation of telephone cables from other telecommunications services, but with combined junction boxes with partitions to preserve adequate separation. Special fittings for both bring the duct up to a suitable position for a skirting outlet where such is required.

There are also *sheet-metal duct* systems. They consist of a series of pressed steel channels with a cover which fits over. The system is complete with junction boxes, skirting bend for outlets, etc.

There are three main floor *duct layout* arrangements in common use:

1 A complete grid of ducts running both ways. This is the most satisfactory but is the most expensive. In one direction the spacing might be as little as 2.5 m.

2 Corridor duct, with ducts off it across individual offices to the external wall. This is convenient as only the junction boxes in the corridor need be opened when altering any lines.

3 Perimeter duct running round near the outer wall with spurs up to skirting outlets. This uses the least ducting but alterations will affect many offices, and positions away from the outer wall would have to be run on the surface.

Ducts to outlets should have an area of 1300 mm^2. Corridor ducts or those serving several outlets should have 2000 mm^2.

Ducts can be formed behind *skirtings* or proprietary metal skirting ducts can be used. There are also systems for wall ducts, which may be useful in special buildings where appearance is not an important factor.

BS 1710 requires telephone cables and ducts to be marked ET in black paint on a light orange background.

The GPO issue, free of charge, an excellent booklet called Facilities for Telephones in New Buildings.

Other telecommunication services

Tape-machines and teleprinters also radio and television rediffusion (ie by line as opposed to wireless reception), is done normally through the GPO so that any apparatus can usually be accommodated in switch-rooms, or rooms adjacent, and distributed in a twin duct with telephones.

Broadcast reception by communal aerial, for up to 12 sets, is now becoming common for flats, and saves the unsightliness of many aerials. Radio programme service is sometimes supplied in hotels and flats from a central aerial and receiving set to loudspeakers by an internal network.

ELECTRICITY

Staff location or call systems are available in many forms, by coloured lights or numbers operated from a central position.

Many types of sound distribution systems are available or can be devised to suit special requirements, as calling or 'page-ing' systems in hotels or works, relaying 'music while you work', etc.

All of these need a wiring network of some kind and the previously discussed duct systems are suitable for such installations.

REGULATIONS

New installations must satisfy the electricity supply authority before they can be connected. The Institution of Electrical Engineers publish *Wiring Regulations for Buildings* which covers materials, electrical design and workmanship and is intended to be specified as a standard for building work. It contains specimen forms for electrical contractors to sign, undertaking that they have carried out the installation in accordance with the IEE Regulations.

Electrical Services in Buildings, by Peter Jay and John Hemsley, published by Elsevier, is recommended for further reading.

ELECTRICAL TERMS

Voltage (V): measure of electrical pressure.

Ampere (A): measure of quantity of electricity (current).

Watt (w): measure of electrical power.

Kilowatt (kw) equal to 1000 watts (volts x amps = watts).

Unit: short for the commercial measure of electrical power, the Board of Trade Unit which is equivalent to 1,000 watts (for one hour, or 1 kilowatt hour (kwH).

kVA: measure of apparent electrical power in alternating current systems (line voltage multiplied by line amperage).

Power factor: ratio of actual power (kw) to apparent power in alternating current system. The balance of inductance and capacitance in an electrical installation may cause the voltage and current fluctuations to become out of phase with each other. This causes the actual power to be less than the apparent power and gives a power factor of less than unity. It is desirable that power factors should be as near unity as possible since otherwise the efficiency of the distribution system is reduced and electricity costs are increased. In large installations additional inductance or capacitance may be purposely added to balance the power factor.

Voltage drop: if a long length of wire is used its resistance to the flow of electricity may give rise to a drop in voltage. The permissible amount of this drop in electrical installations is limited.

LIGHTNING PROTECTION

Strokes of lightning coming from thunderclouds to the ground are particularly likely to strike high or isolated buildings, but may strike anywhere. The peak current averages 20 000 amps but may reach 2 000 000 amps. Substantial structural damage can be caused and fires started if lightning strikes buildings.

Protection against lightning may be achieved by conductors, or as the whole installation is properly called, a lightning protective system. Such a system consists of an air terminal, or air terminals above the top of the building to intercept the lightning, a down conductor and an earth terminal to convey the very heavy currents to earth. An air terminal is usually considered to give good, although not perfect, protection against lightning in a zone defined by a cone, the apex of which is the air terminal and the angle of the side of the cone 45° from horizontal. Several terminals may be required to protect a building and large roof areas may require systems of horizontal conductors. All metallic projections above roofs should be bonded to the air terminals. Down conductors are run down the building externally by the most direct route and usually consist of aluminium or copper strip or rod 20 mm x 3 m and 10 mm diameter.* One down conductor is recommended for a plan area of 100 m² with an additional down conductor for every 300 m² of plan area over 100 m². Special joints are necessary

---

*Bends should be kept to a minimum and re-entrant loops totally avoided.

LIGHTNING PROTECTION

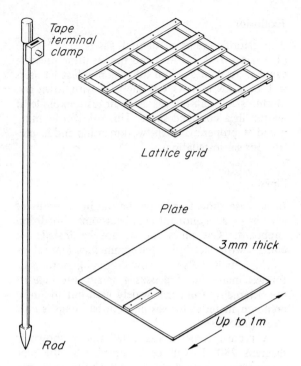

279 Earth Rod and Earth plate for lightning conductor installations

at any junction in a down conductor. At the foot of the down conductor an earth electrode discharges the flows of current to earth. In some metal and reinforced concrete buildings the structural metalwork can be used as a down conductor.

It is not possible to lay down definite rules which control whether lightning protection should be provided in a building. Many factors are involved in a decision including the nature of the use and the consequences of a lightning strike, the type of construction, the nature of the country where the building is sited, the degree of insulation, the height of the building, and the frequency of thunderstorms. CP 326: 1965 not only described lightning protection systems but also gives a series of tables which can be used to give an index of risk and thus aid the decision whether to provide a lightning protection system.

16 Gas

Gas has the virtues of giving almost pollution free products of combustion, entirely acceptable in clean air zones, and of flue requirements which are smaller and often very much simpler than those for solid fuel or oil. Recently the use of gas as a fuel has been revitalised by the discovery of natural gas under the North Sea. This has happened at a time of increasing expense and shortage of other fuels and consequently also of manufactured town gas. As a result all gas supplies in the United Kingdom are being converted to the use of natural gas.

Town and natural

Town gas, originally made from coal, and more recently from oil has a high proportion of hydrogen and a significant proportion of carbon monoxide which made it poisonous. Natural gas is mainly methane and is virtually free of carbon monoxide. While natural gas is, therefore, non toxic, high concentrations in air for breathing can cause asphixiation. The burning velocity (or the rate at which gas ignites after leaving the jet) of natural gas is lower than that of town gas but its calorific value is higher. Consequently different burners are required for town and natural gas. Natural gas is distributed at higher pressure than town gas and this coupled with its higher calorific value per m^3 means that smaller main and distributing pipes in buildings can be used.

Explosion

The calorific value of natural gas is 37 MJ/m^3 (Town gas 18.5 MJ/m^3) and the resulting pressure supplies to domestic and other non-industrial users is 1 to 2 kN/m^2. As was the case with town gas, mixtures of natural gas and air can explode if ignited in a confined space. This risk can be minimised by proper materials, workmanship and inspection for gas installations.

Pipes

In no case should gas pipes be run in unventilated cavities or ducts nor through bedrooms or bedroom cupboards. Gas pipes should not be installed in contact with electrical or telecommunications cables.

Unplasticised PVC is now becoming more usual for gas mains but fire risks generally preclude its use internally. Copper is widely used but the most common material for gas pipes in buildings is mild steel, to BS 1387.

A typical domestic gas installation is shown in diagram 280. It will be observed that the pipe arrangements from the main to the building are not dissimilar to water supply, although the gas pipe

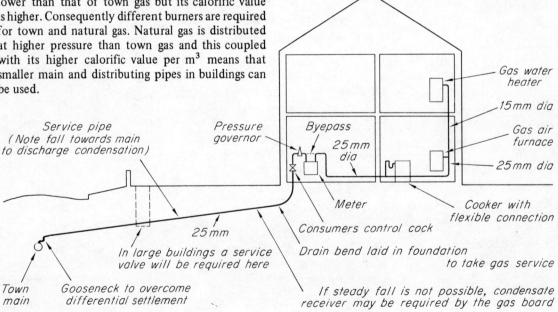

280   A typical domestic gas installation

must fall towards the main so that condensation may drain. Inside the building a main control cock precedes the gas meter which must be sited conveniently for access. Sometimes a pressure governor is fitted before the meter and sometimes, particularly in larger installations a bye pass to the meter is included. For a small domestic installation a meter space 550 mm × 550 mm × 300 mm deep might be appropriate. For large installations meters can be very substantial in size requiring special accommodation and sometimes lifting tackle. The area gas board should be consulted at an early stage in design.

Underground service pipes must be protected by bituminous or other wrapping.* Pipe runs after the meter can be dictated by convenience and neat appearance. It is acceptable to bury gas pipes in floor screeds but mild steel pipes should be wrapped

*During building an entry for the gas pipe should be formed in the foundation. Vitrified Clay pipes may be used and the gaps after installation packed with bituminous filling.

to prevent corrosion if they are buried in plaster. Pipe sleeves should be used wherever gas pipes pass through walls and floors not only for neatness but also to ensure that thermal movement and settlement stresses are prevented.

Most gas appliances are connected directly to the pipework. In the case of cookers and refrigerators, however, it is becoming increasingly usual to provide flexible connections to allow the appliance to be moved for cleaning.

Flues

Very small gas appliances and cookers are permitted to discharge their products of combustion directly into the air of the rooms in which they are sited. All other appliances must have flues but flues for gas appliances are very much simpler than for solid fuel or oil. All gas appliances have to be able to function without a flue, therefore the flue itself is merely a discharge for products of combustion. It does not have to provide a draught to aid combustion. No cleaning is required for gas flues and for single appliances flues formed of hollow blocks set into walls can be used. Products of combustion from gas appliances contain substantial quantities of water vapour and care must be taken that condensation does not cause problems. Vitreous clay and asbestos cement pipes make excellent flues. There are, however, two major devices which simplify gas flue requirements. One is the balanced flue, shown in diagram 97. Where gas appliances can be fixed on an external wall it is possible to use models with a sealed combustion chamber which draws air for combustion from outside and discharges the products of combustion through the same hole in the wall. The vertical flue is totally eliminated. Steam is likely to be produced by any gas flue in cold weather and balanced flues should not be positioned under windows. In multi-storey buildings it is possible to use a single duct running from top to bottom to deliver air for combustion to gas appliances and also take the products of combustion. This arrangement is known as a SE duct and is shown in diagram 281. Sizes of SE ducts are governed by the height of the building and number and type of appliance. The local gas board should be consulted. While the SE duct offers substantial simplification of the requirements it governs the placing of gas appliances and must be considered during the early planning stages of building design.

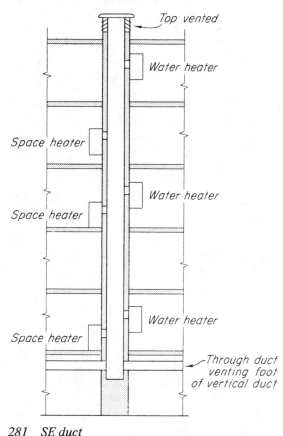

281 SE duct

17 Mechanical conveyors

Mechanical devices for the movement of people and goods are now essential equipment in many types of building. Their spatial and functional requirements and the desire to make the best and most economical use of the installation often exercise considerable influence on construction and planning and they may, as in the case of lifts for multi-storey buildings, make new building forms possible. Mechanical conveyor installations must therefore, in common with many other service installations, be taken into account from the very first stages of design. A bewildering range of devices exists, particularly for the conveyance of goods, but the underlying principles of operation are relatively few and can be found applied both to the movement of people and goods. Table 58 shows the most significant forms of conveyors and typical applications.

MECHANICAL MOVEMENT OF PEOPLE

Four types of installation are normally used in buildings. Lifts, escalators, paternosters and travolators. Lifts are widely employed to make vertical circulation quicker and easier and they make possible buildings rising above the four or five floors which would be a maximum for reasonable access on foot. Escalators occupy considerable floor space and, being inclined raise people through a limited height, but they will deal with large numbers (a 900 mm wide escalator is capable of handling 6000 people per hour). Paternosters are comparatively rare in this country but they are widely used in Europe and several installations have been made here in recent years. They are particularly appropriate where internal circulation is the critical feature rather than inward or outward flow. Travolators assist in the movement of large numbers of people horizontally or up very limited inclines.

Lifts

The essential features of a lift installation are a car to carry passengers supported on a cable running over a pulley (traction sheave) and balanced by a counterweight, as shown in diagram 282. Both the car and counterweight rise and fall between steel guides in a shaft formed in the building, known as the lift well. The traction sheave is housed, together with the motor and control gear, in a motor room over the lift well. The well itself is extended some distance below and above the furthest extent of the car movement to allow for innaccuracies of control. These spaces are known as the pit and

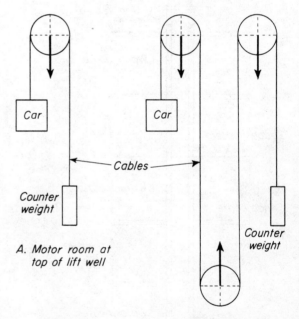

A. Motor room at top of lift well

B. Motor room at foot of lift well

Note: Different cable lengths required
Loads imposed on structure of building
(shown by black arrows)

282 Lift operation

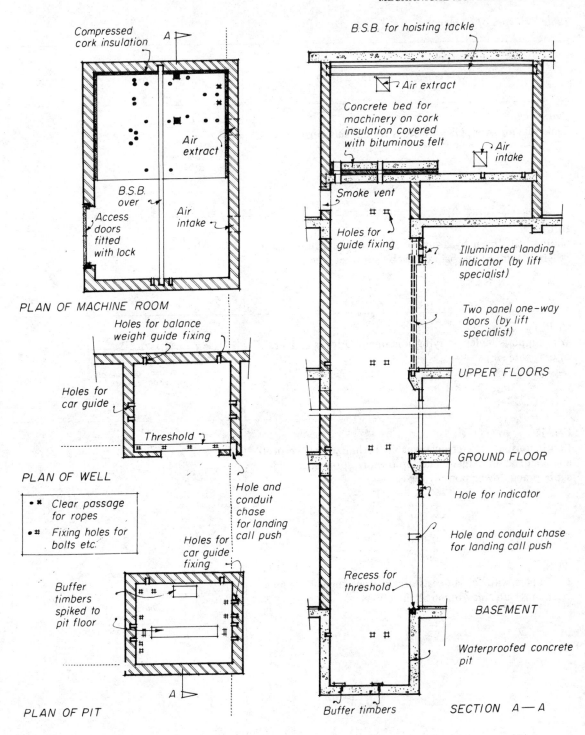

283 *Typical constructional details of lift well and motor room*

MECHANICAL CONVEYORS

Table 77 Types of conveyor

Type of installation	Form
HOIST Cable running over pulley provides vertical movement	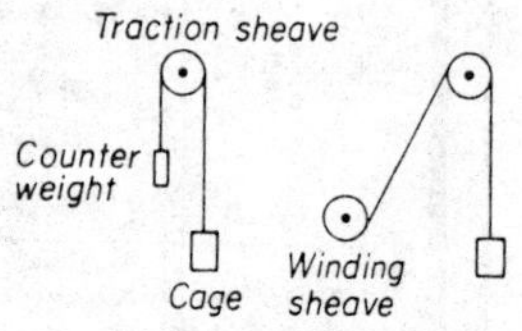 *Traction sheave, Counter weight, Cage, Winding sheave*
BELT Sliding belt driven by drum, or slats or pallets running on rollers and linked by chain driven by sprocket wheels	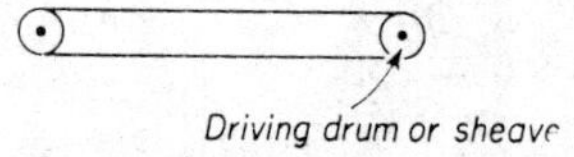 *Driving drum or sheave*
CABLE Wire supports pulley carrying containers. Drive is by inertia or drag wire	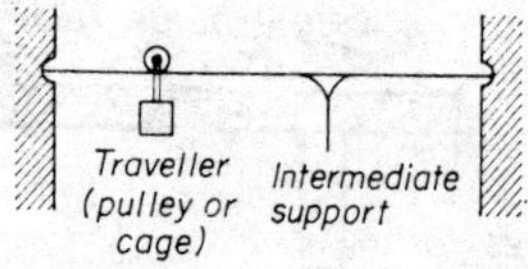 *Traveller (pulley or cage), Intermediate support*
CHAIN Continuous moving chain, capable of horizontal, vertical and inclined movement and of turning corners, provides support and driving power for baskets	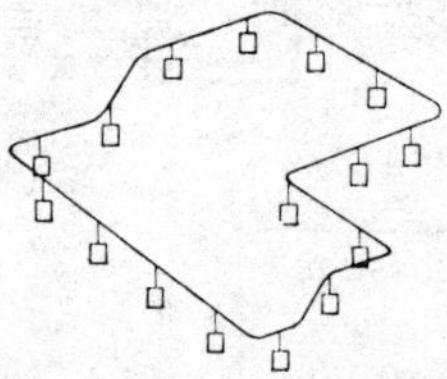
TUBE Containers with felt air seals drawn along tubes by atmospheric pressure. May run in all directions	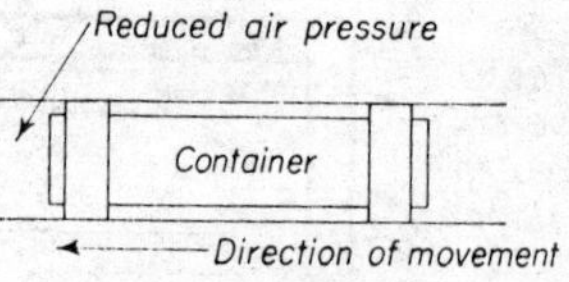 *Reduced air pressure, Container, Direction of movement*

Table 77 Cont/d...

Pattern of movement	Examples of applications		Mobile types
	People	Goods	
Vertical, intermediate stops	Lifts	Lifts Cranes Hoists	Mobile cranes
Horizontal or inclined	Travolators Escalators	Belt conveyors Platform conveyors	Mobile belt conveyors Mobile travolators (airport use)
Horizontal or inclined Intermediate stops	Teleferique	Cash and letter conveyors	
Horizontal or inclined cable does not stop but automatic discharge of baskets at selected points is possible	Paternoster	Basket conveyors Chain conveyors	
Terminal system (Central station, Local station) Ring system Carriers automatically discharged at selected stations	(Atmospheric railway. Historical)	Pneumatic conveyors	

overrun respectively. The top of the lift-shaft is vented to the fresh air. This ensures that in case of fire the lift-well acts as a flue and does not encourage the spread of smoke and fire from one floor to another. Diagram 283 shows these features.

Lift speeds vary from 0.5 m/s for small, slow installations to 6-7 m/s for very high buildings. This higher speed is governed by the capacity of the human ear to adapt to changing atmospheric pressures. It is conventional in this country to use speeds of 1.2 to 1.5 m/s for buildings of 6, 7 or 8 floors and 2.0-2.5 m/s over this height, except for very high buildings where greater speeds may be considered. For low-speed lifts an ac motor driving the traction sheave via a worm gear provides the motive power. In faster installations better control, quieter operation, smoother and more rapid acceleration and other technical advantages can be achieved with a dc motor directly coupled to the sheave and operating on a variable voltage system.

The dc current is usually generated from the mains supply by individual generators for each lift motor. The size for the motor room is therefore increased. Since lift speeds are higher, pit and overrun requirements are also greater.

Control

Control of lift operation is important. Many arrangements are in use but there are five main control possibilities:

1 Operator

This system is rarely employed except in special circumstances such as prestige, or department stores, since other systems can give equal or better service at lower cost. It operates with call buttons at each floor registering in an indicator in the lift. Most other systems can be arranged for dual use so that operators can be employed at times if desired.

2 Automatic

This system operates by means of call buttons on each floor and a set of destination buttons in the lift. The first button pushed sets the lift in motion and no further calls are accepted until the lift is again at rest. Calls are not stored for further action. This system is only suitable for small, lightly trafficked lifts.

3 Collective

Destination buttons for each floor are provided in the lift and two call buttons on each floor, one for upward and one for downward journeys. Calls are registered by the control gear and automatically dealt with in sequence. On a downward journey the lift will stop at all floors where downward calls are waiting or where passengers wish to stop. It will then make an upward journey, stopping for upward calls. If further down calls are made it will then return to collect and deliver these passengers. The push-buttons, both on landings and in the lift, usually give a visual indication showing that the call and destination have been stored by the control gear for operation. It is possible to have simplified versions of this type of control such as collective control on downward journeys only.

4 Group collective

Up to four lifts may be grouped together in a collective control system so that the closest lift will respond or the first lift of the group moving in the desired direction will stop in response to a call.

5 Programmed operation

The performance of a collective, or group collective, control system may be improved by including in the control gear facilities for functional patterns of lift operation to suit particular circumstances, sometimes described as lift logic. It is possible to have lifts 'home' to the ground floor during arrival periods and to positions up the building, closer to possible calls, during departure periods and to have some lifts 'home' to ground and some to intermediate floors during working periods. If the lift is full it is possible to arrange for calls to be ignored until the load has been reduced, thereby saving the time of wasted stops when no more passengers can be taken. When traffic is light some lifts can be taken out of operation, while when it is heavy the journeys of lifts in a group can be timed not to coincide, thereby reducing waiting time.

Motor rooms

Motor rooms are required for all except some special types of lift installation. Normally they are placed over the top of the lift well and require a greater plan area than that of the lift well. It is possible to place motor rooms at the foot of the well or at an intermediate level. The advantages of motor rooms not at the top of the building are easier sound insulation and the avoidance, for the architect, of a perhaps awkward volume at the top of the building. The disadvantages are the increased loads on the structure; the increased length of ropes required and consequent increase in maintenance costs; and the difficulty of achieving adequate ventilation to dissipate the heat generated by the electrical gear. Diagram 282 shows a comparison of the loads imposed on the building and simple roping arrangements for motor rooms at the head and at the foot of a lift well. More complex patterns are used in special circumstances and the ropes for high speed lifts may be wrapped right round the sheave.

Final size and layout of motor rooms depend on many factors. Lift manufacturers publish tables of typical sizes which should be consulted as early as possible for layout and details of sound precautions. Good ventilation is essential in motor rooms to dissipate the heat generated. Rates of the order of thirty air changes per hour may be needed. Some heating may be needed to prevent undue humidity affecting the electrical gear when it is not in operation. A floor strong enough to sustain the loads of machinery and car and counterweight and a beam to support hoisting tackle for lifting the machines must be provided at an appropriate position in the roof. It is usual for the lift installers to use this beam to raise and install their equipment through the lift well itself.

Machine room equipment

The main equipment in the machine room (or lift motor room) will be:

1 Winding machine, comprising the motor, electromechanical brake, worm gear and traction sheave.
2 Controller, mounted on a frame, usually against a wall and contained in a sheet-steel case.
3 The floor selector. This is operated by a steel tape attached to the car and passing over a sheave top and bottom.
4 Overspeed governor, which operates the safety gear on the car in the event of it exceeding a certain speed.

Sound insulation of the machinery is essential Variable voltage motors reduce noise and the gearless traction machine is quieter still. It is best to set the bed-plate of the motor on a heavy block of concrete set on a bed of cork. Even so noise from the lift is difficult to eliminate, the clicking of the contacts is a sharp penetrating noise, so it is best to plan non-living rooms near the lift well. In some cases the lift well forms an independent tower quite independently of the frame or floors of the strucutre, with separations of strips of cork.

Lift well and pit equipment

The lift car is supported on sets of steel ropes, usually four, clamped to the car sling and to the balance weight. Two guides at the point of balance of the car on each side, are fixed to the lift-well wall or to special steelwork. These fixings are at about floor levels. Two other guides are required for the balance weight. These guides are T-section steel accurately machined and finished to 0.05 mm limits.

The lift shaft must extend down below the lowest landing into what is known as the lift pit. In this pit are fixed the buffers, spring-type for slower speeds, oil loaded for high-speed lifts. The depth of this pit varies from 1 m for 0.5 m/s lifts to 1.6 m for 1.5 m/s lifts; with gearless machines this will be 2.5 m for 2.0 m/s and 2.8 m for 2.5 m/s. Similarly at the top, head clearance is necessary for overrun. This is a minimum of 4 m up to 4.6 m for 1.5 m/s lifts, but 5.5 m and 5.8 m for gearless lifts.

Plan size of the well depends on the size and shape of car, the type of doors or gates used and can be obtained from manufacturers' tables. Diagram 284 shows some typical sizes related to one particular arrangement, having the balance weight at the side and a single panel sliding door.*

Where two or more lifts are fitted in one well an additional allowance of 100 mm is made between them for the guide supports.

It is very desirable that the inside of the lift-well be smooth and free from ledges, recesses, etc. The floor trimming should be very accurate. In a framed building it may be appropriate to surround the lift with light partitioning built off the floor. This should be set true to the lift-well dimensions, so that there is no ledge at floor level.

*BS 2653, Part 3, gives outline dimensions for 'light traffic passenger lifts' (up to 8 persons, speeds up to 0.75 m/s), 'general purpose passenger lifts' (up to 20 persons, speeds up to 1.5 m/s) and 'general duty goods lifts' (up to 2000 kg load).

MECHANICAL CONVEYORS

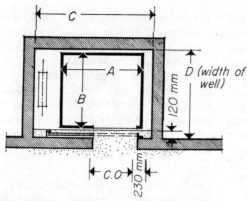

PLAN OF LIFT CAR AND WELL

0.5–1 metre per second	Number of passengers			
	4	5	6	8
Load kg	270	340	400	540
Dimension A	1 m	1.1 m	1.2 m	1.5 m
B	0.8 m	0.9 m	0.9 m	1 m
C	1.7 m	1.7 m	1.9 m	1.9 m
D	1.2 m	1.3 m	1.3 m	1.5 m
E Width of machine room x	2.4 m	2.4 m	2.4 m	2.7 m
y			2.7 m	2.7 m
z				3 m
F x	2.1 m	2.7 m	2.7 m	3 m
y	2.7 m	3 m	3 m	3 m
z				4 m
C.O.	0.7 m	0.7 m	0.8 m	0.8 m

Dimension x for speed of 0.5 metres per second
Dimension y for speed of 0.75–1 metres per second
Dimension z for speed of 1 metre per second
If doors manually operated C.O. is 50 mm less

SECTION

ALTERNATIVES with two-panel doors

GEARED LIFT   balance weight at side · single panel sliding door power operated

284 Typical sizes for lift car, well and machine room

At the front or door side a ledge will, however, be necessary for the landing doors or gates. There is no specific height for this projection, which is often a beam casing, but the underside should be splayed as shown.

The car sizes in diagram 284 are internal sizes and not overall platform sizes and may have to be varied slightly, depending upon the car finish.

Lift doors and gates

Sliding doors of sheet steel are the most satisfactory. They can be made very safe, are easily suited to the closing mechanism and can be made to have a fire resistance of a long enough period to satisfy most authorities.

Generally the doors are 30 mm thick flush, of sheet steel on a light frame, all spot welded, but a finish of special metal like anodised aluminium can be supplied, or they can be made of wood.

The door should have a rigid frame of steel angles, so that it can be fixed as one complete element to ensure accurate positioning of the tracks.

The hangers for the doors are usually fitted with plastic rollers, of large size, on ballbearings for reasonably quiet operation.

All landing doors and gates are fitted with electro-mechanical safety locking gear, with a special emergency lock release operated by a secret key.

Single-panel sliding doors are suitable for small offices, hotels and flats. They are particularly suitable for council flats, as they are simple and safe. Two-panel one-way doors give a wider entrance in proportion to the total width of well. The leading door panel moves at twice the speed of the second panel. Two-panel centre-opening is the best arrangement for busy lifts and in openings over 1.5 m four-panels centre-opening may be used. Single hinged doors of wood are quite suitable for lifts for residences, small hotels, etc. They should have spring closing and check mechanism.

Collapsible steel gates are suitable for slow speed attendant-operated lifts and are commonly specified for goods lifts, but they are noisy and provide no smoke check, which the fire authorities usually require. They should not be power operated.

Flush-leaf sliding gates are virtually the same type of gate with flat 1.6 mm metal sheaths fitted by one edge to the outside of the steel pickets so that the sheaths slide over one another and pack away at one side. To get a clear opening of 1 m, this type of gate will need a landing opening of 1.35 m. These gates normally satisfy the fire authorities and look quite neat.

Shutter gates operate rather similarly, but the 150 mm wide 1.6 mm steel plates are hinged together and secured to the pickets alternately so as to open with a concertina effect. They can be made to conform to 2-hour fire test of the Fire Offices Committee. To get a 1 m clear opening the landing opening must be 1.25 m.

The power operation for sliding doors is usually by a small electric motor operating an arm through a worm gear and spur wheel. This is arranged so that the door moves fast in the middle of the travel but slows down to the open and closed position. A sensitive strip can be fitted to the leading edge of a car door so that if an obstruction is met both doors will stop and re-open. Power operation is possible for shutter or flush-leaf sliding gates.

The lift car

This has a basic framework of steel angles and channels which is called the sling, incorporating the fixing for the guide shoes and the *safety gear*. There are two kinds of safety gear, the dead-grip type giving instantaneous action for the slow and medium speed lifts, and the wedge-clamp type giving gradual action for high-speed lifts. When the lifting ropes break or stretch unduly, or, when the governor operates, if the lift exceeds a fixed speed, the safety gear is brought into action by an independent steel rope.

In the dead grip type two pairs of cams with serrated surfaces grip the guides and stop the lift in a very short distance. The wedge-clamp type, controlled by governor, is operated by wedges which are forced against the clamps which steadily close upon the guides and bring the car gradually to rest within a prescribed distance.

The car can be made of wood panels or of sheet metal on light framing. There is a considerable range of decorative finish available, such as aluminium sheet with matt, burnished or anodised finish, metal-faced plywood, plywood or blockboard with wood veneer or plastic veneer, or linoleum or rubber sheet. Car surfaces get considerable wear and should be robust and easily cleaned and renovated. This particularly applies where luggage or goods are likely to be carried, as in flats, hotels, etc. Floors can be wood, lino, rubber, plastic, cork, or with fitted carpet.

MECHANICAL CONVEYORS

Vision panels glazed with wired glass or armour plate are necessary where the lift is manually operated, but large panels in high-speed lifts only tend to cause alarm among the passengers, who otherwise are not, in a good lift, aware of their speed of travel. It is important that there is good ventilation in the car, either by simple grilles or by concealed louvres in the roof; these are often combined in the lighting fitting which can well be set in a recess in the ceiling with air gaps on all sides. It is important in all passenger lifts, except those in private houses, that the lift is on all the time and operated by a key only.

Planning for lift installations

In very large buildings separate vertical circulation towers may be justified and in staircase access flats separate sets of lifts are essential. In the majority of buildings, however, the most critical planning consideration is the grouping of lifts together. This ensures that if any lift is available in the building a person entering may immediately use it, whereas if the lifts were scattered it is possible to have both idle lifts and waiting passengers. The way in which the lifts are grouped is important. Two or three lifts may satisfactorily be placed side by side but if more than this are placed in a row it is not easy for passengers to note the arrival of a lift, or if the warning of arrival system is good, to get in before the doors close. Lifts should never be separated by corridors or stairs since this not only makes worse the difficulties of view and access described above but introduces conflicting circulation arrangements and makes control interconnection and maintenance more difficult. Diagram 285 shows how grouping lifts to face each other gives better visibility and quicker access than arranging them in one bank.

Ideally groups of lifts should be placed at the centres of gravity of the circulation systems that they serve. This is not always convenient on the entrance floor since the lifts may then be remote from the entrance and may also constrict the planning of the entrance floor which often has to serve a purpose different from that of the floors above. If placed too close to the entrance general circulation may be restricted and waiting rendered unpleasant by people passing. In general, therefore, a lift lobby of adequate size to accommodate the numbers likely to be waiting should be provided where it will not be affected by or affect the other movement through the entrance but be easily seen and readily reached by people coming into the entrance. This arrangement is very compatible with the satisfaction of users who generally appear to be more prepared to walk short distances on upper floors than on the entrance level. The waiting areas for lifts should be clear of other circulation, otherwise unsatisfactory working and general dissatisfaction with both circulation and waiting can result. In very tall buildings it may be economical to divide the building into two from the point of view of lifts. One lift lobby and group of lifts serving ground to mid-height and the other passing from the ground floor directly to mid-height and stopping in the normal way from mid-height to the top. In domestic buildings ease of access for deliveries and adequate manoeuvring space for perambulators must be borne in mind.

Size and number of lifts

Some basic considerations govern the minimum size and number of lifts. Where vertical circulation is dependent on the lift and walking upstairs is impracticable it is desirable to have two lifts to allow for the possibility of one being out of action due to defect or routine maintenance. Well-maintained lifts are very reliable and in large cities where routine maintenance can be carried out at night and repair teams can be very rapidly available, thought might be given to relaxing this requirement where it conflicts with economy and where doing so has no critical consequences. In some circumstances minimum size is critical. In multi-storey housing lifts must be large enough to take perambulators and furniture. It is not usual, however, for them to be

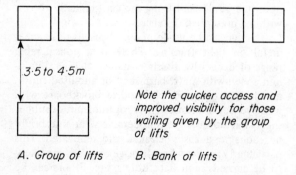

285 Grouping of lifts for best service

MECHANICAL MOVEMENT OF PEOPLE

Possible four-lift installations

Speed (m/s) motor	Capacity	Round-trip time	Waiting time
1.25 variable voltage control	13 persons	133 sec	33 sec
1.5 variable voltage control	12 persons	124 sec	31 sec
2 gearless	11 persons	109 sec	27 sec

large enough to take coffins, although some local authorities insist on adequate space to take a stretcher. Where firemen's lifts are required they have to be of 8-person capacity, which is also the standard for perambulator lifts. In other special cases particular minimum size considerations may apply.

The installation appropriate to a particular building depends on the traffic to be served and the speed, size and number of lifts. Speed is usually considered in terms of round-trip time which includes the time taken for entering the car, doors closing, acceleration, running at standard speed and deceleration, doors opening and closing, passengers entering and leaving, and acceleration for the number of stops that the lift will make. The traffic to be served in office buildings is normally assumed to be 75 per cent of the population on the second floor and above, to be taken up from the ground floor in a period of 30 minutes. In flats the usual standard is 6 per cent of the population on second floor and above to be moved in 5 minutes.

If in a particular case the size of the lift is assumed and the round trip time estimated from the manufacturer's data on lift performance and the probable number of stops (from theoretical lift data) it is easy to determine the number of lifts that will be required.

$$\frac{30 \text{ minutes} \times \text{lift capacity}}{\text{Round trip time}} = \text{Number of people discharged by one lift}$$

$$\frac{\text{Total number of people}}{\text{Number discharged by one lift}} = \text{Number of lifts required}$$

If very large lifts have been chosen the round trip time will be long and arrivals may have to wait some time before they can enter. The average waiting interval can be assessed by dividing the round trip time by the number of lifts. In office buildings this value should not be greater than 30 seconds, while in flats 90 seconds is appropriate. If the waiting interval is too long a greater number of smaller lifts should be considered.

As an example, consider an office building of ten floors of 780 m² each lettable area. On the basis of 7 m² per person in offices, this gives 111 persons on each floor, or 888 persons on eight floors above the first. The transportable population within the half hour is 75 per cent of this, or 666 persons. If four lifts are used it means that each lift must be capable of taking 166 persons in 30 minutes. From tables in a lift manufacturer's handbook there is a choice of:

The small advantage of the gearless lift in this case would probably not warrant the considerable extra cost unless a specially smooth service was wanted. If three lifts were considered they would all have to be 2.5 m/s gearless, and each would have to take 17 persons. The waiting time works out at 49 seconds, which is too long.

Diagram 286 shows some dimensions which enable an approximate idea of lift car and well sizes to be obtained. In proportion lift cars should

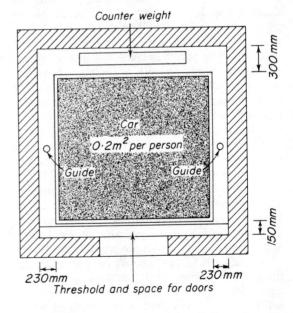

286 *Approximate lift car and well dimensions for preliminary planning purposes*

be square, or with the door in the longer side if they are rectangular. Deep narrow lift cars make exits difficult and time-consuming.

Goods and other lifts

The type and size of *goods lift* is usually determined by the type of goods transported, their bulk and weight, or more probably the dimensions and loaded weight of the trucks used for their transport. In the absence of any specific figure, 340 kg per m^2 of car is allowed. Speeds of goods lifts are not usually an important factor, loading and unloading taking up much more time than the travel time; 0.25 or 0.5 m/s are usual.

Service lifts, such as food lifts for restaurants, have been referred to in the section on controls. These can be specially designed to suit any particular set of circumstances, but the following particulars would suit a lift to take 50 kg (balance weight at the back):

> Car size: 560 mm square x 690 mm high
> Well size: 780 mm wide x 780 mm deep
> Hatch size: 560 mm wide x 690 mm high
> Headroom at top: 2.7 m.

The motor and control gear can usually be accommodated in the same plan area as the well and within the headroom mentioned, provided that doors are set in each wall so that access can be got to any part. Doors to the hatches can be hinged or, better, sideways parting, in which case more space would be needed in the well than given above, which only allows for vertical sliding or bi-parting vertically sliding (called *rise and fall shutters*). The latter give the best service. Lifts for special purposes include *hospital bed lifts.* Code of Practice 407.101 gives much information on these. Accurate levelling is important, so slow speed is favoured unless it is likely to be very busy. A minimum depth of 2.3 m is advisable to take a wheeled bed and minimum width of 1.2 m, but this should be increased to 1.6 m wherever possible. Height should be 2.2 m. If the doors are power operated, means must be provided of keeping them open when required. Open and close buttons in the car are probably the best.

A special lift for *firemen* may be required by the fire authorities, as in London in buildings over 24 m high. This should have direct access from the street and have a special switch at that entrance which will cancel all calls and return the car to that level, after which the car pushes only are operative. The lift should take 760 kg, or 8 firemen, and have a speed of 0.7 m/s. The car is to be of steel and 2.5 m high, with shelf 1.8 m up. An electric supply separate from the main supply to the building is necessary.

Basement hoists for goods only, and to take 1 000 kg, consist of a platform with dwarf sides and back running between two channel guides. They are normally operated by lifting ropes, which are attached to the top of one guide and pass under the platform and up to the other guide, and then down to a winding drum in a small machine enclosure at the side. A platform 1.4 m^2 needs a well 1.7 m x 1.5 m and a pit 1 m deep. The machine enclosure at the side may be 2.4 m x 1.5 m. The speed is very slow.

Hand power lifts

These are useful for small serveries or food lifts in small restaurants, or even houses, as goods lifts where the use is infrequent, or for basement hoists or ash hoists. They are covered by Code of Practice 407.301.

The lighter types of hand-power lift are made of timber, including guides, the heavier, ie over 1000 kg, are usually framed in steel angle with wood panels. Hand-power lifts for over 100 kg are not recommended for frequent use. Dimensions are not much different from those for the electric service lift given below.

The operation of the lift is by hauling rope which passes outside the enclosure. The self-sustaining gear is automatic in action, locking in any position as soon as the hauling rope is not pulled.

The ash hoist is, of course, made of steel and is not normally balanced. It is under-driven like the electric basement hoist, but by a chain which is wound through a self-sustaining chain winch.

Escalators

Escalators can move very large numbers of passengers. To achieve a similar service a very large number of lifts, occupying more floor space, would be needed. They can be reversed in their direction of operation thereby lending themselves to 'tidal flow' situations. Typical applications where large numbers of people have to be moved over limited heights and where tidal flow is marked are departmental stores, underground railway stations and exhibitions. Where numbers of passengers justify it two or even three escalators may be used offering various possibilities of inward, outward and two-way flow. The speed of escalators is normally 0.45 to 0.5 m/s; faster travel is possible but apparently produces little increase in the capacity of the escalator. Typical widths and discharge capacities are shown in Table 78.

Diagram 287 shows other typical dimensions for small escalators. 0.6 m wide steps will permit one laden adult or an adult and child to stand on the step, while 1.0 m width permits passing even when carrying parcels.

It is possible to use a series of escalators to serve several floors. A corkscrew or scissors arrangement gives almost continuous travel for passengers and a double corkscrew may be arranged to give a choice of both up and down directions from each floor.

The large floor openings required by escalators pose a problem of fire spread and it may be necessary to provide a separate hall on each floor separated from the rest of the building by walls with openings protected by automatic fire doors. Automatic shutters closing the openings and water curtains may be considered in some cases.

The need to plan for escalators is clear and also the need to carry the loads that they may impose on the structure. It is also important to remember that escalators are normally delivered to the building as a unit ready for installation and adequate access is important.

Table 78 Discharge capacities of escalators

Width of step m	Nominal width m	Overall width m	Discharge capacity persons per hour
0.6	0.8	1.3	5000
0.8	1.0	1.5	7000
1.0	1.2	1.7	8000

Paternosters

A paternoster consists of a number of open-fronted cars supported on endless chains running over

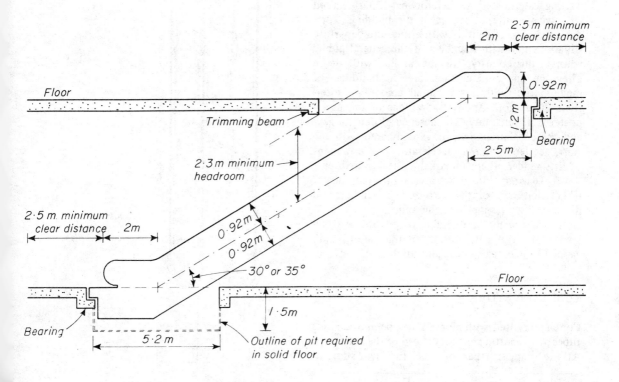

287 Typical dimensions for a small escalator

MECHANICAL CONVEYORS

sprocket wheels at head and foot. The cars move continuously, providing both up and down movement. Support is only at the top of the car and consequently the cars remain vertical and the same way up at all times. No doors are fitted to car or opening, although aprons are fitted above and below the car to prevent anyone entering the wrong area. Safety devices to stop the motor are fitted and leading edges are hinged to prevent injury if improper use is made. The safety record of paternosters is good. The cars are shallow and capable of taking two people side by side. Passengers have to judge the motion of the car, usually about 0.35 m/s, and step in as it passes. The maximum discharge capacity of a single paternoster is stated by manufacturers as being about 600 persons per hour, but this type of apparatus is normally selected for the type of service it offers rather than a specific capacity.

The main advantage of the paternoster is that passengers can arrive at any floor, intending to travel up or down, and immediately begin the journey. Normal lift installations would inevitably cause some waiting. Paternosters are therefore particularly suited to serve the internal traffic of the building. They do not deal so effectively with arrival and departure situations nor with members of the general public who, in this country, are not familiar with them. They are appropriate, therefore, in buildings of medium rise, with a substantial amount of inter-floor circulation. Another advantage of the paternoster is its reliability. The absence of control gear and the continuous motion as compared with the start, accelerate, travel, decelerate, stop movement of lifts renders them remarkably free from breakdown. Diagram 288 shows a view of a paternoster installation free of the surrounding building and diagram 289 gives a plan of the well.

It will usually be desirable to provide a lift in association with paternosters for the use of handicapped people, tea trolleys and goods.

TUBE CONVEYORS

Containers fitted with air seals may be moved along tubes by exhausting air from ahead of the container. At the simplest a pair of tubes link two stations.

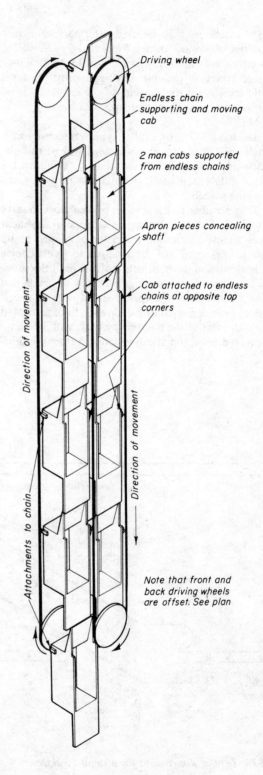

288 Isometric view of paternoster installation

TUBE CONVEYORS

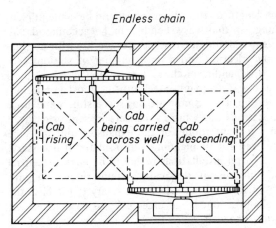

Note method of supporting cab from the two endless chains by means of attachments at opposite top corners

289 *Plan of well for paternoster installation*

Air is exhausted from the end of one tube and containers can therefore be despatched in one direction. Excess air flow is prevented by flaps at entry and exit points and consequently a bypass connection between the ends of the tubes remote from the point of exhaust will produce an air flow away from the exhaust end and carriers can also be sent in this direction. This arrangement is very simple and robust and no sophisticated controls are required. No choice of destination is offered unless further pairs of tubes are installed. The system is, therefore, particularly suitable for cases where all communication is from one central point to a number of outlying stations and where the outlying stations do not communicate with one another. This is the case in department stores with a central cashier's office and this type of installation has been used in this way for many years. The convenience of the system for conveying papers and a variety of small articles has led to its use in other circumstances where outlying stations do communicate. At first this necessitated a central exchange where containers were manually transferred from one tube to another. This arrangement required a great deal of tube and substantial space was required for this near the main station. Improved efficiency and economy can be achieved by a ring-main type of installation with automatic selection of destination and several systems have been devised to give this sort of service. Table 77 shows typical layout patterns of tube systems. Diagram 290 shows a station on a ring main, demonstrating how containers may be received and sent from a point on a ring main. Diagram 291 shows a typical carrier unit.

Both round and rectangular tubes are used for tube installations, sizes varying from just over 50 mm to 150 mm diameter for round tubes and up to 380 x 130 mm for rectangular tubes. As well as cash, bulky documents, drawings, films, diet

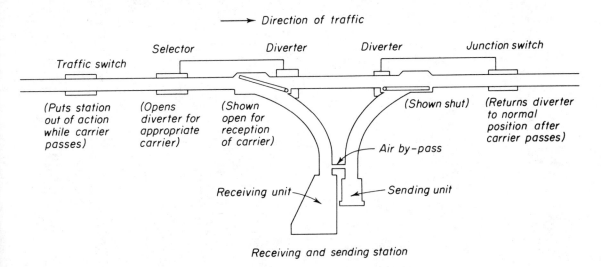

290 *Receiving and sending station on a pneumatic tube ring main system*

MECHANICAL CONVEYORS

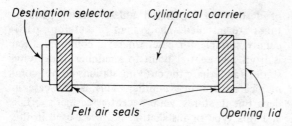

291 *Typical carrier unit for pneumatic tube ring main system*

sheets, drugs, tools and small spare parts may be carried. Movement in all directions, including the vertical, is possible and there is no close limitation on length of run. Provision should be made in planning so that easy bends can be accommodated. Electrical and acoustic devices are used for destination selection and a large number of stations (80-120, depending on the system employed) can be used on one ring main. Steel tube is most commonly used but PVC is now proving successful. Some noise at stations is inevitable as carriers are sent or received and a plant room for the exhauster unit must be allowed for in planning and insulated if necessary. In central installations it should be near the central station but in ring-main systems discussion with the manufacturers will be necessary to establish the right position. In all cases exhaust to the fresh air will be needed.

18 Firefighting equipment

Modern buildings require not only means of escape, access for the fire brigade and structural protection, but also first aid equipment for occupants to use on small fires while waiting for the arrival of the brigade and in some cases fixed installations to help contain the fire until the arrival of the fire brigade. In very large and high buildings special installations may also have to be provided for the fire brigade. Expert advice on the appropriate provisions and the maintenance desirable for particular cases may be obtained from local fire brigades which normally maintain an office for this purpose.

Fires are extinguished by removing the fuel, excluding oxygen or cooling the materials involved to below ignition temperature. Removal of combustible materials is carried out manually in case of need. Firefighting equipment acts by excluding oxygen or cooling. Water is cheap, readily available and acts extremely efficiently as a cooling agent. It is therefore used, except in cases where electricity or oil or other special features which render it hazardous are present or where its use would cause damage which could be avoided by other means.

HAND EXTINGUISHERS

Buckets: water

Buckets were widely used before chemical extinguishers and hose-reels were developed. They are cheap and their use is well understood. On the other hand they are liable to be empty unless well maintained and are expended at a throw. Projecting the water to any point above floor level is difficult and the main application of buckets will be to fires, including rubbish, occurring at floor level. Fire buckets should be painted red, have round bottoms to prevent use for other purposes, be fitted with covers and be fixed at a convenient height. A better method is to have galvanised steel tanks filled with water containing a number of fire buckets.

A normal provision of fire buckets is 3 to each 250 m² of floor area with a minimum of 6 buckets.

Buckets: sand

Sand is non-conducting and can be employed for small fires associated with electrical apparatus where water would not be possible and also for small liquid fires which are covered or absorbed by the sand. The technique can only be applied on flat surfaces. In tanks the sand will merely sink to the bottom.

Soda-acid

Bicarbonate of soda is dissolved in water in an enclosed container. Sulphuric acid in a bottle in the container is released into the water either by

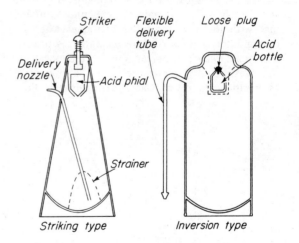

292 *Types of soda-acid extinguishers*

breaking the bottle by means of a metal striker or by inverting the extinguisher. The chemical reaction in the container generates carbon dioxide gas which pressurises the container and forces the water through a nozzle. Diagram 292 shows two forms of soda-acid extinguisher. It will be noted that it is important to keep the extinguisher the

329

FIREFIGHTING EQUIPMENT

correct way up and that this is different for the two extinguishers shown, one of which remains upright while the other must be inverted. Reversing an extinguisher will allow the carbon-dioxide to escape while leaving the liquid undischarged.

For firefighting purposes this type of apparatus acts as a water extinguisher. The chemicals merely provide the power to produce the jet. The advantages of soda-acid extinguishers compared to buckets are the freedom from evaporation and risk of inappropriate use as well as the jet delivery which enables the water to be carefully directed at a fire from some distance away. Soda-acid extinguishers must be protected against frost.

A normal standard of provision is 10 litres of extinguisher capacity for each 250 m^2 of floor area with a minimum of 20 litres on each floor.

A similar performance is given by an extinguisher operated by a compressed gas cartridge. This type has the additional advantage that antifreeze can be added to the water.

Foam extinguishers

Diagram 293 shows a foam extinguisher. Chemicals dissolved in the water in the inner container mix with those in the outer container when the extinguisher is inverted and generate gas pressure and foam which is forced out of the nozzle by the pressure. A gas pressure type is also available as in the case of soda-acid extinguishers. Foam extinguishes fires by excluding oxygen. They can be used in most cases where water is appropriate but are especially suitable for burning liquids since the foam can float upon the surface.

The usual scale of provision is the same as that for soda-acid extinguishers.

Vaporising liquids

A number of liquids which vaporise rapidly when applied to a fire are used in extinguishers. They act by excluding the oxygen. They are non-conducting, do not spoil most materials on to which they might be sprayed and disperse by vaporisation after use. They are widely used for electrical apparatus and motor vehicles. There are some serious disadvantages: they are to varying degrees toxic, either before or after use. When used externally winds may disperse the vapour and allow the fire to re-establish itself.

Carbon dioxide

Carbon dioxide gas can be compressed to a liquid and stored in pressure cylinders. When the pressure is released the liquid vaporises and expands very rapidly and the gas will extinguish fires by excluding oxygen. It causes no damage to materials on which it is used and for this reason is often used for machines, libraries, museums and similar situations. It is non-conducting and well suited for use on electrical equipment, and it is effective on small liquid fires. It is not toxic but can cause suffocation by excluding oxygen from the lungs. The discharge, which is made through a nozzle (shown in diagram 240), has an effective range of only 1 to 2.5 m. Many CO_2 extinguishers have a control valve so that complete discharge is not inevitably made immediately the extinguisher is used. An extinguisher containing 3 kg of carbon dioxide is regarded as the equivalent in extinguishing capacity to 3 water buckets.

Dry powder

These extinguishers deliver a cloud of inert powder expelled by carbon dioxide or nitrogen. The powder cools the flames, separates them from the combustible material and to some extent excludes oxygen. The powder is non-conducting, so that use on electrical equipment is possible and it is also effective on small fires of liquids. In appro-

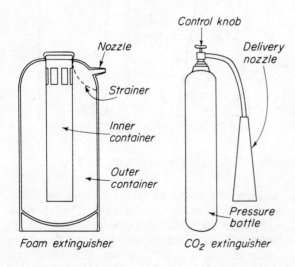

293   Foam and carbon dioxide extinguishers

FIXED APPARATUS

priate circumstances 1.8 kg of powder may be regarded as the equivalent to 1 water bucket and 9 kg of powder may also in appropriate cases be regarded as the equivalent of a 9-litre foam extinguisher.

Blankets

Asbestos, or glass fibre, blankets can be used to wrap around people whose clothes are on fire and also to cover small fires or fires in small open containers. They are often used where welding is taking place and in kitchens.

FIXED APPARATUS

Hose-reels

Small diameter rubber hoses for use by the occupants of buildings provide a very much more effective measure than buckets or soda-acid extinguishers and in many circumstances will be competitive in price. Rubber hose is used, usually of 18 or 25 mm internal diameter. Up to 22 mm

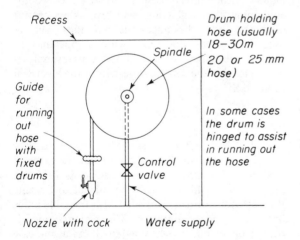

294 Hose reel in recess

of unreinforced hose or up to 36 m of reinforced hose is wound on to a drum. The hose is connected to a water supply serving the spindle of the drum and fitted with a small-diameter nozzle with control cock. Diagram 294 shows such a reel installed, as is usual, in a recess in corridor or landing wall. In case of need the control valve is turned on, then the person using the hose takes the nozzle end of the hose and moves towards the fire, pulling out the hose as he goes. Diagram 294 shows the guide which is used to make this possible with fixed drum hose-reels. A superior arrangement is the use of a swinging arm hose-reel where the drum automatically swings to follow the line of movement of the hose.

The hydraulic requirements for hose-reels is that they should be able to deliver 22 litres per minute at a distance of 6 m from the nozzle and that three should be capable of operating at any one time. A pressure of 207 kN/m² (21 m head) at the connection is usually considered appropriate for each reel. Many water authorities will allow connection of hose-reels direct to the main and when the mains pressure is adequate this provides a simple supply arrangement. In some cases break-tanks are called for, usually of 1150 litres minimum capacity. If the break-tanks can be high enough to supply the hose-reels by gravity no other special provision is needed. If they are not or if in the case of direct connection the building is too high for the hose-reels to operate, duplicate pumps capable of appropriate water delivery must be provided. They are usually installed in the basement and automatic means for bringing them into operation, such as flow or pipeline switches, or by operation of the reel, must be provided.

Hose-reel installations must be taken into account when buildings are being planned. They should be sited near means of escape so that they are readily available to people leaving the building and if the floor is full of smoke after the hose-reel has been in use the hose itself can guide its users to safety. There should be sufficient hoses, of appropriate lengths, so positioned that the hose can

(a) enter every room
(b) when fully extended reach to within 6 m of every part of the floor.

Sprinklers

There is clearly great advantage, particularly for buildings involving special fire risks or only intermittently supervised, such as very large or high buildings, underground car parks, warehouses or stores, for an automatic fire-fighting installation which does not depend for its operation on the presence of occupants. Sprinkler installations fulfil such a purpose. They consist of a grid of water pipes fixed under the ceiling with delivery heads normally on a 3 m square grid. Water is prevented from emerging by a glass or quartzoid bulb containing liquid. Diagram 295 shows a typical sprinkler head. When the temperature rises the liquid expands, breaking the bulb which is preventing the water passing. Bulbs are available which break at temperatures ranging from 68°C to 180°C. A jet of water then impinges on the shaped deflector plate which results in a spray of water being delivered over an area of about 10 m². The flow of water is arranged so that it works a turbine-operated fire alarm. The sprinkler installation therefore both sounds the alarm and helps to minimise and contain the fire. In an unheated building there would be danger of freezing in the winter. It is possible to overcome this by means of the dry pipe system where special valve arrangements enable the delivery pipes to be filled with air under pressure. When this pressure drops, due to the opening of a sprinkler head, the water is admitted to the delivery pipes.

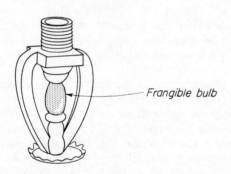

295 *Typical sprinkler head*

Sprinkler systems require central control and test gear. This is often arranged in the basement, but there is much to be said to having this gear, or at least the alarm and stop valve, in a prominent position near the entrance to the building. This will enable the water to be shut off quickly before too much damage is done as soon as the fire is under control. Two independent water supplies are generally thought desirable for sprinkler installations. In towns it is sometimes possible to have supplies from water mains served by two independent trunk mains. When this is not the case storage on the site is necessary. High-level cisterns capable of holding 22.5 m³ to 34 m³ used to be the standard provision. Pressurised cylinders housed in the basement are becoming universal. 22.5 to 50 m³ capacity is usual.

The regular grid required for sprinklers for which complex standards exist depending on the fire protection of the building and the nature of the activities, and the need for control gear and storage space mean that installations must be planned for in the early stages of building design. Rules are laid down for sprinkler supply pipes. The minimum size for them is 25 mm diameter, rising in stages until a 50 mm diameter pipe can serve 18 sprinklers (or 10 in the case of some special risks). Installations with more than 150 sprinklers will require a 150 mm diameter pipe.

A special type of sprinkler can be used for oil fires. Although a jet of water would do more harm than good, a fine spray falling on the surface of burning liquid cools the surface and also, by turning into steam, excludes oxygen.

It is also possible to obtain valves which open on the bursting of a sprinkler bulb. They then admit water to a series of sprinklers. This technique is sometimes used in the case of rooms containing electrical equipment. Sprinklers cannot discharge over the apparatus but they can be used to protect door and window openings to prevent the spread of fire if the water supply is controlled by the temperature within the electrical room itself.

Drenchers

Are devices similar to sprinklers which deliver a curtain spray usually used to protect the external face of a building from some adjacent fire risk. Since they would inevitably be subject to freezing they are not normally provided with frangible bulbs but have open waterways. The water supply is turned on manually when required. As in the case of sprinklers, the minimum size of supply pipe which can serve two drenchers is 25 mm diameter. Ten drenchers require a 50 mm diameter pipe, 36 a 75 mm diameter, and over 100 a 150 diameter pipe. Not more than 12 drenchers may be fixed in any one horizontal branch. Periodical flushing of the pipework is required. The Fire Offices Committee have recently published Rules for Automatic Sprinkler Installations which allow for more rational design of pipework.

Other extinguishing media

Both hose reels and automatic systems using fixed pipes can be provided for delivering carbon dioxide, foam or, in rare cases, powder from a central reservoir.

Fusible links

Two strips of brass soldered together with low-melting-point solder can be used as a link in a straining wire system. The solder will melt and actuate the system as do sprinkler bulbs. A wide variety of situations can be dealt with by this method. Projection room shutters are held open by a wire running over the projectors with a fusible link over each; the valve controlling entry of oil to a boiler room may be operated by a lever and weight and held open by a wire passing over the boilers with a fusible link over each. Automatic fire-doors, fly-tower roof-lights and fire dampers in ventilation ducting are other devices which may be operated in a similar way.

FIRE ALARMS

The simplest form of fire alarm is a series of manually-operated switches, usually installed behind glass which has to be broken in order to gain access. The operation of any one switch sounds alarms on all floors and records at a central indicator which alarm was operated. Automatic devices which will give warning in case of need can be used. They operate either by being affected by sharp rises in temperature or by being able to detect smoke. In both cases the detector units are placed at strategic points, usually each covering an area of between 20 to 100 m² and wired back to a central indicator. Each type of alarm can, if desired, be arranged to repeat at the local fire station.

EQUIPMENT FOR USE BY FIRE BRIGADE

In buildings of modest size hard access for fire appliances to a reasonable proportion of perimeter is required but no special installation for use by the fire brigade is normally needed. Water for fire-fighting purposes is taken from hydrants in the highway.

Fire brigade hydrants

If the building is extensive or sited a long way from the road, one or more fire brigade hydrants may be needed. Fire brigade hydrants are 65 mm in diameter with a valve and a special type of coupling for hoses. If the water serving the hydrants can flow only in one direction, which is the case in communication pipes to buildings, then a minimum 100 mm diameter is needed and the section of the service pipe up to the hydrant will have to be increased to this diameter. If the water supply to the building is metered a problem arises since it might often be inappropriate to install a 100 mm diameter for comparatively small flow. Diagram 105 shows a way in which the difficulty may be overcome by interrupting the large diameter pipe with a valve under the control of the fire brigade to be turned on only in case of fire, the supply to the building being maintained by a small diameter by-pass including the meter.

Access by fire brigade appliances

For firefighting and emergency escape to take place effectively fire brigade appliances such as pumps and escape ladders must be able to be positioned close to buildings. It is usual for the local authority to insist that an appropriate proportion of the perimeter of the building is accessible in this way. See *MBC: Structure and Fabric*. In buildings remote from the highway with only access roads on the site it may be necessary to construct special roadways or hard-standing to meet this requirement.

High buildings

Until comparatively recently the height of buildings in England was limited to 24 m plus a further 6 m in the roof space. This requirement related to firefighting and escape. The move towards high buildings necessitated the provision within the building of adequate means of vertical access for firefighting which would not be interrupted even when several floors were on fire. The lobby approach staircase or more colloquially the 'fire tower', equipped with wet or dry riser and firemen's lift, has been developed to meet this need.

In buildings over 24 m in height it is usual for one of these towers to be required for every

1000 m² of plan area. The tower consists of a protected stair separated by fire-resisting and self-closing doors at each floor level from a ventilated lobby which itself is cut off from the rest of the floor by similar doors. The tower must be provided with a lift (minimum size 8 persons) which, while it may form part of the normal passenger or goods lift provision of the building, must be powered separately and be controlled in such a way that the fire brigade may override normal use. This lift requires ready access to the outside at ground level.

The stairs and lobby have to be vented to prevent smoke accumulating in case of fire. In a tall building with internal stairs considerable cost and space were involved. Recent high buildings have been provided with pressurised stairs and lobbies. Provided an adequate pressure can be maintained smoke will not be able to enter the pressurised area The cost of fans and input ducts is less than the value of space occupied by the natural draught vents and the mechanical system appears to give a better performance. The most popular arrangement seems to be one where the stair is moderately pressurised at all times and this is automatically increased when any fire detector registers an outbreak.

Means of conveying water for fire-fighting up the building are needed and this is achieved by permanent pipe installations with fire brigade hydrant connections at each floor level. In buildings up to 45 m high a 100 mm diameter 'dry riser' is required. This is equipped with an inlet at ground level to which the fire brigade pumps may be connected. An automatic air vent is needed at roof level and provision for draining down without flooding at ground level is desirable. Between 45 m and 60 m, a 150 mm diameter riser is used, while over 60 m in height.'wet risers' are required. These have their own water reserves on site and are supplied by duplicate pumps in the building itself. An alternative power supply is needed. A standby generator or diesel power for one pump is normal. In wet riser installations the fire brigade may call for a reserve of water which may be 50 m³. In buildings higher than 69 m, pumping in stages up the building may become necessary.

Emergency control of services

It should be possible for the fire brigade to turn off the main electrical and gas intakes at a readily accessible spot so that fire fighting may be easier and safer.

Advice from fire brigade

Local fire brigades are normally very pleased to be asked for advice about firefighting provisions for proposed buildings and it is sensible to take advantage of this at an early stage in design.

19 Ducted distribution of services

In all but the smallest buildings it is usual for the runs of the various services to be grouped together for convenience. Special passages through the building, called service ducts, are normally planned and constructed to accommodate the service runs and to allow for installation, maintenance and replacement. The purposes achieved by these service ducts may be summarised.

1 Concealment of services. (This involves not only appearance but also heat, noise, condensation, etc).

2 Installation of individual services can be independent of the building operation. This enables the service installations to be installed completely in one continuous process which is economical, minimises faults and makes initial testing and adjustment much simpler.

3 Inspection and maintanance are made simpler.

4 Alterations and additions are facilitated.

5 Services are protected.

The most usual pattern of ductwork in buildings involves three types of ducts.

1 *Main ducts* giving basic horizontal distribution services. They usually occur below the ground floor or for air ducts above the roof of the top floor.

2 *Vertical ducts* which rise at suitable points to distribute services to a particular area on each floor.

3 *Lateral ducts* which give local horizontal distribution of services (Suspended ceilings are very often used as lateral ducts.)

Main horizontal ducts below ground may take the form of floor trenches as shown in diagram 296. In very heavily serviced buildings or buildings, such as hospitals, where it would be inappropriate to raise floor panels for access, crawlways, diagram 297, or if the number of services justify it, walkways as shown in diagram 298 and 299 may be used.

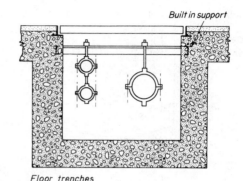

296 *Floor trench*

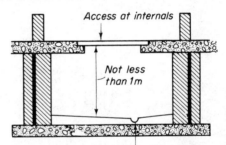

297 *Crawlway for services distribution*

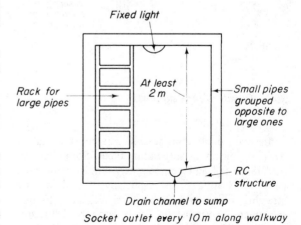

298 *Walkway for services distribution*

335

DUCTED DISTRIBUTION OF SERVICES

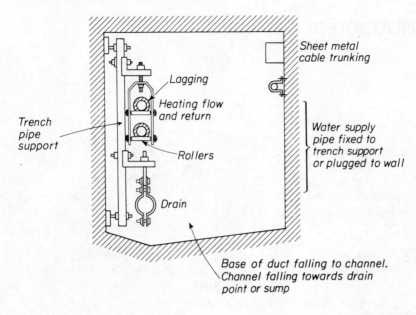

299 Typical main duct running below ground floor level with access for inspection and maintenance

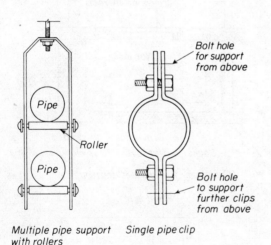

300 Details of pipe supports

In main ducts considerable care must be taken with the means of suspending pipes. Some have to be laid to fall and consequently require adjustable hangers. Heating pipes, especially if laid in long straight runs, will exhibit significant thermal movement and must be firmly supported but not restrained for longitudinal movement. Diagram 300 gives details of the roller pipe support and the adjustable level pipe clip which are shown in diagram 299. Where long straight lengths of pipe occur it may be necessary to have loop-or bellows-type expansion joints for pipes other than heating and hot water. In underground ducts care must be taken to avoid damp penetration (see *MBC: Structure and Fabric* for constructional details).

All major underground ducts should be provided with channels to carry away seepage and leakage. It is usual to arrange these channels to drain to the sump in the boiler room.

Diagram 301 shows a vertical duct. In this case the duct is simply a recess in a corridor wall covered by a ply facing. A hole is left in the concrete floor to allow the services to be installed and this is subsequently filled in as a fire precaution. Access to the service runs is very easy and on upper floors the interference to circulation while work is going on is not likely to be critical. If this arrangement is not acceptable and work on the services must be able to be carried out without access from the corridor, a walk-in type of duct, running continuously up the building with open gratings at each working stage and ladders for access, can be provided. It is very much more expensive and space-consuming than the type previously described and fire precautions may be necessary to prevent spread of flame from one floor to another.

DUCTED DISTRIBUTION OF SERVICES

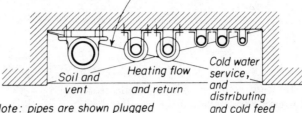

301 Typical vertical duct

Note: pipes are shown plugged to masonary wall. Where this is not practical angle iron supports could be provided and pipes fixed by U-bolts or saddle bands

302 Ventilation trunking suspended from structural flow and concealed by suspended ceiling

303 Typical corridor suspended ceiling showing services distribution above the ceiling

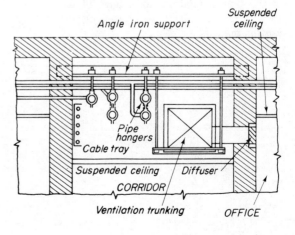

337

DUCTED DISTRIBUTION OF SERVICES

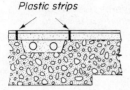 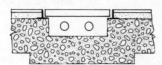

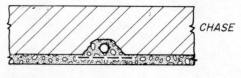
PLAN

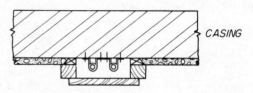
PLAN OF SECTION

304   Floor chases

Positions for vertical ducts should be selected with care. Correction to lateral ducts may often be difficult to contain within ceiling spaces as they leave the vertical duct and distribution may be complicated if the duct is adjacent to a flue, lift or stairwell which prevents lateral distribution in that direction.

Lateral ducts are very often formed in ceiling spaces. Diagram 302 shows a basic method of suspending a ventilating trunking at high level one a corridor and diagram 303 shows a typical corridor ceiling space used for services.

No specific rules can be given for sizes since cases vary so widely. A general method of approach is to decide on the number and type of pipes and casings to be accommodated (including any space reserved for additional installations), estimate their size and the spacing required and design the duct accordingly.

Most of the normally encountered services can be accommodated together in a single duct, provided that in some cases the right relationship between services is maintained (eg cold water pipe liable to condensation should not be fixed above electrical trunking). CP 413, *Design and Construction of Ducts,* for services gives a table of the mutual relations of a range of services one to another, and deals with problems of fire prevention in ducts.

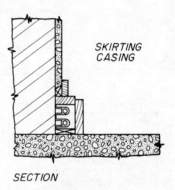
SECTION

305   Wall chase and pipes casings

Final runs of services are often exposed on wall surfaces. Where it is desired to conceal them pipe casings and chases may be used. Diagrams 304 and 305 show typical examples.

Electrical conduits and underfloor trunkings and skirting ducts are dealt with in chapter 15.

CI/SfB

The following tables are based on the *Construction Indexing Manual* published by the RIBA (1968). References are given only to the chapters within which some aspect of the appropriate symbol will be found.

Abbreviations *ES, M, SF* (1) and (2) and *CF* are given for *Mitchell's Building Construction: Environment and Services; Materials; Structure and Fabric Parts 1* and *2*, and *Components and Finishes*.

Table 1   Elements

(1)   **Substructure**
(10)   *Site**
(11)   Excavations, land drainage *SF (1)* 4, 8, 11: *SF (2)* 3, 11
(12)
(13)   Floor beds *SF (1)* 4, 8; *SF (2)* 3
(14)
(15)
(16)   Foundations, retaining structures *SF (1)* 4; *SF (2)* 3, 4
(17)   Pile foundations *SF (1)* 4; *SF (2)* 3, 11
(18)
(19)   *Building**

(2)   **Primary elements**
(20)   *Site**
(21)   External walls, walls in general, and chimneys *SF (1)* 1, 5, 9; *SF (2)* 4, 5, 7, 10
(22)   Internal walls, partitions *SF (1)* 5; *SF (2)* 4, 10; *CF* 9
(23)   Floors, galleries *SF (1)* 8; *SF (2)* 6, 10
(24)   Stairs, ramps *SF (1)* 10; *SF (2)* 8, 10
(25)
(26)
(27)   Roofs *SF (1)* 1, 7; *SF (2)* 9, 10
(28)   Frames *SF (1)* 1, 6; *SF (2)* 5, 10
(29)   *Building**

(3)   **Secondary elements** if described separately from primary elements
(30)   *Site**
(31)   Secondary elements in external walls, external doors, windows *SF (1)* 5; *SF (2)* 10; *CF* 3, 4, 5, 7
(32)   Secondary elements in internal walls, doors in general *SF (2)* 10; *CF* 3, 7
(33)   Secondary elements in or on floors *SF (2)* 10
(34)   Balustrades *CF* 8
(35)   Ceilings, suspended *CF* 10
(36)
(37)   Secondary elements in or on roof, roof lights, etc. *SF (2)* 10; *CF* 6
(28)
(39)   *Building**

(4)   **Finishes** if described separately
(40)   *Site**
(41)   External wall finishes *SF (2)* 4, 10; *CF* 14, 15, 16
(42)   Internal wall finishes *CF* 13, 15
(43)   Floor finishes *CF* 12
(44)   Stair finishes *CF* 12
(45)   Ceiling finishes *CF* 13

* These classes are not used in general documentation but have special application in project documentation.

339

CI/SfB

(46)
(47) Roof finishes *SF (2)* 10; *CF* 18
(48)
(49) *Building**

(5) **Services** (mainly piped, ducted)
(50) *Site**
(51) Refuse disposal *ES* 13
(52) Drainage *ES* 11, 12
(53) Hot and cold water *ES* 9, 10; *SF (1)* 9; *SF (2)* 6, 10
(54) Gas, compressed air
(55) Refrigeration
(56) Space heating *ES* 7; *SF (1)* 9; *SF (2)* 6, 10
(57) Ventilation and air-conditioning *ES* 7; *SF (2)* 10
(58)
(59) *Building**

(6) **Installations** (mainly electrical, mechanical)
(60) *Site**

(61)
(62) Power *ES* 14
(63) Lighting *ES* 8
(64) Communications *ES* 14
(65)
(66) Transport *ES* 15
(67)
(68) Security
(69) *Building**

(7) **Fixtures**
(70) *Site**
(71) Circulation fixtures
(72) General room fixtures
(73) Culinary fixtures *CF* 2
(74) Sanitary fixtures *ES* 10
(75) Cleaning fixtures
(76) Storage fixtures *CF* 2
(77)
(78)
(79) *Building**

Tables 2/3 Construction Form/Materials
Table 2 is never used without Table 3
Table 2 **Construction form**

E	cast in situ *M* 8; *SF (1)* 4, 7, 8; *SF (2)* 3, 4, 5, 6, 8, 9
F	Bricks, blocks *M* 2, 4, 6, 12: *SF (1)* 5, 9: *SF (2)* 4, 6, 7
G	Structural units *SF (1)* 6, 7, 8, 10; *SF (2)* 4, 5, 6, 8, 9
H	Sections, bars *M* 2, *SF (1)* 5, 6, 7, 8; *SF (2)* 5, 6
I	Tubes, pipes *SF (1)* 9; *SF (2)* 7
J	Wires, mesh
K	Quilts
L	Foils, papers (except finishing papers) *M* 9, 13
M	Foldable sheets *M* 9
N	Overlap sheets, tiles *SF (2)* 4; *CF* 18
P	Thick coatings *M* 10, 11; *SF (2)* 4; *CF* 12, 13, 18
R	Rigid sheets, sheets in general *M* 3, 12, 13; *SF (2)* 4
S	Rigid tiles, tiles in general *M* 4, 12, 13; *CF* 12, 15
T	Flexible sheets, tiles *M* 3, 9; *CF* 17
U	Finishing papers, fabrics *CF* 17
V	Thin coatings *CF* 17
X	Components *SF (1)* 5, 6, 7, 8, 10; *SF (2)* 4; *CF* 2, 3, 4, 5, 6, 7, 8
Y	Products

CI/SfB

Table 3 **Materials**

In formed products

e Natural stone *M* 4, *SF (1)* 5, 10; *SF (2)* 4
f Formed (precast) concrete, asbestos based materials, gypsum, magnesium based materials *M* 8; *SF (1)* 5, 7, 8, 9, 10; *SF (2)* 4, 5, 6, 7, 8, 9; *CF* 13
g Clay *M* 5; *SF (1)* 5, 9, 10; *SF (2)* 4, 6, 7
h Metal *M* 9; *SF (1)* 6, 7; *SF (2)* 4, 5, 7
i Wood *M* 2, 3: *SF (1)* 5, 6, 7, 8, 10; *SF (2)* 4, 9; *CF* 2
j Natural fibres and chips, cork, leather *M* 3
m Mineral fibres *M* 10; *SF (2)* 4, 7
n Rubbers, plastics, asphalt (preformed), linoleum *M* 11, 12, 13; *SF (2)* 4, 9; *CF* 12, 18
o Glass *M* 12, *SF (1)* 5; *CF* 5

In formless products

p Loose fill, aggregates *M* 8, 10, 15
q Cement, mortar, concrete, asbestos based materials *SF (1)* 4, 7, 8; *SF (2)* 3, 4, 5, 6, 8, 9
r Gypsum, special mortars, magnesium based materials *M* 15; *SF (2)* 4; *CF* 13
s Bituminous materials *M* 11; *SF (2)* 4

Agents, chemicals, etc.

t Fixing, jointing agents, fastenings, ironmongery *M* 14; *CF* 7
u Protective materials, admixtures *M* 1, 2, 8, 9; *CF* 17
v Paint materials *CF* 17
w Other chemicals

x **Plants**
y **Materials in general** *M* 1

Table 4 **Activities Requirements**

Activities
(Af) Administration, management in general
(Ag) Communications in general
(Ah) Preparation of documentation in general *CF* 11
(Ai) Public relations in general
(Aj) Controls in general
(Ak) Organizations in general
(Am) Personnel, roles in general
(An) Education in general
(Ao) Research, development in general
(Ap) Standardization, rationalization in general *CF* 1, 11
(Aq) Testing, evaluating in general *CF* 1

(A1) **Management** (offices, projects)
(A2) **Financing, accounting**
(A3) **Design, physical planning** *SF (1)* 4; *SF (2)* 3
(A4) **Cost planning, cost control, tenders, contracts**
(A5) **Production planning, progress control** *SF (2)* 1, 2: *CF* 1
(A6) **Buying, delivery**
(A7) **Inspection, quality control** *CF* 1
(A8) **Handing over, feedback, appraisal**
(A9) **Arbitration, insurance**

(B) **Construction plant** *SF* 11; *SF (2)* 2, 11

(C) **Labour***

* *These classes are not used in general documentation but have special application in project documentation.*

CI/SfB

(D) Construction operations *SF (1)* 11; *SF (2)* 2, 11

Requirements, properties
(E1) Construction requirements *SF (1)* 1, 2; *SF (2)* 1; *CF* 1

(E2) User requirements *ES* 1, 2, 3, 4, 5, 6; *CF* 1

(E3) Types of user

(E4) Physical features *CF* 1

(E6) Environment in general, amenities *ES* 1, 2, 3, 4, 5, 6,
(E7) External environment *ES* 1, 2, 3, 4, 5, 6
(E8) Internal environment *ES* 1, 2, 3, 4, 5, 6
(F) Layout, shape, dimensions, tolerances, metric *SF (1)* 2; *CF* 1
(G) Appearance, aesthetics, art

(H) Physical, chemical, biological factors, technology *CF* 1
(I) Air, water *ES* 2, 3; *SF (1)* 5
(J) Heat, cold *ES* 5
(K) Strength, statics, stability *SF (1)* 3
(L) Mechanics, dynamics *SF (1)* 4; *SF (2)* 3
(M) Sound, quiet *ES* 6
(N) Light, dark *ES* 4, 8
(Q) Radiation, electrical
(R) Fire *M* 1; *SF (2)* 10
(S) Durability, weathering defects, failures, damage *M* 1, *SF (1)* 4; *SF (2)* 3, 4, 5
(U) Special requirements, efficiency, working characteristics

(V) **Effect on surroundings**

(W) **Maintenance, alterations**

(Y) **Economic, time requirements** *SF (1)* 2; *SF (2)* 3, 4, 5, 6, 9

Index

Absorption, sound   110
Absorption, sound, coefficients   122
Acoustics, room   120
Activated sludge   289
Admittance factor   93
Admittance method   92
Air: change rates   38, 165
    conditioning plant   156
    handling plant   156
    handling plant, sizes   168
    intake and exhaust, positions   35
    lift, water   194
    moisture content   26
    movement   32
    movement, comfort criteria   37
    movement, patterns   32, 34
    space, thermal resistance   78
    temperature, outdoor air   105
    trunkings   158
    velocities, comfort   33
    velocity, in trunkings   166
Airbourne sound insulation   111
Aircraft noise   119
Alarms, fire   333
Alkathene tube   232
Alternating solar gain factor   93, 106
Aluminium RWPs   263
Ampere, difinition   10
Anti-syphon pipe   256
Anti-syphon trap   258
Apostilb   173
Artesian borehole   192
Artificial sky   55
Asbestos cement drains   267
Asbestos cement pipes   230
Asbestos cement RWPs   264
Automatic flushing cisterns   220
Axial flow fan   156

BRS ball valve   227
    sky component protractors   42
BZ classification   174
Backdrop manhole   282
Bacteria, ventilation criterion   37
Balanced flue   154
Ball valves   227
Barriers, vapour   23
Basement drainage   285
Bathrooms, electrical provisions   292, 304, 297
Baths   221
Beaufort Scale   33
Beddings, drain   270
Bedford, Thomas   68

Bidet   200, 201, 221
Bilham's formula   21
Bins, refuse   291
Biological filter   288
Blankets, fire   331
Block spacing, daylight criteria   50, 52
Bodman ratio   174
Body heat balance   71
Body odour, ventilation criterion   37
Boiler room, schematic diagram   145
Boiler room, sizes   167
Booster, pneumatic, water supply   207
Boreholes   192
Bottle trap   257
Bowl urinal   221
Branch bends   273
Brick manholes   272
British Zonal classification   174
Buckets, firefighting   329
Building Regulations, thermal   76
Building, heat blance   72
Building materials, thermal properties   79
Bus bars, electric   299

CIE sky   39
Cable tap system, electrical   300
Cable trunking, electrical   299
Calculation sheet for Waldram Diagram p48   51
Calculation sheet for Waldram Diagram p 53 with
    circuits for daylight criteria   54
Calidity   68
Calorific values of fuels   131
Candela, definition   10
Capillary solder joints   231
Carbon dioxide extinguishers   330
Casing, pipe   338
Cast iron drains   268
Cast iron pipes   232
Cast iron sanitary appliances   216
Caulking ferrule   259
Cavity resistance, thermal   78
Ceiling ducts   337
Cement and sand joints   266
Central heating   139
Centrifugal fans   156
Cesspools   286
Chain conveyor   316
Channels, manhole   273
Channels, waste, flow   260
Chases, electrical conduit   298
Chases, pipe   338
Chimneys, metal, stacks   36
Chimneys, positioning to discharge effluent gasses   35

343

INDEX

Chute, refuse 293
Circuit, basic electric 296
Circuit breaker, electric 297
Circuits, low pressure hot water space heating 142
Cistern, cold water storage 200
Cistern, duplicate 206
Cistern, insulation 206
Cistern sizes 235
Cistern type water heater 153
Clo values 70
Cold bridges 80
Cold water distribution 199
Combined drainage 275
Comfort indices, thermal 68
Common wastes 260
Communication pipe, water 198
Compaction, refuse 295
Comparative sizes of pipes and trunkings for heat distribution 141
Compression joints 230
Concrete drains 267
Concrete manholes 274
Condensation 23
Condensation prediction, analogue 31
   graphical 26
   numerical 28
Conduction, thermal 76
Conductivity, thermal 79
Conduit, electrical 298
Conservancy system, sewage disposal 286
Control systems, heating 149
Control systems, lifts 318
Convection, thermal 76
Connectors, electric and gass 136
Convectors, water 146
Conversion factors, imperial to SI 12
Conveyors, mechanical 314
Copper pipes 230, 234
Corbelled closets 218
Connected effective temperature 68
Cost, comparative of heating 128
Covers, manhole 274
Crawlway, services 335
Cylinders, hot water 212
Cylinder, 'Primatic' 211
Cylinder, sizing, hot water storage 237

Damp proof course 22
Damp proof membrane 22
Daylight, criteria block spacing, 50, 52
Daylight factor 40
Daylight indicator 52
Daylight protractors, BRS 41
Daylighting, general 39
Daylighting, model analysis 55
Daylighting standards 40
Dead leg, hot water distributing 200, 210
Decibel 110
Declination, of sun 58
Design process 19
Detergents, effects in drainage pipework 250

Dew point temperature 25
Diaphragm ball valve 228
Diffusers, air 159
Dilution of sewage 286
Discharge capacities of drains 281
Discounted cash flow 128
Distribution circuits, electric 302, 305
Distribution, cold water 199
Distribution, hot water 203
Distribution systems, air 155
Distribution systems, heat 141
Diurnal temperature cycle 71
Double glazing, acoustic 119
Drain bedding 270
   construction 265
   discharge capacities 281
   entry to 264, 275
   joints 266
   materials 266
   sizing 283
   subsoil 291
   trench width 271
Drainage basement 285
   design of 251
   estimation of flows and sizes 245
   flow in pipes 247
   hydraulic design 245
   installations 245
   land 290
   minimum cover 276
   minimum falls 251
   paved areas 265
   performance standards 245, 252
   planning and layout 265
   regulations 246
   rodding point 278
   systems 275
   through walls 271
   typical layout 277
   underground 265
   ventilation 276
   velocities of flow 281
Draincock 199
Draught stabilizer 153
Drawings, electrical layout 305
Drenchers 332
Drinking water 209
Drinking water cistersn 210
Driving rain 22
Droop line diagram for surpath diagram 60, 61
Droop line diagram for Waldram diagram 48, 49
Drop manhole 282
Drop system, heating 143
Dry bulb temperatures, recommended 69
Dry riser, fire fighting 334
Dual duct system of air conditioning 160
Duct, drainage 260
Duct, electrical 299
Duct, noise from 119
Ducts, services 335
Ductube 300
Dustbins 292

INDEX

Ear, response to sound 109
Earthenware, sanitary appliances 216
Earthing, electrical 296
Eastern closet 218
Eaves gutter 262
Effluent disposal, sewage 289
Electric, basic circuit 296
    bus bars 299
    cable trunking 299
    circuit breaker 297
    distribution 302, 305
    earth 297
    emergency supply 304
    fixed apparatus 304
    flow trunking 301
    fuse 297
    intake 302
    insulation 297
    lamps 175
    lighting 172
    lighting, heat gain from 176
    overhead distribution 301
    polarity 297
    radiant heaters 136
    ring main 304
    single phase supply 296
    skirting trunking 301
    socket outlets 305
    switches 303
    terms 310
    thermal storage heaters 137
    three phase supply 296
    tubular heaters 136
    underflow heating 136
    water heaters 213
    wiring systems 298
Electrical, drawings 305
    schedules 305
Electricity general principles 296
    mains distribution 296
    emergency supply 304
Emergency supply, electric 304
Energy to heat, transfer mechanism 130
Energy sources 129
Environmental design, general 12
    factors 12
    temperature 93
Effective temperature 68
Equatorial comfort index 68
Equivalent temperature 67, 68
Equivalent warmth 68
Escalators 324
Estimation air handling plant sizes 168
    boiler room sizes 167
    drain sizes and falls 281
    ducts, air 170
    flue sizes 170
    hot water storage 235, 237
    lift sizes and numbers 322
    refrigeration plant 169
    water supply pipes 236
    eupatheoscope 67

excess heat, criteria for ventilation 37
exhaust position, ventilation 35
expansion joints, pipe 336
explosion doors, flue 153
exposure 74
external reflected component of daylight factor 40
external surface resistance 77
extinguishers, fire 329
extract devices, ventilation 158

Falls, drain 251
Fan convectors 147
Fanger 68
Fans, axial flow 156
    centrifugal 156
    induced draught 154
    propellor 155
Ferrules, caulking 259
Filters, air conditioning 156
    biological, sewage 288
    water 196
Fire, alarms 333
    brigade equipment 333
    extinguishers 329
    fighting equipment 329
    high building equipment 333
Fireclay 216
Firefighting equipment 329
Fireman's lift 334
Fixed apparatus, electrical 304
Floating floors 118
Floor: chases, pipe 338
    channel, waste 260
    floating 118
    noise insulation 116
    trench, services 335
    trunking, electrical 301
    U-values 84
Flow in: drains 247, 281
    gutters 263
    RWPs 263
    stacks 248
    water pipes 249
Flues 145, 151
Flues, balanced 154
Flues, sizes 170
Flushing cistern, automatic 220
    manual 219
Foam extinguishers 330
Free outlet water heaters 213
Fuels, calorific values efficiency of production and distribution 130
Fumes, comfort criteria 37
Furring, pipes and boilers 196, 212
Fuse, electric 297
Fusible links 333

GPO telephones 307
GRP manholes 274
Garchey system, refuse disposal 204

INDEX

Gas, balanced flue 154
   calorific value 312
   heaters 135
   installations 312
   pipes 312
   systems for heating 130
   water heaters 211
Gate valve 226
Geysers 210
Glare 41
Globe thermometer 68, 70
Globe valve 226
Gnomonic projection 66
Goods lift 324
Granular bedding, drains 270
Graphical condensation prediction 26
Grilles, ventilation 158
Grinders, refuse disposal 294
Ground water 291
Gulley discharge of waste over 253
   rainwater, trapped 264
   waste connection 259
Gutters rainwater, general 261
   materials 263
   sizing 262

Hand lifts 324
Hard water 195
Health closet 217
Heat: general 67
   balance of buildings 72
   balance of human body 71
   distribution systems 142
   distribution media 139
   emitters, low pressure hot water 146
   gains and requirements 73
   gain, solar 57
   insulation 75
   loss, calculations 86
      maximum rate 86
      seasonal 87
   output by young males 107
   pump 134
   pumps, recovery from exhaust air 163
   recovery wheel 164
   transfer 76
   transfer media 139
Heating installations, general 127
   control 149
   costs 128
Heating system selection 127
Hertz, definition 10
High buildings, firefighting 333
   ventilation 36
   water supply 207
High pressure hot water space heating 144
High velocity ventilation systems 159
Hose reels 331
Hot water: cylinder 212
   distribution 203
   heaters 213
   storage 236
   supply 209
   space heating, low pressure 142
Housemaid's closet 219
Humpheys 69
Humus chamber 289
Hydraulic: jump 248
   ram 194

Illumination, electric 172
   natural 39
Immersion heaters 212
Impact sound insulation 111
Incinerators, refuse 295
Indirect cylinder, patent 211
Induced draught fans 154
Induced syphonage 254
Induction system, ventilation 159
Input devices, ventilation 158
Inspection chambers 271
Instantaneous water heaters, electric 214
   gas 211
Insulation, electric 297
   noise 113
   thermal 75
Intake, electrical 302
   position, ventilation 35
Interceptor, sewer gas 275
Intermediate switches, electric 303
Internal: lavatories, ventilation 164
   surface resistance 77
   surface temperature 92
   telephones 308
Internally reflected component of daylight factor 40
Interstitial condensation 26

Joints drain 266
   pipe 230
Joule, definition 10

Kata thermometer 67
Kelvin, definition 10
Kilner jar trap 257
Kilogramme, definition 10

Ladder system, space heating 143
Lamps, electric 132
Land drainage 290
Lateral ducts 335
Lavatory: basins 224
   ventilation 164
Lead caulked joints for cast iron 268
Lead pipes 229
Lifts general 314
   car 321
   construction 315
   control 318
   doors and gates 321
   fireman's 334
   goods 324
   hand 324
   motor room 319
   planning fan 332

INDEX

Lifts – *continued*
    typical sizes and numbers   320, 323
    ventilation   315, 319
    well   319
Light guague copper pipes   230
Lighting: calculations day   42
    electric   177
    criteria   50, 173
    costs   177
    day   39
    electric   172
    fittings   175
    fittings, extract through   161, 177
        heat gain   176
    levels   40, 173
    quality   172
Lightning protection   310
Lobby, ventilated, fire   333
Low pressure hot water space heating   142
Lumen, definition   10
    method   197
Luminance   174
Luminance design   177
Lux, definition   10

McAlpine trap   258
Machine room, lift   319
Mains distribution, electricity   296
    water supply   197
Mains head, water supply   197
Manhole covers   274
Manholes   271
Manholes, back drop   282
Marley: collar bass   255
    rodding point drainage   278
Marscar drainage system   278
Materials: building, thermal properties   79
    drain   266
    gutters   263
    pipes   229
    rainwater pipes   263
    sanitary appliance   216
    thermal properties   79
Maximum rate, heat loss   86
Mechanical conveyors   314
Metabolic rates   70
Meter, water   198
Metre, definition   10
Mild steel pipe   231
Mineral insulated copper covered cable   299
Mixing ratio, vapour   25
Mixing valve, heating   150
    water supply, thermostatic   228
Model analysis, daylighting   55
Moisture: general   21
    content of air for various building types   26
    content of air, prediction   28
    migration   23
Mole drains   291
Moon and Spenser sky   39
Motor room, lift   319
Mushroom extract, ventilation   158

Natural: lighting   39
    ventilation   36
Neutral, electric   296
Newton, definition   10
Noise: constructional precautions   113
    control   113
    floor standards   116
    indices   119
    party wall standards   113
    planning for   113
    pollution level (LNP)   119
    services installations   119
Non-concussive taps   226
Nomograms: daylight, internal velectal component   46, 47
    drain sizing   280
    water pipe sizing   242
Nozzles, ventilation   159

Ogee joints   267
Oil: calorific value   131
    heaters, direct   130, 136
        central system   130
    storage   131, 132
One pipe system: heating   142
    waste and soil   253
One wag switches, electrical   303
Orientation   74
Outgo, sewage   290
Outlet, electric socket   303
Overcast sky   39
Overflow: bath   223
    lavatory basin   224
Overhead distribution, electric   301

P-trap   257
PVC drains   268
    pipes   233
Packaged air conditioning   171
Paladin   292
Panels, embedded heating   147
Paper sacks, refuse   292
Party wall standards, noise   113, 115
Patent indirect cylinder   211
Paternoster   325
Paved areas, drainage   265
Perceived noise decibels (PNdB)   119
Percolating filter   288
Permanent supplementary artificial lighting   173
Pillar tap   226
Pipe: casings   338
    flanges   234
    fixing wall   234
    hangars   336
    materials and jointing   229
    rollers   336
    sleeves   234
    supports, in ducts   336
    sizing: comparative heating   141
        drains   281
        hot and cold water   235
    trenches   335

347

INDEX

Pitch fibre drains 267
Planning: drains 265
    heat loss 74
    lifts 322
    noise 113
Plant sizes: air handling 168
    boiler 167
    refrigeration 169
Plant spaces, air conditioning 158
Plastics: drains 268
    pipes 232
    sanitary appliances 216
    sheathed wiring electric 298
Plugs, formation in drain pipes 248
Plumbing duct 260
Pneumatic: booster, 207
    tube conveyor 326
Polarity, electric 297
Polythene pipe 232
Portsmouth ball valve 227
Powder fire extinguisher 330
Precast concrete manholes 272
Precipitation 21
Predicted mean vote 68
Pressed steel sanitary appliances 216
Pressure: fluctuation in drain pipe 247
    vessel for high pressure hot water space heating 144
    water heaters 213
'Primatic' cylinder 211
Propellor fan 156
Protected stair 333
Psychometric chart 24
Pumped water supply 207
Pumps: sump 285
    water supply 192

Radian, definition 11
Radiant heaters 136
Radiation, thermal 76
Radiators 146
Rain 21
Rainfall intensity 21
Rainwater: gulley 264
    pipe materials 261, 263
    pipe and gutters, sizing 263
    roof outlet 261
    shoe 264
Rates of heat loss, comparative 75
Rates, ventilation 38, 165
Recovery rate, hot water supply 236
Reflected sound 121
Reflection factor, light 173
Refrigeration plant sizes 169
Refuse chute 293
Refuse collector and storage 292
Relative humidity 25
Relative humidity, criterion for ventilation 37
Resealing trap 258
Resistance, thermal 28, 77
Resistance, vapour 28
Resistivity, thermal and vapour 28
Resultant temperature 68

Reverberation time 120
Ring main, electric 304
Rising damp 22
Rodding point drainage, Marley 278
Rollers, pipe 336
Roof outlet, rainwater 261
Roof U-values 82
Room acoustics 120
Rose, shower 223
Rubber ring joint 233, 266, 267
Run-around coil 163

S-trap 257
SE duct 313
SI units 9
Sabine's formula 122
Sacks, paper 292
Sanitary appliances 216
Schedules, electrical 305
Seals, trap 256
Seasonal heat loss 87
Second, definition 10
Selection of heating systems 127
Self syphonage 249
Separate drainage 275
Septic tank 287
Service ducts 335
Service installations, noise in 119
Service pipe, water 200
Service reservoir 197
Sewage: disposal 286
    treatment 287
Sewer: connection to 274
    gas interceptor 275
Sewerage systems 274
Showers 221
Shunt duct system 165
Simultaneous demand units 238
Simultaneous discharge units 281
Single phase supply electric 296
Single pipe system, heating 142
Single stack system 254
Sinks 225
Sizing: drain 283
    flues 170
    heating pipes and trunkings conparative 141
    heating and air conditioning plant 167-169
    lifts 322
    manholes 272
rainwater pipes and gutters 263
ventilation trunkings 166
    waste and soil pipes 261
    water pipes 239
Skirting heaters 146
    trunking, electrical 299[
Sky component 40
Sleeve coupling, drains 266
    waste 233
Slop hopper 219
Sludge, activated 289
Snow 22, 265
Soakaway 290

INDEX

Socket outlet, electric 305
Soda/acid extinguisher 329
Softening, water 196
Solar: gain factor 93
    heat collector 133
    heat gain 57
    heat gain diagram 62
    intensities 103
    radiation 57
    ray cone 58
Solid fuel: calorific value 131
    heating systems 130
    storage 131, 132, 133
    stoves 135
Solvent weld joints 233
Sound absorption 110
    absorption coefficents 122
    insulation 111
    propogation 109
    transmission 111
Sparge pipes 220
Speech interference levels 110
Springs, water 192
Sprinklers 331
Squatting WC 218
Stack effect 36
Stacks, flow in 247
Standing waves 123
Steel pipes 231
Steradian, definition 11
Sterelization, water 195
Stopcock, water 198
Storage: cold water 199
    cistern 200
    cistern sizes 235
    fuel 131
    hot water 236
    hot water cylinder 236
Stove, solid fuel 135
Subsoil drains 291
Sub-surface irrigation 289
Sump pump 285
Sun, movement of 57
Sunlight general 55
    criteria 66
    penetration standards 56
Sunpath diagram for 52°N lat. 60
Sunpath diagrams for use with Waldram diagram 63-65
Surface resistance, thermal 77
Surface temperature 92
Surface water disposal 290
Swan neck 262
Switch board GPO 308
Switching, electric 303
Syphonage, of taps, induced 247, 250
    self 249, 255
Syphonic WC 217

TNO meter 66
TRS wiring, electric 298
Taps 226
Telecommunications 307

Telephones 306
Temperature difference, design 86, 91
    diurnal variations 71
    surface 92
Ten percent level (L_{10}) 119
Terrazzo, sanitary appliances 216
Thermal: comfort 67
    indices 68
    installation 127
    insulation 75
    properties of buildings 71
    properties of building materials 79
    resistance and resistivity 28
    storage heaters 137
    transmittance 77
Thermostatic mixing valve: heating 150
    water supply 223
Thin wall copper pipes 234
Three phase supply, electric 296
Trade refuse 295
Traffic noise 119
Traffic noise index (TNI) 119
Transmission, sound 110
Transmittance, thermal 77
Traps: unsealing 249
    types and seals 256
Treatment: sewage 287
    water 195
Trenches, drain 269
    drain, width 271
    pipe 335
Trunking, cable 299
    floor 301
    skirting 301
    ventilation 166
    ventilation, noise in 163
Tube conveyors 326
Tubular heaters, electric 137
Two-pipe system: heating 142
    waste and soil 253
Two-way switches 303

U-values: calculation 80
    floors 84
    roof 82
    statutory 76
    walls 81
    windows 85
Umbrella spray, shower 223
Underflow heating, electric 136
    water 149
Underground drainage 265
Undersink water heater 213
Upfeed system, heating 143
Urinals 219
Utility services 189

Valves 226
Vapour barriers 23
    pressure 25
    resistance and resistivity 28
    sources 25

INDEX

Vapourising liquid extinguishers 330
Variable volume ventilation systems 161
Ventilated lobby, fire 333
Ventilation: criteria 37
   drain 276
   intake and exhaust positions 35
   internal lavatories 164
   lift 315, 319
   loss of heat due to 75
   natural 36
   plant 158
   plant size 168
   rates 36, 104, 38, 165
   systems 155
   trunkings general 158
      fixing 337
         sizes 166
Venting, trap and brand waste 250
Vertical ducts 335
Vitreous china 216
Vitrified clay drain pipes 267

WCs: generally 217
   ground floor and drain connection 258
   seats 218
Waldram diagrams: 45
   for CIE sky and vertical glazing 48
   drop line diagram for above 49
   calculation sheet for above 51
   for CIE sky, no glazing (for block spacing) 53
   calculation sheet for above 54
   sunpath diagrams for use with above: 52°N 63
         54°N 64
         56°N 65
Walkway services 335
Wall: chase 338
   U-value 81
Warmth, standards of 67
Warning pipe 201, 219
Wash: basin 224
   down WC 217
   fountain 224

Waste and soil systems 253
Waste and soil pipe size 261
Wastes, common 260
   connection to gulley 259
   flow in 249
   length of 254
Water: buckets, fine 329
   consumption 191
   distribution cold 199
   distribution hot 210
   filter 196
   hardners 195
   heaters 210
   mains: head 197
        supply 197
   meter 198
   softening 196
   storage cold 199
       hot 236
   supply: high buildings 207
      hot 209
      installations, noise 200
      service reservoir 197
      table 22, 291
      tower 197
      treatment 195
      waste preventer 219
Watt, definition 10
Wells 192
Wet riser, fire fighting 334
Wheel, heat recovery 164
Wild heat 73
Wind 32
Windows: solar gain 91
   sound insulation 118
   U-value 85
Wiped joints 229
Wiring systems, electric 298
Working plane, lighting 41
Wrought iron pipe 231

Zinc RWPs and gutters 264